HARVARD EAST ASIAN MONOGRAPHS

89

Studies in the Modernization of
The Republic of Korea: 1945–1975

Rural Development

North Korea

126°   127°   128°   129°

Kŭmhwa   Sokch'o

East Sea

38°   38°

KYŎNGGI   Ch'unch'ŏn
        Ŭijŏngbu   KANGWŎN
                                    Kangnŭng
                                        Mukho
        Seoul   Pukp'yŏng   Samch'ŏk
Inch'ŏn   Yŏju   Wŏnju   Chŏngsŏn
        Suwŏn
        Ansŏng   Ch'ungju

37°   37°
                NORTH
Ch'ŏnan   CH'UNGCH'ŎNG
                Ch'ŏngju   Ŭmsŏng

SOUTH   NORTH KYŎNGSANG
CH'UNGCH'ŎNG
        Taejŏn
        Yŏnmu
Changhang   Kimch'ŏn   P'ohang
36° Kunsan   36°
Okku   I-ri   Taegu   Kyŏngju
        Chŏnju
Yellow Sea   NORTH CHŎLLA

Pŏpsŏngp'o   SOUTH KYŎNGSANG

                                        Chinhae
                Kwangju   Chinju
35°   Samch'ŏnp'o   Pusan   35°
                SOUTH CHŎLLA   Changsŭngp'o
        Sunch'ŏn
Mokp'o   Yŏsu
                        Mijo-ri

        Sin'gŭm-ni

128°   129°

THE REPUBLIC OF KOREA

0   20 40 60 80 100 120 140 160
KILOMETERS

0   20   40   60   80   100
MILES

34°

CHEJU ISLAND   Cheju
        Sŏgwip'o
126°   127°

Studies in the Modernization of
The Republic of Korea: 1945–1975

Rural Development

SUNG HWAN BAN
PAL YONG MOON
DWIGHT H. PERKINS
with contributions by
VINCENT BRANDT
ALBERT KEIDEL
JOHN E. SLOBODA

PUBLISHED BY
COUNCIL ON EAST ASIAN STUDIES
HARVARD UNIVERSITY

Distributed by
Harvard University Press
Cambridge, Massachusetts and London, England
1982

The Council on East Asian Studies at Harvard University publishes a monograph series
and, through the Fairbank Center for East Asian Research and the Japan Institute,
administers research projects designed to further scholarly understanding of
China, Japan, Korea, Vietnam, Inner Asia, and adjacent areas.

The Harvard Institute for International Development
is Harvard University's center for interdisciplinary research, teaching, and technical
assistance on the problems of modernization in less developed countries.

The Korea Development Institute
is an economic research center, supported in part by the Korean government,
that undertakes studies of the critical development issues and prospects of Korea.

Library of Congress Cataloging in Publication Data

Ban, Sung Hwan.
Rural Development.

(Studies in the modernization of the Republic of
Korea, 1945-1975) (Harvard East Asian monographs; 89)
Bibliography: p.
Includes index.
1. Agriculture—Economic aspects—Korea—Addresses,
essays, lectures. 2. Rural development—Korea—
Addresses, essays, lectures. 3. Korea—Rural condi-
tions—Addresses, essays, lectures. I. Moon, Pal
Yong, joint author. II. Perkins, Dwight Heald, joint
author. III. Title. IV. Series. V. Series:
Harvard East Asian monographs; 89.
HD2095.5.B37   338.1'09519'5   79-28625
ISBN 0-674-78042-6

Foreword

This is one of the studies on the economic and social modernization of Korea undertaken jointly by the Harvard Institute for International Development and the Korea Development Institute. The undertaking has twin objectives; to examine the elements underlying the remarkable growth of the Korean economy and the distribution of the fruits of that growth, together with the associated changes in society and government; and to evaluate the importance of foreign economic assistance, particularly American assistance, in promoting these changes. The rapid rate of growth of the Korean economy, matched in the less developed world (apart from the oil exporters) only by similar rates of growth in the neighboring East Asian economies of Taiwan, Hong Kong, and Singapore, has not escaped the notice of economists and other observers. Indeed there has been fairly extensive analysis of the Korean case. This analysis, has been mainly limited to macroeconomic phenomena; to the behavior of monetary, fiscal, and foreign-exchange magnitudes and to the underlying policies affecting these magnitudes. But there are elements other than these that need to be taken into account to explain what has happened. The development of Korean entrepreneurship has been remarkable; Korea has an industrious and disciplined labor force; the contribution of agricultural development both to overall growth and to the distribution of income requires assessment; the level of literacy and the expansion of secondary and higher education have made their mark; and the combination and interdependence of government and private initiative and administration have been remarkably productive. These aspects together with the growth of urban areas, changes in the mortality and fertility of the population and in public health, are the primary objects of study. It is hoped that they will provide the building blocks from which an overall assessment of modernization in Korea can be constructed.

Economic assistance from the United States and, to a lesser extent, from other countries, has made a sizable but as yet unevaluated contribution to Korean development. A desire to have an assessment undertaken of this contribution, with whatever successes or failures have accompanied the U.S. involvement, was one of the motives for these studies, which have been financed in part by the U.S. Agency for International Development and, in part, by the Korea Development Institute. From 1945 to date, U.S. AID has contributed more than $6 billion to the Korean economy. There has also been a substantial fallout from the $7 billion of U.S. military assistance. Most of the economic assistance was contributed during the period before 1965, and most of it was in the form of grants. In later years the amount of economic assistance has declined rapidly and most of it, though concessional, has been in the form of loans. Currently, except for a minor trickle, U.S. economic assistance has ceased. The period of rapid economic growth in Korea has been since 1963, and in Korea, as well as in other countries receiving foreign assistance, it is a commonplace that it is the receiving country that is overwhelmingly responsible for what growth, or absence of growth, takes place. Nevertheless, economic assistance to Korea was exceptionally large, and whatever contribution was in fact made by outsiders needs to be assessed. One of the studies, *The Developmental Role of the Foreign Sector and Aid,* deals with foreign assistance in macroeconomic terms. The contribution of economic assistance to particular sectors is considered in the other studies.

All the studies in this series have involved American and Korean collaboration. For some studies the collaboration has been close; for others less so. All the American participants have spent some time in Korea in the course of their research, and a number of Korean participants have visited the United States. Only a few of the American participants have been able to read and speak Korean and, in consequence, the collaboration of their colleagues in making Korean materials available has been invaluable. This has truly been a joint enterprise.

The printed volumes in this series will include studies on the growth and structural transformation of the Korean economy, the foreign sector and aid, urbanization, rural development, the role of entrepreneurship, population policy and demographic transition, and education. Studies focusing on several other topics—the financial system, the fiscal system, labor economics and industrial relations, health and social development—will eventually be available either in printed or mimeographed form. The project will culminate in a final summary volume on the economic and social development of Korea.

Edward S. Mason

Edward S. Mason
Harvard Institute
for International Development

Mahn Je Kim

Mahn Je Kim
President,
Korea Development Institute

A Note on Romanization

In romanizing Korean, we have used the McCune-Reischauer system and have generally followed the stylistic guidelines set forth by the Library of Congress. In romanizing the names of Koreans in the McCune-Reischauer system, we have put a hyphen between the two personal names, the second of which has not been capitalized. For the names of historical or political figures, well-known place names, and the trade names of companies, we have tried to follow the most widely used romanization. For works written in Korean, the author's name appears in McCune-Reischauer romanization, sometimes followed by the author's preferred romanization if he or she has published in English. For works by Korean authors in English, the author's name is written as it appears in the original publication, sometimes followed by the author's name in McCune-Reischauer romanization, especially if the author has published in Korean also. In ordering the elements of persons' names, we have adopted a Western sequence—family name first in all alphabetized lists, but last elsewhere. This is a sequence used by some, but by no means all, Koreans who write in English. To avoid confusion, however, we have imposed an arbitrary consistency upon varying practices. Two notable exceptions occur in references to President Park Chung Hee, and Chang Myon, for whom the use of the family name first seems to be established by custom and preference. Commonly recurring Korean words such as si (city) have not been italicized. Korean words in the plural are not followed by the letter "s." Finally, complete information on authors' names or companies' trade names was not always available; in these cases we have simply tried to be as accurate as possible.

Geographic Terms (from largest to smallest)

Urban
   si – city
   pu – old term for a city
   ku – borough
   tong – precinct; (see rural tong)
   t'ong – sub-precinct
   pan – neighborhood

Rural
   to – province
   kun – county
   ŭp – town (formerly the county seat)
   myŏn – township
   tong – group of villages
   i (~ri, ~ni) – village
   purak – hamlet

Contents

Contents

Tables

Appendix Tables

Figures

Figures

Abbreviations

ADB	Asian Development Bank
AID	Agency for International Development
BOK	Bank of Korea
CEB	Combined Economic Board
CFFC	Central Federation of Fisheries Cooperatives
ECA	Economic Cooperation Administration
ED	enumeration district(s), in agricultural census
EPB	Economic Planning Board
FAO	Food and Agriculture Organization
FFYP	First Five-Year Plan
FOA	Foreign Operations Administration
GDCF	gross domestic capital formation
GDP	gross domestic product
GMSA	Grain Management Special Account
GNP	gross national product
IBRD	International Bank for Reconstruction and Development (The World Bank)
ICA	International Cooperation Administration
KASS–POP	Korean Agricultural Sector Study—Population
MAF	Ministry of Agriculture and Forestry (through 1972) Ministry of Agriculture and Fisheries (1973–)
MHA	Ministry of Home Affairs
MOC	Ministry of Construction
MOF	Ministry of Finance
M/T	metric tons
NACF	National Agricultural Cooperative Federation

NAERI	National Agricultural Economic Research Institute
NCM	New Community Movement (Saemaul Undong)
OECF	Overseas Economic Cooperation Fund
NDP	net domestic product
ORD	Office of Rural Development
PL 480	Public Law 480
RGO	Rural Guidance Office (s)
UNKRA	United Nations Korea Reconstruction Agency
USOM/K	United States Operations Mission to Korea
YBAFS	*Yearbook of Agriculture and Forestry Statistics*

Preface

Many people have contributed to this study. Vincent Brandt, Albert Keidel, and John Sloboda have each provided an entire chapter, as noted in the Table of Contents. The remaining chapters are the work of the three principal co-authors, Sung Hwan Ban, Pal Yong Moon, and Dwight H. Perkins. Although they were mutually helpful in a variety of ways, each had special responsibility for certain portions of the book. The key Chapter 3 and Appendix B were written by Dr. Ban; Chapters 6, 7, and 8, and Appendix A, by Dr. Moon; Chapters 1, 2, 4, 10, 11, and 13, and Appendix C, by Dr. Perkins.

PART ONE

Agriculture's Role

ONE

Introduction

Three decades ago, on the eve of Korea's liberation from Japanese rule, most of the people of Korea were farmers working tiny plots of land and paying a large share of the harvest to their landlords. In 1975, a majority of these same people lived in cities or held non-agricultural jobs in the countryside. Those who remained on the farm had achieved substantial gains in output and income and no longer had to share those gains, for the most part, with absentee owners of the land. This book is about these changes and how they were achieved.

Korean agricultural development did not begin in 1945 nor did it end in 1975. Even in the late nineteenth century, Korean farmers were poor in terms of income but rich in skills. In the words of one European traveler in the 1890s:

> I was much surprised with the neatness of the cultivation ... The crops of wheat and barley were usually superb, and remarkably

free from weeds—in fact, the cleanliness would do credit to 'high farming' in the Lothians.[1]

Korean farmers, of course, had to be skilled. In the late nineteenth century some 15 million people were crowded onto less than half the acreage available to the hundred thousand farmers (plus their families) of Iowa, and Korean land was far poorer than that of Iowa besides. Survival, therefore, depended on a high degree of skill.

Where advances could be accomplished by a farmer working on his own, they were often achieved in pre-modern Korea. But many kinds of progress in farming are beyond the means of the independent farm family working alone. Many require the skills found only in scientific laboratories or the ability to mobilize large numbers of people to build dikes and canals. The government of late Yi dynasty Korea, however, was incapable of providing either and was, instead, preoccupied almost exclusively with its own survival.

The Japanese Colonial Government did bring change to rural Korea. Desirous of ensuring an adequate supply of rice for Japan's own cities, Japan began to create institutions to promote an increase in agricultural output.[2] An extension service was founded, agricultural credit was made available, and Korean agricultural output did grow. These changes were later to provide a base of experience on which post-1945 efforts to improve farming could be built, but there were few direct benefits to Korean farmers at the time. Education in rural areas expanded and the death rate declined, but careful estimates by economists suggest that the standard of living of most Korean farmers actually fell.[3] Large amounts of rice were exported to Japan, while Koreans themselves shifted to eating a larger proportion of the coarser grains. The main cause was a rising level of tenancy as more and more Korean farmers lost their land to both Korean and Japanese landlords.

The changes that have been instituted since 1945 have involved far more than efforts to raise agricultural production, although production has risen substantially. Equal, if not more important, have been the measures that have redistributed the benefits of increased farm output. This study, therefore, is about rural development, not just agricultural production. We shall be as concerned with the impact of land reform and the quality of local government as we are with the rate of growth of agricultural outputs and inputs.

Our approach is historical—we shall attempt to describe and explain the causes and effects of various developments as they actually occurred through time. By analyzing how farmers responded to problems and opportunities over time and how various government measures helped and hurt, one can often learn more than by an in-depth attempt to analyze farmer behavior and agricultural policy at a given point in time. Both approaches, however, have their uses, and our only claim is that there are valuable insights for both the present and future from an understanding of how the Korean countryside got to where it was in 1975.

Rural Korea, needless to say, did not develop in a vacuum but in the context of overall Korean economic development since 1945. Rural Korea's role in overall development, however, differs markedly from that of many other developing countries. A Korean agricultural revolution did not precede or lead development in the other sectors of the economy. There were no substantial net flows of savings or tax dollars from the rural to the urban sector. There was a flow of agricultural produce in exchange for manufactures, and, most important, a massive migration of labor from the farms into the factories, transport, and trade. For the most part, however, it was agriculture that benefited from the industrial and export boom rather than the reverse. This boom provided a rapidly expanding market for farm output and for rural "surplus" labor, together with an increasing supply of such key inputs as chemical fertilizer and farm machinery. How agriculture, the dominant economic sector in 1945, was able to receive more than it gave during the next three decades is the subject of Chapter 2.

If rural development involves more than increases in agricultural production, it is still difficult to conceive of sustained rural progress without steady increases in farm output. For Korea, the already mentioned shortage of arable land made increases in output particularly difficult to achieve, and yet Korea's overall rate of agricultural growth was comparable to that in other developing nations with richer endowments of land. Part of the explanation for this respectable performance is that Korean farmers stressed products that were less land-intensive than grain, products such as vegetables, fruits, and mulberry (for silk). This changing product mix is the subject of Chapter 3.

The other part of the explanation for the pace of farm output growth is the steady rise in inputs. One input that did not rise, however, was labor. In fact, the amount of labor available per farm or per hectare fell steadily as young people left for the city and had to be replaced by the mechanization of activities such as threshing and the pumping of water, activities that occur during the peak period of labor demand. By the mid-1950s Korea was no longer a labor-surplus economy even in the limited sense where the marginal product of labor is above zero but below the going rural wage rate.[4] Another input that rose little was land, despite aid-supported efforts to open up new land in the early 1960s.

A key input that did increase rapidly was chemical fertilizer, and there was also a steady expansion in the percentage of paddy land irrigated. Korean farmers' experience with chemical fertilizers began in the Japanese occupation period, much earlier than in most currently developing nations, and in the late 1940s and 1950s United States aid enabled Korea to continue large-scale imports of fertilizer. In the 1960s, Korea began the rapid development of its own chemical fertilizer industry and, by the end of the period covered in this study, had saturated the domestic fertilizer market and was on the verge of becoming an exporter. Together with improved irrigation and other complementary inputs such as improved plant varieties (*t'ongil* rice, for example), Korea by the 1970s had achieved grain yields nearly comparable to those of such an advanced agricultural nation as Japan.

The analysis of the role of agricultural inputs in this study is carried out on two levels. In the latter part of Chapter 3, a formal but not very successful attempt is made to estimate an aggregate Korean agricultural production. In Chapter 4, the major inputs and the role each has played are described and analyzed in detail. A major problem in analyzing the sources of growth in Korean agriculture is how to explain the rise in productivity that accounted for half the rise in farm output. One major source of this increase in agricultural productivity was the improvement in marketing and rural transport that occurred over the period. As the analysis in Chapter 5 indicates, it is the relative degree of isolation that explains much of the difference in regional income. Reducing this isolation, therefore, is one major way of achieving a rise in productivity.

Improvements in transport, marketing, and other inputs, of course, do not appear out of thin air. Some, notably unskilled labor, are "produced" by the farm household itself. But many others become available only if the government takes steps to provide them. During the late 1950s and particularly the 1960s, the Korean government began to build a rural infrastructure capable of stimulating sustained growth. The Office of Rural Development, which houses the extension service, and the agricultural cooperatives, really arms of the government handling credit and the supply of certain inputs, were created and strengthened. There was also a series of five-year plans that provided the framework within which government investment in agricultural projects took place.

Although agricultural investments accounted for a significant fraction of total government investments, the amounts were a tiny share of total national investment, both public and private. During the 1960s, government economic policy was primarily concerned with the industrialization program and the export drive. Nowhere was this more apparent than with the government's agricultural procurement price policy. During the 1950s and 1960s, with the help of large imports of PL 480 grain (American subsidized exports of grain under Public Law 480), the government held farm prices down in an effort to stem inflationary pressures. But in the late 1960s and even more in the 1970s, the government began to turn its attention to agriculture. Farm price policies were reversed and agriculture's terms of trade with industry improved markedly. By 1974 and 1975, the level of government investment was also increased by a large amount in real terms. And, perhaps most important, through the New Community Movement (Saemaul Undong), the government succeeded in mobilizing local government officials to promote and cooperate positively with village "self-help" efforts, efforts that involved attempts to improve the quality of rural life as much as or more than to push production.

Agricultural investment policies and the five-year plans are the subject of Chapter 6. Chapters 7 and 8 deal with government credit and price policies respectively, and in Chapter 9 an attempt is made to estimate the degree of success achieved in mobilizing local government to promote the New Community Movement.

The final part of this study turns to the question of what has happened during the past thirty years to the quality of rural life, with particular emphasis on how the benefits of agricultural growth and Korean economic development in general have been distributed to the rural population. In the late 1940s through the Korean War, Korea carried out one of the twentieth century's most thoroughgoing land reforms. This reform, the subject of Chapter 10, left only 16 percent of the land in the hands of tenants, and the effect on the incomes of former tenants was dramatic. Because former landlords received very little compensation for their land, those without alternative sources of income were wiped out, while the income of former tenants must have risen about 33 percent on the average.

Since the mid-1950s there has been no further major redistribution of rural income, with the exception of the rise in farm procurement prices in the 1970s. While the size distribution of income within agriculture has changed little since the early 1960s, rural incomes have risen along with output. As pointed out in Chapter 11, rising incomes have led to better diets, improved housing, and more and more education for rural children, most of whom are now going on to middle school and even high school. Improvements in rural health, however, have lagged badly.

Further improvements in rural income distribution have been stymied in part by the persistence of substantial differences in income per capita among regions, the subject of Chapter 5. Remote areas have been opened up with a greatly improved road network during the past decade, but it has been farmers living closer to the cities who have been able to take the greatest advantage of the opportunity to supply urban residents with vegetables. In time, migration out of the rural areas may bring about a greater equality between regions as it has in some other countries. Certainly, as pointed out in Chapter 12, that migration is occurring at an extraordinarily rapid rate.

There have been many failures and false starts in Korea's rural development programs, but the overall picture is one of substantial progress. Some of that progress has been due to efforts of the government; more has stemmed from the efforts of farmers themselves responding to the opportunities of a rapidly increasing urban market. The story of Korean rural development is not so much one of a long, weary struggle to raise more grain on too little land, although grain output has risen, as have grain imports. It is more a story of farmers making the best use of their resource endowment by putting greater emphasis on the cash crops. It is also a story of deliberate efforts by the government to redistribute income and improve the quality of life of the farmer at the beginning and the end of the three decades covered by this study. In the decade and a half in the middle of the period, however, Korean government efforts to help the rural sector were comparatively feeble, and those that did exist were heavily dependent on U.S. aid.

There is at least one fundamental lesson in Korea's experience for other developing countries. In the 1970s, the focus of most international aid-giving agencies has been on ways to assist the rural poor. To many aid-givers the obvious solution to this problem is to see that the assistance given is applied directly to the poorest segments of the farm population. The "trickle down" effects of more general development efforts, it is argued, take too long to reach the poor, even when they are not diverted altogether to the rich. Korea, however, is a country where "trickle down" did work. In a relatively short period of time, rapid urban and industrial development did lead to marked improvements in rural living conditions.

A study of Korea's experience also suggests that direct efforts to help the rural poor may be difficult to achieve prior to a major and general development effort. In the 1940s and 1950s, with the notable exception of land reform, the Korean government's ability to implement rural development programs was extremely limited. It was not only that general development efforts brought with them an enhanced ability to build roads, establish fertilizer plants, and the like. A case can also be made that it was overall development that made it possible to improve the quality of the agricultural extension services and to provide the government with the capacity to implement programs, such as the New Community Movement, aimed directly at rural areas. Perhaps Korea would have found the trained personnel to implement these programs even in the absence of rapid general development, and perhaps the government would have acquired the will and the capacity to support the efforts of these personnel. But there are grounds for skepticism. A more plausible view is that experience implementing successful export-led industrial growth paid off over time in an increased will and ability to do something comparable for the rural sector.

TWO

Agriculture's Role in Korean Economic Development

There are two parts to the story of agriculture's place in the overall development of the Korean economy. On the one hand, the contribution of the agricultural sector to the extraordinary rise of 10 percent a year in national product over the 1960s and the first half of the 1970s has been modest. The story of Korea's extraordinary growth performance during these fifteen years is a tale of industry and industrial exports. Agriculture's growth rate has lagged far behind the average of other sectors, and the farm population has contributed little but its "surplus" labor to the more modern and rapidly growing sectors.

The other side of the story, however, is that the farm population has itself benefited greatly from the rapid growth of the non-farm parts of the economy. Rapid urbanization and industrialization have created a large increase in demand for agricultural production, while at the same time providing employment opportunities outside agriculture for rural youth. Large jumps in foreign-exchange earnings have made it possible to import first more fertilizer and then the feedstock for domestic fertilizer plants (and the plants themselves).

In many models of economic development, agriculture in the early stages of growth is seen as playing an essential role. Not only does rising labor productivity in agriculture free workers for the factories, but rural savings are a major source of investment funds for the non-agricultural sector. Rising rural labor productivity also contributes to an overall increase in farm output and thus prevents a worsening of the urban sector's terms of trade with agriculture, a worsening that, if it occurred, would cut into industrial profits and hence investment and growth.[1]

This model fits poorly with the last thirty years' performance of Korean agriculture. Farmers have saved little and invested little if any of their savings until recently in the non-farm sector. The terms of trade did not turn in favor of agriculture and against industry in the 1960s almost as much because of rising food and particularly grain imports as because of increased agricultural productivity. When the terms of trade did shift in favor of agriculture in the 1970s, it was a result of a deliberate change in government policy, not a worsening performance of agricultural production.

Agriculture, therefore, has not in any way led Korean economic development. Korea is, in contrast, a model of how an industrial revolution can precede and help bring about an agricultural revolution rather than the reverse.

THE CONTRIBUTION OF AGRICULTURE
TO OVERALL GROWTH

There are various ways of demonstrating the modest role agriculture has played in Korean economic development. To begin with, there are the estimates of agriculture's share in gross national product (GNP). In the mid-1950s (and earlier) Korea was a fairly typical less-developed country with half its GNP originating on the farm (see Tables 1 and 2). Until the early 1960s, the decline in agriculture's share was not large but, when industry and exports took off after 1964, the share of agriculture plummeted. Within ten years that share had fallen from around 40 percent to below 20 percent of GNP. Of a total increase in GNP of 2,665 billion wǒn between 1964 and 1975 (in constant 1970 prices), agriculture contributed only 285 billion wǒn or just over 10 percent. Manufacturing plus public utilities, construction, and transport, which together accounted for only 23.7 percent of GNP in 1964, experienced a sixfold increase over the next eleven years. The 1840-billion-wǒn increase in value added in these sectors accounted for nearly 70 percent of the increase in GNP. The remaining 20 percent of the 1964–1975 rise came mainly from the service sector, particularly trade, a sector that grew largely in order to handle the increased volume of manufactures. In comparison with other countries at comparable levels of per capita national product, Korea's agricultural performance was not typical. Agriculture's (plus mining's) share in the average "large" country fell considerably more slowly than was the case in Korea (see Table 2). Only in such neighbors as Taiwan and Japan was the performance of the farm sector similar to that in Korea.

It is important from the outset to understand that agriculture's small and declining contribution was not the result of an unusually slow growth rate in agriculture. Nor can one build an argument around the theme that Korean farmers were so backward by international standards that a slow agricultural growth rate was inevitable. As the growth rates in Table 3 indicate, Korea's agricultural performance over nearly the past two decades was above the world average and comparable to the overall Asian average. It was better than that of the land-rich nations of Africa and Latin America, not to mention North America.

Far from being backward, Korea, judged by the productivity

of its land, was one of the most advanced farming nations (Table 4). Korean rice yields were more than double those of south and southeast Asia and nearly a match for Japan's heavily

TABLE 1 Agriculture's Share in GNP and Population

| Year | Share of Agriculture and Forestry in GNP (%) | | Share of Farm Population in Total (%) | Farm Population (millions) |
	Current Prices	Constant 1970 Prices		
1953	46.4	45.6	61.0	13.15
1954	39.1	46.7	61.2	13.17
1955	43.9	45.5	61.9	13.30
1956	46.0	42.1	60.3	13.45
1957	44.2	42.5	59.2	13.59
1958	40.2	43.1	58.2	13.75
1959	33.6	40.9	58.2	14.13
1960	35.9	39.9	58.3	14.56
1961	39.5	42.5	56.5	14.51
1962	35.6	38.6	57.1	15.10
1963	41.2	38.3	56.6	15.27
1964	44.5	40.7	56.2	15.55
1965	36.7	37.6	55.8	15.81
1966	33.8	37.1	54.1	15.78
1967	29.5	32.4	54.4	16.08
1968	26.6	29.2	52.7	15.91
1969	27.0	28.7	50.7	15.59
1970	26.2	26.2	45.9	14.42
1971	27.0	24.5	46.2	14.71
1972	26.5	22.8	45.4	14.68
1973	23.5	20.1	44.5	14.64
1974	23.4	19.2	40.2	13.46
1975	24.0	19.2	38.2	13.24

Source: BOK, *National Income in Korea, 1975,* pp. 144–147, and *Economic Statistics Yearbook 1976.*

subsidized rice farmers. One of the main reasons for this high land productivity, of course, was that Korea has very little land and a considerable supply of farm workers. In terms of acreage per person in the farm sector, in fact, Korea's arable land endowment was probably the smallest in the entire world, if one excludes city-states like Singapore (see Table 5). A theme to which we shall return in subsequent chapters is that many of the

TABLE 2 International Comparisons of Agriculture's Share in Domestic Product

Country	1953–1955	1960–1962	1964–1966	1972–1974
Korea				
GDP per capita (U.S. 1961 $)	81	–	–	206
(U.S. 1964 $)	74	80	–	175
Share of agriculture in GDP	.503	–	–	.262
Share of primary product in GDP	.513	.467	–	.273
Chenery-Syrquin predicted primary shares for large countries	.522	.510	–	.354
Taiwan				
GNP per capita (U.S. 1961 $)	126	–	202	338
Share of agriculture in NDP	.338	–	.268	.158
	1905–1907	1920–1922		1938–1940
Japan				
GNP per capita (U.S. 1961 $)	162	244	–	356
Share of agriculture in NDP	.280	.216	–	.138

Source: Kwang Suk Kim and Michael Roemer, "Growth and Macro Change in the Structure of National Product, 1945–1975," Korea Development Institute Interim Report 7601, pp. 8–12, 8–17, 8–23.

TABLE 3 Agricultural Growth Rates—International
Comparisons

Country or Area	1952/1961	1961/1971	1952/1971
Korea (FAO)	3.1[a]	3.7	3.5
(this study) 1965 prices	3.1	3.7	3.4
World	2.8	2.6	2.7
North America	1.1	2.1	1.6
Latin America	3.5	2.4	2.9
Africa	2.8	2.9	2.9
Asia (east and south)	4.2	2.6	3.4
India	3.6	2.2	2.9
Japan	2.4	2.0	2.2
Thailand	5.2	3.6	4.4
Taiwan	4.1	3.9	4.0

Source: FAO *Production Yearbook 1968* and *1971.* FAO methods of estimating
agricultural output differ slightly from the other estimates used in this study.
Note: [a]1954–1961 because 1952 and 1953 were during the Korean War.

TABLE 4 Paddy Rice Yields—International Comparisons

Country	Year	Yield (kg./hectare of sown acreage)
Korea (Republic of)	1952–1956	3340
	1961–1965	4110
	1970	4550
Japan	1952–1956	4340
	1961–1965	5020
	1970	5640
Taiwan	1952–1956	2810
	1961–1965	3670
	1970	4160
India	1952–1956	1280
	1961–1965	1480
	1970	1700
Thailand	1952–1956	1350
	1961–1965	1760
	1970	1970

Source: FAO, *Production Yearbook 1969* and *1971.*

TABLE 5 Arable Land Per Capita—International Comparisons
(1968–1970)

Country	*(1)* *Arable* *Land* *(1,000 ha)*	*(2)* *Total* *Population* *(millions)*	*(3)* *Agricul-* *tural* *Population* *(millions)*	*(4)* *(1)/(2)*	*(5)* *(1)/(3)*
Korea (Rep. of)	2,311	32.422	17.300	.071	.134
Japan	5,510	103.540	21.329	.053	.258
China (PRC)	105,000	800.000	640.000	.131	.164
India	164,610	550.376	372.605	.299	.442
Thailand	11,415	35.814	27.398	.319	.417
United States	176,440	205.395	8.216	.859	21.475

Source: FAO, *Production Yearbook 1971,* except for the estimates for China, which were made by the author.

trends in Korean agriculture can only be understood as reflecting this land-short factor endowment. As will be demonstrated at length subsequently, Korean agriculture did not contribute more to overall economic growth, not because farmers failed to respond to stimuli or even because of government neglect, although prior to the 1970s there was some of that, but because of a basically unfavorable factor endowment. In agriculture there are limits to the degree that one can substitute capital and labor for land, whereas in manufacturing there are few such limits.

TRANSFERS OUT OF AGRICULTURE

Not only did the agricultural sector itself contribute little to the rise in GNP; as already indicated, except for labor, the transfer of resources out of the rural sector to the rapidly growing sectors was also modest.

Data on farm-household savings and direct taxes are presented in Tables 6 through 9. Ideally, one should also include an estimate of the indirect tax burden on agriculture as well but,

TABLE 6 Farm Household Savings and Taxes, 1958–1975
(per household)

Year	Farm Household Income (wŏn)	Farm Household Expenditure (living expenditure)	Savings (wŏn)	Savings % of income	Taxes and Public Charges (wŏn)	Taxes and Public Charges % of income
1958	42,910	45,352	−2,442	−5.7	1,513	3.5
1959	40,198	39,643	555	1.4	1,580	3.9
1960	44,750	45,499	−749	−1.7	1,386	3.1
1962	59,286	55,739	3,547	6.0	1,476	2.5
1963	82,799	77,464	5,335	6.4	1,935	2.3
1964	107,913	101,118	6,795	6.3	2,748	2.5
1965	109,839	100,492	9,347	8.5	3,062	2.8
1966	118,349	109,878	8,471	7.2	3,366	2.8
1967	138,718	127,667	11,051	8.0	2,615	1.9
1968	161,096	143,104	17,992	11.2	3,032	1.9
1969	186,852	171,371	15,481	8.3	4,324	2.3
1970	230,170	207,766	22,404	9.7	3,283	1.4
1971	317,236	244,463	72,773	22.9	4,137	1.3
1972	359,120	309,665	49,455	13.8	4,282	1.2
1973	426,756	337,350	89,406	21.0	6,399	1.5
1974	465,794	435,490	30,304	6.5	8,720	1.9
1975	722,716	616,280	106,436	14.7	12,687	1.8

Sources: MAF, *Report on the Results of Farm Household Economic Surveys* for 1962–1974, and BOK, *Farm Household Survey* for 1958–1960. Income excludes inventory, animals, and tree appreciation. 1958–1960 data have been converted from hwan to wŏn (1 wŏn = 10 hwan).

TABLE 7   An Estimate of Farm Household Savings in
Current Prices

	(1) Number of Farm Households (1,000s)	(2) Savings per Household (wŏn)	(3) Total Savings of Farm Households (1) X (2) (million wŏn)
1958	2,218	–2,442	–5,416
1959	2,267	555	1,258
1960	2,350	–749	–1,760
1961	2,327	–	–
1962	2,469	3,547	8,758
1963	2,416	5,335	12,889
1964	2,450	6,795	16,648
1965	2,507	9,347	23,433
1966	2,540	8,471	21,516
1967	2,587	11,051	28,589
1968	2,579	17,992	46,401
1969	2,546	15,481	39,415
1970	2,488	22,404	55,741
1971	2,482	72,773	180,623
1972	2,452	49,455	121,264
1973	2,450	89,406	219,045
1974	2,381	30,304	72,154
1975	2,379	106,436	253,211

except for a few general remarks, that is beyond the scope of
this study.

Savings estimates are notoriously inaccurate, largely because
savings are usually taken to be equal to the residual after all
sources of income and expenditure have been totaled. Thus,
errors on either the income or expenditure side will affect the
savings estimate, and it is an act of faith to believe that the
errors on one side cancel out those on the other. Korean rural

TABLE 8 The Contribution of Farm Savings to Gross Domestic Capital Formation

Year	(1) Gross Domestic Capital Formation as % of GNP (current prices)	(2) Household and Non-Profit Institution Saving (as % of GNP)	(3) Household and Non-Profit Institution Saving (as % of GDCF)	(4) Farm Household Saving (as % of GNP)	(5) Farm Household Saving (as % of GDCF)
1958	13.0	3.8	29.4	-2.6	-20.3
1959	10.7	1.4	13.1	0.6	5.3
1960	10.9	-1.2	-11.0	-0.7	- 6.6
1961	13.1	0.9	7.2	–	–
1962	13.0	-2.9	-22.4	2.5	19.3
1963	18.5	0.6	3.3	2.6	14.3
1964	14.6	2.0	13.6	2.4	16.3
1965	15.2	0.2	1.5	2.9	19.2
1966	21.7	4.1	19.0	2.1	9.6
1967	22.1	2.0	8.9	1.5	10.2
1968	26.8	1.4	5.4	2.2	10.8
1969	29.8	6.0	20.1	1.6	6.4
1970	27.2	3.4	12.5	2.2	7.9
1971	25.6	2.9	11.3	6.4	11.4
1972	20.9	3.8	18.1	4.0	15.1
1973	26.2	8.2	31.1	6.2	17.0
1974	31.4	7.5	23.9	1.9	3.4
1975	27.1	–	–	6.2	23.4

Sources: (1), (2), and (3), BOK, *Economic Statistics Yearbook 1976*; (4) and (5), derived from Table.

TABLE 9 Farm Household Cash Receipts and Disbursements
(per household)

Year	(1) Cash Receipts (current prices)	(2) Cash Disbursements (current prices)	(3) Net (1) − (2)	(3) % of farm income (incl. taxes)
	(yen)	(yen)	(yen)	
1933	198.8	182.9	15.9	4.9
1938	304.1	266.3	37.8	6.7
	(wŏn)	(wŏn)	(wŏn)	
1954	5,977	6,778	−801	
1955	11,974	12,774	−800	
1956	16,492	18,034	−1,542	−5.8
1957	18,199	19,578	−1,379	−3.5
1958	21,378	23,884	−2,506	−4.9
1959	21,071	22,464	−1,393	
1960	23,688	25,902	−2,214	
1962	31,834	31,400	434	0.7
1963	37,570	41,291	−3,721	−4.5
1964	54,729	57,831	−3,102	−2.9
1965	62,265	64,716	−2,451	−2.2
1966	72,159	74,553	−2,394	−2.8
1967	86,770	87,138	−368	−0.3
1968	101,171	100,743	428	0.3
1969	121,456	117,665	3,791	2.0
1970	157,174	146,411	10,763	4.7
1971	189,154	167,090	22,064	7.8
1972	228,591	210,702	17,889	5.1
1973	288,594	243,331	45,263	11.3
1974	391,792	319,563	72,229	14.1

savings estimates are further complicated by the inclusion of farm-household inventory appreciation in money terms in income and hence in savings. For reasons given elsewhere, inventory appreciation is systematically removed from the estimates of income used in this study. Although the resulting estimates of rural savings are considerably more plausible than if inventories were retained, plausibility is not the same thing as a high degree of reliability. The rural savings estimates from the Farm Household Surveys, for example, are in general consistent with the level of household savings in the national accounts, but the match-up on a year-to-year basis is not very close.

The main conclusion that can be drawn from these tables is that, while farmers saved a respectable proportion of their income in the 1960s (but not in the late 1950s), relatively little of this income found its way into the modern sector. For savings to be invested in sizable amounts outside the agricultural sector, it is reasonable to assume that those savings would usually have to be in the form of cash that could be deposited in banks or *kye* (financing clubs) or lent to firms directly. But in Korea in the 1950s and 1960s, farmers generally appear to have had a negative cash flow. How they financed this steady outflow is a subject for a later chapter, but, if the data are accurate, farmers were left without any surplus cash for investment elsewhere. Farm savings, therefore, must mainly · have gone into household improvements, farm equipment, and the like.

Beginning in the 1970s, this picture changes significantly. For the first time since the 1930s, the Korean rural population had substantial savings in cash. In the 1930s this positive net cash flow probably reflected the strong cash position of landlords and other high-income groups. In the 1970s, however, it probably reflected the general rise in farm incomes in that period, but more on this subject later. It is reasonable to argue, although impossible to prove, that this increase in farm household cash savings played an important part in the rise of nationwide household savings in 1973 and 1974. Whether one accepts this hypothesis depends very much on whether cash or total farm household savings best matches the concept of household savings as used in the national accounts.

Even with the increase in savings rates in the 1970s, it is still a fair generalization to state that rural savings played only a small part in the rapid acceleration in gross domestic capital formation after 1960. Throughout most of the post-1960 period, particularly after 1965, rural savings financed only about 10 percent of total capital formation, a share about equal to the percentage contribution of agriculture to the increase in national product.

Direct taxes and public charges made even less of a contribution to gross domestic capital formation than savings, even if one were to make the unrealistic assumption that all rural direct taxes and public charges were used to make possible a net increase in capital rather than current government expenditures. The addition of indirect taxes would not appear to change this picture much. The largest indirect taxes are on alcohol, "commodities," and petroleum products (nearly 80 percent of total indirect taxes in the 1970s). It is unlikely that farmers pay any significant part of the petroleum tax, and, because of lower incomes and a high percentage of self-supplied consumption goods, it is probable that the rural share of taxes in the other two categories is substantially less than the rural share of population. Indirect taxes on agriculture may take twice or even three times as much revenue out of agriculture as the very small direct taxes (mainly the agricultural land tax), but indirect and direct taxes on agriculture together probably amounted to less than total government expenditures on agricultural investment alone, at least in the 1970s. Thus, even if taxes on the rural sector made a contribution to gross capital formation, the government budget was not a vehicle for transferring substantial sums of money out of agriculture to other sectors.

Labor, of course, is a different story. There were massive transfers of workers out of agriculture into industry and other urban occupations. Furthermore, as will be demonstrated in the chapter on rural-urban migration, there is evidence to suggest that these migrants came disproportionately from the more educated portions of the rural population. Thus, the farm sector contributed its investment in training as well as the strong arms and backs of many of its workers. The precise size of this transfer is in large part a question of definition. The rural population in 1975 was virtually the same as what it was in 1953–1955. If, instead of holding steady, the rural population had maintained its share in the total population at 1953–1955 levels, there would have been nearly 22 million rural people instead of 13 million. The share of the 9 million difference who entered the non-agricultural labor force, together with the investment made by the rural areas in their upbringing and education, represents the farm sector's contribution to Korea's industrialization. An attempt at a more precise calculation of the size of this contribution would serve little purpose in this study.

THE CONTRIBUTION TO AGRICULTURE

If the contribution of agriculture to Korean industrialization was modest in all areas other than labor, the impact in the opposite direction, from industry and foreign trade to agriculture, was anything but modest.

The shift in Korea's population from rural to urban areas created a large increase in the demand for food without creating any matching ability to supply that food. The result was a rapidly expanding market for the produce of those who remained on the farm. A rough estimate of the magnitude of this increase in demand is presented in Table 10. The Urban Household Surveys on which these estimates were based are not very reliable, but they are adequate for the use to which they are put here.

The rise in urban demand for food came mainly from the increase in numbers of the urban population, not from a rise in food consumption per person or per urban household. The

income elasticity of demand for food is low in Korea as elsewhere in the world.

TABLE 10 An Estimate of Non-Farm Household Demand
for Food, 1955–1975

Year	*(1)* Number of Non-Farm Households *(1,000s)*	*(2)* Per-Household Consumption of All Foods per Month (constant 1970 wŏn)	*(3)* Total Non-Farm Demand for Food (million 1970 wŏn)	
			(a) *(per mo.)*	*(b)* *(per yr.)*
1955	1,548	10,000	15,480	185,760
1960	1,969	10,000	19,690	236,280
1965	2,338	10,054	23,506	282,075
1970	3,374	12,120	40,890	490,680
1974	3,774	12,297	46,409	556,907
1975	4,032	12,245	49,372	592,462

Sources: (1) Obtained from EPB, *Korea Statistical Yearbooks,* and BOK, *Economic Statistics Yearbooks,* various years, except for 1974–1975 data, which were obtained by dividing non-farm population by 5.3 to yield farm household estimate.
(2) Obtained by dividing the expenditure per urban household from the Urban Household Survey by the food component of the all urban consumer price index, except for 1955 and 1960, which were assumed.
(3) (a) (1) X (2).
    (b) (2) X 12.

The rise in urban demand for food was, nevertheless, substantial; in real terms the rise was more than 200 percent over the
entire twenty-year post-Korean War period. The 592 billion wŏn
estimate for non-farm food demand in 1975 in constant 1970
prices (Table 10) was equivalent to over 70 percent of the total
gross value of crop production in that year (also in constant
1970 prices). The rise in rural demand for food was, in contrast,
very small. Since the rural population was not increasing during
this twenty-year period (or more accurately, it first rose and
then fell), any increase in rural food demand had to come from
rising rural per-household income. But because, as already

mentioned, the income elasticity of demand for food is low, the rise in demand from this source was also not very great. In fact, if the Farm Household Survey estimates (deflated by an index of prices paid by farmers for food) can be believed, farm household expenditures on food (including self-supplied food) between 1963 and 1974 actually declined slightly, despite a 20 percent rise in real per household income over the same period. This result seems improbable and very likely reflects biases in the sample, but the basic point is clear. Increases in rural food demand due either to a rise in income or a rise in the number of households were very small.

Because of a lack of consistency in the way some relevant data are reported (foreign trade figures are in U.S. dollars rather than wŏn, for example), it is difficult to calculate a full balance sheet giving the total supply and demand for food in particular years. But rough estimates are possible and are revealing. In 1970, food expenditure per farm household was just under 100,000 wŏn, which would make total farm household food demand a little below 250 billion wŏn. It is unlikely that this total in 1970 prices would be much different in either 1975 (income was up but the number of households was down) or in 1965 (income was down but the number of households was up, albeit only slightly). In 1955, lower incomes and a smaller number of households might have kept rural food demand down to a level as low as 150 billion wŏn. Thus, of a total increase in domestic food demand of roughly 500 billion wŏn (in 1970 prices) between 1955 and 1975, rural demand accounted for about one-fifth and urban (and other non-rural demand), four-fifths. Put differently, without the increase in urban demand, the total domestic demand for food in 1975 would have risen by 30 percent instead of by nearly 150 percent.[2] Of course, in reality, if the urban population had stayed home on the farm, food demand would have risen by more than 30 percent, but the basic point is clear. Rising urban income made possible a large increase in the demand for food.

Not all of this increase in food demand was satisfied by

supplies from Korean farms, however. A significant portion was provided for by imports from abroad. Data on agricultural exports and imports are presented in Table 11. The figures include a number of non-food items (notably cotton on the import side), but the trend in net imports of food is not much affected whether non-food items are kept in or left out. In any case, one is really interested in total agricultural product supply and demand, not just food.

In terms of percentage shares, exports of agricultural products fell substantially and steadily in the 1960s and 1970s, and imports also fell, although only slightly. But percentage shares obscure the effects of Korea's dramatic growth of foreign trade in general and exports in particular. From less than 4 billion wŏn in 1956–1960 (in 1970 prices), agricultural exports rose to around 70 billion wŏn (also in 1970 prices) in 1971–1975.[3] Imports of agricultural products over the same period rose from 20 to 130 billion wŏn for a net import balance in 1971–1975 of 60 billion wŏn.

This rise in imports cut into the demand for domestic agricultural produce, but it does not automatically follow that imports thereby cut into a potential increase in farm income either from a rise in farm output to meet this demand or from an improvement in the rural-urban terms of trade. Nor would a greater rise in agricultural exports necessarily have benefited the Korean farmer. Whether the impact of a lower level of imports or a higher level of exports would have helped or hurt farmers depends on how these changes would have been achieved.

In the 1930s, for example, agriculture provided the lion's share of export earnings (Table 11), and over 40 percent of Korea's rice production was exported to Japan (Table 12), but the Korean farmer was certainly not the beneficiary of this process. These rice exports were supplied mainly from the rents paid to landlords. High exports, therefore, depended on a high level of tenancy and a related very unequal distribution of income. Land reform, to be discussed in a later chapter, eliminated most rents and with them the main source of exportable agricultural

TABLE 11 Exports and Imports of Agricultural Products,
1911–1975
(annual averages)

Year	Agricultural and Food Product Exports		Agricultural and Food Product Imports	
	million yen	*% of total X*	*million yen*	*% of total M*
1911–1915	19.3	64.6	11.0	17.4
1916–1920	76.0	55.3	27.6	16.1
1921–1925	161.2	59.5	56.7	20.3
1926–1930	199.6	59.1	94.1	24.1
1931–1935	216.1	55.3	73.0	16.9
	million U.S. dollars	*%*	*million U.S. dollars*	*%*
1956–1960	4.5	19.3	62.0	16.7
1961–1965	21.1	22.1	72.4	16.7
1966–1970	58.3	11.7	201.8	14.4
1971–1975	299.1	9.7	664.6	14.3

Sources: 1911–1935 data from Wontack Hong, "Trade and Subsidy Policy and Employment Growth in Korea" (mimeographed, Korea Development Institute, 1976). The 1956–1975 data are from BOK, *Economic Statistics Yearbook 1961, 1967,* and *1976* and include beverages, tobacco, and animal and vegetable oils, as well as food.

TABLE 12 Korea's Grain Exports and Imports, 1910–1939

Year	Rice Production	Rice Exports	Net Rice Supply	Imports of Millet
	(annual averages in 1,000 M/T)			
1910–1914	1,459	47	1,412	28
1915–1919	2,095	268	1,827	41
1920–1924	2,163	471	1,692	127
1925–1929	2,223	840	1,383	294
1930–1934	2,589	1,105	1,484	204
1935–1939	3,078	1,199	1,879	n.a.

Source: Wontack Hong, "Trade and Subsidy Policy and Employment Growth in Korea."

commodities. This reduction in farm exports was clearly to the benefit of the typical Korean farmer.

After 1945, Korea became a net importer of grain; by the 1960s and 1970s, grain imports were sizable and, together with cotton, accounted for the bulk of Korea's agricultural imports (see Table 13). Initially the bulk of these imports were paid for with American aid (PL 480) but, as Korean foreign exchange earnings grew, the Koreans themselves took over payment for steadily increasing levels of farm product imports (see Table 14).

TABLE 13 Imports of Grain, 1955–1975
(annual averages in 1,000 M/T)

Years	Wheat	Rice	Barley	Other
1955	42.3	0.8	3.5	14.3
1956–1960	277.3	41.1	171.9	26.5
1961–1965	491.2	23.5	149.9	66.7
1966–1970	879.2	360.9	56.6	96.9
1971–1975	1,623.0	563.0	383.9	58.7

Sources: BOK, *Economic Statistics Yearbook 1960, 1964, 1967, 1976,* and EPB, *Korea Statistical Yearbook 1969, 1973.*

It is worthwhile speculating to what degree PL 480 imports cut into domestic grain demand and worsened agriculture's terms of trade and hence lowered farm incomes. Certainly a case can be made for this point of view (and will be made in a later chapter). But the grain imports of the late 1960s and the first half of the 1970s are another matter. Imports of grain in the late 1960s amounted to over 12 percent of domestic production and, in the first half of the 1970s, to over 20 percent. Any attempt to extract this large an amount of grain out of the rural areas for the city population would have been difficult to bring about. For reasons that will become apparent in later chapters, it is unlikely that grain output would have expanded to meet this demand. And the shift in terms of trade toward agriculture would have had to be very large to get this increase in supplies

TABLE 14 The PL 480 Role in Agricultural Imports
($ U.S. millions)

Year	Total Grain Imports	Total PL 480 Aid	PL 480 Grain Imports	Total Cotton Imports	PL 480 Cotton Imports	Other PL 480 Imports
1955	6.4	–	–	20.1	–	–
1956	31.2	33.0	20.0	25.0	8.2	4.8
1957	84.3	45.5	33.0	26.9	1.8	10.7
1958	51.1	47.9	46.9	31.7	0.5	0.5
1959	17.5	11.4	4.3	30.8	7.0	0.1
1960	20.6	19.9	19.2	28.6	0.8	0
1961	n.a.	44.9	22.6	n.a.	21.5	0.8
1962	40.1	67.3	34.4	34.2	31.3	1.6
1963	107.2	96.8	62.6	38.2	31.8	2.5
1964	60.8	61.0	28.0	37.3	30.5	2.5
1965	54.4	59.5	29.7	40.8	29.7	0.1
1966	61.3	38.0	11.2	42.8	26.7	0
1967	76.6	44.4	7.9	49.3	34.0	2.4
1968	129.3	55.9	27.3	49.1	24.7	4.0
1969	250.3	74.8	31.6	52.0	39.0	4.2
1970	244.8	61.7	33.0	62.7	27.4	1.3
1971	304.0	33.7	18.0	84.2	15.7	0
1972	282.7	–	–	85.5	–	–
1973	444.1	–	–	112.4	–	–

Sources: BOK, *Economic Statistics Yearbooks,* and EPB, *Korea Statistical Yearbooks.*

out of existing output. The price elasticity of rural demand for grain is bound to be low under such circumstances. Faced with similar circumstances, nations that have desired to avoid rising grain imports have frequently resorted to imposing compulsory grain delivery quotas on farmers. (The Soviet Union in the 1930s is the classic example of this policy.) Compulsory quotas in effect lowered rural incomes and created barriers to future increases in output by lowering farmer incentives, since the typical farmer could not be sure whether an increase in output would raise his income or simply his delivery quota.

Industrialization, therefore, by contributing to a rapid rise in foreign exchange earnings, helped Korea in a fundamental sense to avoid policies that could have seriously cut into farmer incomes and incentives. Furthermore, instead of a desperate push to produce more grain, Korean farmers could concentrate on raising the output of cash crops, an activity that brought higher returns to the labor-intensive factor endowment of the rural areas than did specialization in grain.

Rising foreign exchange earnings, of course, contributed to Korean agriculture in other ways as well. The import of fertilizers and fertilizer plants has already been mentioned. There have also been small amounts of imports of farm machinery (U.S. $5.7 million in 1975) and insecticides. World Bank loans which must be paid back out of future foreign-exchange earnings have been used to finance irrigation works. And there are many other examples.

The main point, however, is the one made at the beginning of this chapter. Korea's industrialization, urbanization, and export drive have all contributed in a major way to the rise in agricultural output and farm incomes in the late 1960s and the first half of the 1970s by increasing farm product demand, by making possible the import of key inputs, and by allowing the government to avoid policies that would have hurt farmer incentives. Agriculture has been a major beneficiary of Korean economic growth, but not a major cause of that growth.

PART TWO

Agricultural Inputs and Outputs

THREE

The Growth of Agricultural Output and Productivity [1]

The main objectives in this chapter are to estimate the rate of growth of: 1) gross agricultural output; 2) factor inputs; 3) total productivity and the partial productivities with respect to land and labor over the three-decade period. A related objective is to estimate the rate of growth of commodities as well as commodity groups and changes in the share of various commodity groups in total agricultural production.

TRENDS IN AGRICULTURAL OUTPUT

Total agricultural production in Korea has increased by slightly less than 2.7 times over the past twenty-nine years from 1945 to 1974, with an occasional decline due to unfavorable weather or the Korean conflict. The indexes of total agricultural production and gross agricultural output together with gross value added in agriculture[2] are depicted in Figure 1, and their annual average compound growth rates in selected periods are shown in Table 15. The three series move in very similar patterns, although the growth rate of gross output is slightly less than that of gross production and the growth rate of gross value added is slightly less than that of gross output. Over the whole period, gross agricultural production grew at a compound growth rate of 3.41 percent per annum. During the same period, gross agricultural output grew at an annual rate of 3.20 percent and gross value added in agriculture at 2.90 percent per annum. The difference in growth rates among the three different concepts of production indicates the increasing use of intermediate products within agriculture and the increasing dependence of Korean agriculture upon purchased inputs. In this respect, Korean agricultural progress is similar to what has occurred in the agricultural sectors of many advanced economies.

We can observe in Figure 1 that agricultural output has been growing steadily over the entire period except during the Korean War. In spite of this steady growth, some distinct growth phases are observable during the period. The period from 1946 to 1952 is characterized by ups and downs in agricultural output and a low growth rate. Total agricultural output grew at a high rate of slightly less than 7 percent per year from 1946 to 1949. This high growth rate is mainly due to the recovery of production from disturbances in World War II when output declined markedly, falling at the rate of 3.46 percent per year from 1939 to 1945.[3] By 1949, agriculture had nearly recovered to its prewar peak level. The outbreak of the Korean War in 1950, however, brought agricultural output down again to an annual rate of 4.87 percent per year from 1949 to 1952. As a result, over the entire 1946–1952 period, agricultural output grew at only 0.88 percent per year. By the time the cease-fire talks began in 1952, agricultural output had fallen below the level of 1945. By 1954, however, total agricultural production had passed the peak production level of the pre-Korean War period thanks to an

FIGURE 1 Indexes of Gross Production, Gross Output, and
Gross Value Added in Agriculture
(three-year moving averages, semi-log scale)

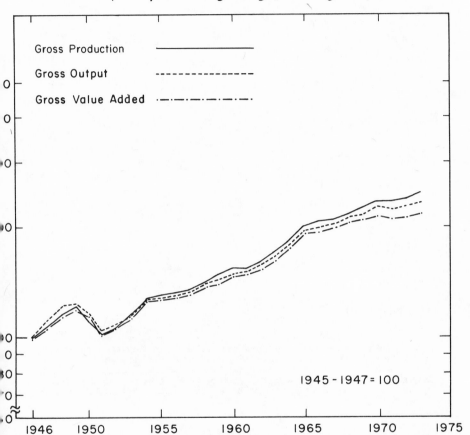

TABLE 15 Growth Rates of Total Production, Total Output,
and Gross Value Added in Agriculture
(%)

Period	Total Production		Total Output		Gross Value Added	
1946–1952	0.83		0.88		0.70	
1946–1949		6.82		6.97		6.79
1949–1952		–4.83		–4.87		–5.04
1952–1954	10.87		10.54		10.48	
1954–1973	3.48		3.19		2.83	
1954–1965	3.99		3.89		3.80	
1954–1960		2.75		2.53		2.35
1960–1965		5.49		5.55		5.57
1965–1973	2.79		2.23		1.51	
1965–1970		3.13		2.84		2.14
1970–1973		2.24		1.21		0.48
Whole period 1946–1973	3.41		3.20		2.90	

Note: Growth rates, here and in subsequent tables, are annual compound rates of increase between three-year averages of the data centered for the years shown. They are calculated from indexes of physical quantities of the specified items, or of value aggregates of them at constant (1970) prices, or of ratios of such magnitudes. Definitions, sources of data, and methods of estimation are explained in detail in Appendix B.

average annual growth rate of 10.54 percent between 1952 and 1954, made possible by the recovery of productive facilities from wartime devastation.

Since 1954, Korean agriculture has entered a long-run steady and sustained growth path. Total agricultural production, output, and gross value added in agriculture grew at an annual rate of 3.48 percent, 3.19 percent, and 2.83 percent respectively from 1954 to 1973. During the post-1954 period, two different subperiods are observable. Between 1954 and 1965, the annual growth rate of total agricultural output was 3.89 percent.

Further, during this period, the growth rate was accelerating from 2.53 percent between 1954 and 1960 to 5.55 percent between 1960 and 1965. The rate of growth has been decelerating since 1965 from an annual growth rate of 2.84 percent between 1965 and 1970 to 1.21 percent between 1970 and 1973.

COMPARISONS OF COMMODITY GROWTH

The major agricultural products of Korea are grouped into the following categories: food-grains, fruits, vegetables, special crops, monopoly crops (that is, leaf tobacco and ginseng, silkworm cocoons, and livestock and livestock products). The food-grains include rice, the barleys (common and naked barley and wheat), miscellaneous grains (millet, corn, sorghum, buckwheat, and so on), the pulses (beans, including soybeans and peas), and potatoes (both sweet and white). The chief fruits are apples, pears, peaches, grapes, and persimmons. More recently the production of oranges has been increasing rapidly on the southern island of Cheju. Many kinds of vegetables are grown in Korea; among the major ones are Chinese or celery cabbage, radishes, garlic, red peppers, onions, watermelons, and sweet melons. Special crops include fiber crops such as cotton and hemp, oil crops such as sesame and perilla, and others. The main classes of livestock are Korean native cattle (used both as draft animals and for meat), hogs, and poultry. In recent years the number of dairy cattle has increased rapidly although the total cannot be compared with that for Korean native cattle.

In Table 16 annual compound growth rates of increase at 1970 constant prices are presented. Figures 2, 3, and 4 depict changes over time of various farm products and groups of products.

Although we can observe significant differences in growth rates between product groups and between individual products, the growth pattern of total agricultural production is quite

TABLE 16 Annual Growth Rates of Agricultural Products,
1945–1973

(%)

	Periods					
Products	1946–1949	1949–1952	1946–1952	1952–1954	1954–1960	1960–1965
Rice	4.98	–8.10	–1.78	13.31	2.14	3.33
Barleys	11.04	–4.94	2.74	15.87	4.34	4.43
Miscellaneous Grains	6.88	10.38	8.62	–13.06	–0.08	4.88
Pulses	6.15	–4.77	0.54	8.59	–0.29	3.55
Potatoes	1.09	–1.87	–0.40	15.59	1.58	22.06
Total Food Grains	5.70	–6.90	–0.80	13.19	2.38	4.96
Fruits	19.02	7.30	13.01	5.31	6.25	12.32
Vegetables	16.08	9.21	12.59	–6.23	0.01	9.55
Special Crops	6.24	–4.36	0.80	3.84	–6.31	–4.10
Monopoly Crops	8.21	–33.60	–15.23	25.89	6.62	15.98
Crop by-products	7.03	–4.91	0.88	9.08	1.95	5.82
Total Crops	7.03	–4.91	0.88	9.08	1.95	5.82
Silkworm Cocoon	2.63	–4.99	–1.26	5.41	–3.13	9.14
Livestock	5.50	–4.23	0.44	32.25	7.86	2.33
Livestock Products	1.37	1.79	–0.13	65.23	23.71	9.57
Livestock & Its product	4.71	–3.68	0.43	33.73	9.26	3.25
Total Agric. Production	6.82	–4.83	0.83	10.87	2.75	5.49

TABLE 16 (continued)

| Products | Periods | | | | | Whole period |
	1954–1965	1965–1970	1970–1973	1965–1973	1954–1973	1946–1973
Rice	2.68	1.12	1.61	1.30	2.10	2.01
Barleys	4.38	1.82	–4.47	–0.59	2.26	3.32
Miscellaneous Grains	2.14	–0.75	–8.08	–3.57	–0.30	0.59
Pulses	1.44	6.94	3.52	5.64	3.19	2.98
Potatoes	10.43	–4.73	–7.03	–5.60	3.37	3.37
Total Food Grains	3.54	0.87	–0.21	0.46	2.23	2.32
Fruits	8.97	8.29	12.38	9.81	9.32	9.82
Vegetables	4.24	7.87	3.58	6.24	5.07	5.80
Special Crops	5.31	9.01	–0.66	5.30	–0.98	–0.80
Monopoly Crops	10.78	2.89	22.84	9.97	10.43	5.15
Crop by-products	3.69	2.43	1.88	2.22	3.07	3.01
Total Crops	3.69	2.43	1.88	2.22	3.07	3.01
Silkworm Cocoon	2.27	23.54	12.40	19.24	9.10	6.43
Livestock	5.27	4.07	1.81	3.04	4.34	5.28
Livestock Products	16.86	16.81	8.21	12.64	15.07	14.53
Livestock & Its product	6.46	6.44	3.14	5.21	5.95	6.52
Total Agric. Production	3.99	3.13	2.24	2.79	3.48	3.41

Note: Growth rates are annual compound rates of increase between three-year averages of the data centered within the years shown.

FIGURE 2 Indexes of Production of Agricultural Commodity
Groups
(1970 constant prices, three-year moving
averages, semi-log scale)

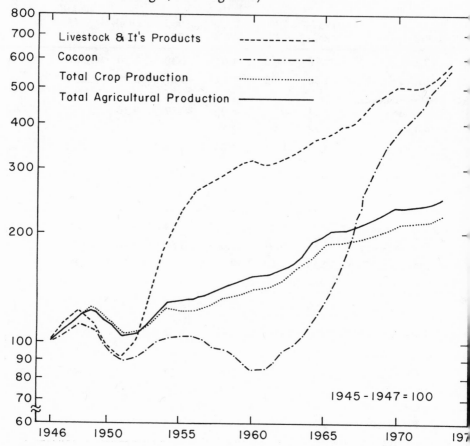

FIGURE 3 Indexes of Crop Production
(1970 constant prices, three-year moving
averages, semi-log scale)

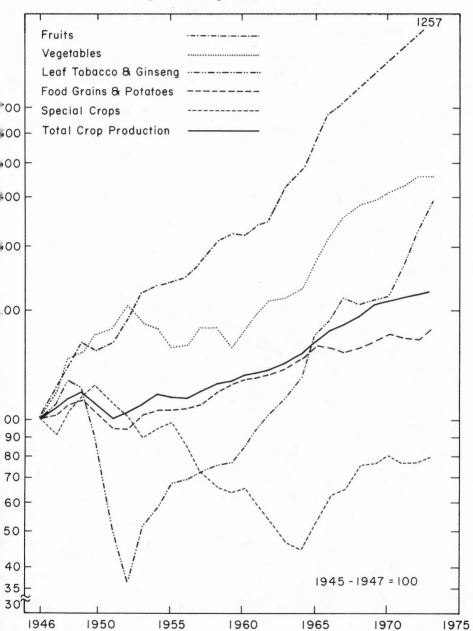

FIGURE 4 Indexes of Food-Grain Production
(1970 constant prices, three-year moving
averages, semi-log scale)

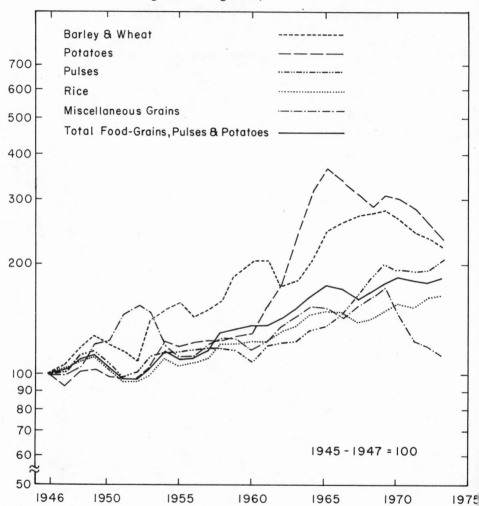

similar to that of total crop production because the share of crops in total agricultural production is dominant. For the whole period from 1946 to 1973, total agricultural production has grown at an annual compound rate of 3.41 percent, while total crops grew at 3.01 percent per annum. Silkworm cocoon and livestock and its products, in contrast, grew at 6.43 and 6.52 percent respectively during the same period, or at more than twice the rate of crop production.

The crop production growth rate declined from 3.69 percent per year during the 1954–1965 period to 2.22 percent during the 1965–1973 period. Furthermore, the growth of crop production, which had been accelerating during the pre-1965 period from 1.95 percent per annum for the 1954–1960 period to 5.82 percent per year for the 1960–1965 period, declined from 2.43 percent for the 1965–1970 period to 1.88 percent during the 1970–1973 period. There have been significant differences in the growth performances of the various crops. The production of food grains including potatoes, for example, grew rather moderately compared to the rate for fruits and vegetables.

Production of rice, the most important farm product in Korea, increased fairly steadily from 1954 on, although it decreased somewhat from 1956 to 1968, due mainly to a decrease in yield. The annual growth rate of rice production for 1954–1973 was 2.10 percent, slightly less than the population growth rate for the period. Prior to 1965, the increase in rice production reflected both increased crop area planted to rice and increased yield. The area planted to rice increased from 1,089,000 hectares in 1955 to 1,228,000 hectares in 1965. The yield per 10a (one-tenth of a hectare) also increased from 272 kilograms (in polished rice) to 311 kilograms from 1955 to 1965. For the post-1965 period, however, the main source of growth in rice production occurred through an increase in yield. In fact, the area planted to rice declined from 1,228,000 hectares in 1965 to 1,204,000 hectares in 1974. But the yield per 10a had increased to 369 kilograms by 1974. The gain in yield was mainly due to the wide adoption of high-yielding rice varieties such as *t'ongil* and the increased application of commercial fertilizer and other chemicals, and the improvement of irrigation facilities, a subject discussed at greater length in the next chapter.

Production of the barleys, the second major type of food grain in Korea, increased faster than rice, at 3.32 percent per annum for the whole period. However, most of this growth occurred prior to 1966. The annual growth rate fell from 4.38 percent for 1954–1965 to 1.82 percent for 1965–1970 and to –4.47 percent for 1970–1973. In fact, production of the barleys declined from 2,136,000 metric tons in 1965 to 1,851,000 metric tons in 1974. The reduced production was mainly due to the reduction of crop area planted to barleys from 1,201,000 hectares to 935,000 hectares, with only a very moderate rise in yields. A particularly notable decline in crop area occurred in common barley, wheat, and rye with crop area of naked barley remaining almost constant. Area planted to common barley declined from 541,000 hectares in 1965 to 375,000 hectares in 1974, wheat from 152,000 to 67,000 hectares, and rye from 26,000 to 8,000 hectares. The decline in production of the barleys in recent years can be attributed to a lack of incentives for farmers, resulting from a fall in their relative price and stagnant productivity. The market price of barley rose only 3.88 times from 1965 to 1974, compared to 4.54 times for rice, and 4.40 times for all grains including potatoes.

Production of miscellaneous cereals had the lowest rate of increase (0.59 percent per year) among commodity groups over the period as a whole. This low rate was due to a reduction in area planted to these crops from 219,000 hectares in 1955 to 81,000 hectares in 1974, because of their low level of land productivity compared to other crops. In spite of a substantial gain in yield of these cereals (from 39 kilograms per 10a in 1955–1957 to 115 kilograms in 1972–1974), the yield level is still much lower than other cereals in both quantity and value terms.

The production of pulses showed moderate growth over the whole period, with ups and downs in their growth rate in different periods. The main source of production increase was the increased yield, with crop area remaining at a nearly constant level. The yield has increased from 54 kilograms per 10a in 1955 to 105 kilograms in 1974.

Potatoes are a food supplemental to grain in the Korean diet, especially among rural and low-income urban families. Potato production increased during the period prior to 1965, due largely to an increase in crop area, from 326,000 metric tons (in polished weight) in 1960 to 1,045,000 metric tons in 1965. During the same period, the area planted to potatoes increased from 107,000 hectares to 213,000 hectares. Since 1965 both production and crop area have decreased as yields stagnated. Production had decreased to 543,000 metric tons and crop area to 123,000 hectares by 1974. The production of potatoes is sensitive to the nation's food situation, increasing rapidly when the nation is faced with a general food shortage.

The production of fruits has increased rapidly through the period studied. The demand for fruits and fruit prices have risen faster than for other field crops. Because income elasticity of demand for fruits is higher than for other crops, fruit demand has benefited most from the rise in per capita income.

Vegetable production has also increased rapidly. In the past, vegetables were grown chiefly for home consumption, on the open field during the later spring and autumn seasons. Recently the expanding production of high-quality vegetables for the market has become an important source of farmers' cash income. Newly developed vinyl greenhouses have made it possible to furnish fresh vegetables the year round.

Silkworm-cocoon production has been highly volatile. Production increased after World War II and dropped during the Korean War. During World War II, cocoon production had been discouraged by the Japanese Government General (in Korea), which encouraged the production of food-grains to meet increasing wartime demand. Cocoon production in all Korea declined from

22,713 metric tons in 1940 to 9,098 metric tons in 1944.[4] After the war, production had begun to increase, but with the outbreak of the Korean War production declined again. Another setback to production occurred in the 1955–1962 period within a few years of recovery from the Korean War, mainly reflecting the appearance of synthetic fiber substitutes for silk in the domestic market. The post-war nadir in cocoon production was 1960. Since then the production of cocoon and silk has spurted ahead dramatically, growing at an annual compound growth rate of 16.6 percent for the period from 1961 to 1973. This rapid growth in cocoon production was mainly a reflection of the increased foreign demand for Korean silk. Korea first exported raw silk amounting to about U.S. $4 million in 1961; and exports of cocoon, raw silk, and related products amounted to U.S. $7.4 million in 1965, U.S. $36.6 million in 1970, and U.S. $62.8 million in 1974, accounting for a large share of total agricultural exports. The exports of raw silk (here and hereafter including cocoon, silk waste, and silk yarn) accounted for 40.2 percent of total agricultural exports in 1965, 54.1 percent in 1970, and 38.4 percent in 1974.

In the early stages of silk export, the main importer was the United States, which accounted for 81 percent of total Korean raw silk exports in 1965. When Japan shifted from net exporter to net importer of raw silk in the 1960s, Korean raw silk partially filled the vacuum left in the international market, and in the 1970s Japan itself became a major customer for Korean raw silk.

For the whole study period, the production of livestock and livestock products grew at 6.52 percent per year. This growth rate is more than twice as high as that for crop production. A particularly dramatic increase in the production occurred just after the Korean War, with an annual compound growth rate of over 33.7 percent for the period from 1952 to 1954, reflecting recovery from wartime conditions. A 9.26 percent per year rate was achieved from 1954 to 1960 and an overall rate of slightly less than 6 percent per year for the whole 1954–1974 period.

It is notable that the production of livestock products has increased much faster than livestock production. The production of livestock products increased at 14.53 percent per year for the whole three-decade period, while livestock production increased at 5.28 percent per annum. This high growth rate in the production of livestock and its products reflected the rapid increase in demand for animal protein by domestic consumers. As per capita income increases, consumers spend relatively more of their food bill on protein, indicating a higher income elasticity of demand.

Some structural change in the major livestock in Korea is occurring. Although the native Korean breed is still predominant, the number of dairy cattle and the production of milk have been increasing rapidly in recent years, and a shift in production from native cattle to dairy and beef cattle is further expected as per capita income increases and mechanization of agriculture progresses. Dairy and beef cattle production, as well as that of hogs and poultry, depends in part upon imported feed in addition to by-products from domestic crop production.

The differences in growth rates among various farm products over the years have altered the relative compositions of total agricultural production, as shown in Table 17. All crops accounted for 88 percent of the value of gross agricultural production in 1955–1957, but for only 82 percent in 1972–1974. Food crops, including potatoes, accounted for 71 percent of the value of gross agricultural production in 1955–1957, but for only 60 percent in 1972–1974. Meanwhile the shares of other commodity groups increased: monopoly crops from 1.5 to 4.2 percent, fruits from 1.8 to 3.3 percent, livestock and livestock products from 11.8 to 15.1 percent, and cocoon from 0.6 to 2.5 percent.

TABLE 17 Percentage Composition of Agricultural Production
by Commodity Groups, 1955–1974
(current prices)

Commodities	Year			
	1955–1957	1962–1964	1969–1971	1972–1974
All Crops	87.54	89.96	83.53	82.48
Food Grains	71.30	77.05	61.81	59.92
Rice	49.46	49.20	40.56	41.59
Barleys	12.83	16.16	12.67	10.43
Miscellaneous crops	0.87	1.12	0.56	0.44
Potatoes	5.37	8.38	5.49	4.89
Pulses	2.77	2.19	2.53	2.57
Monopoly crops	1.46	1.83	2.47	4.19
Fruits	1.84	1.83	2.98	3.32
Vegetables	9.39	7.21	13.80	9.69
Special crops	2.42	0.87	1.39	1.70
Crop by-products	1.14	1.17	1.08	3.66
Livestock & its products	11.77	9.52	14.94	15.05
Livestock	9.98	7.56	10.97	11.32
Livestock products	1.78	1.96	3.97	3.73
Cocoon	0.57	0.36	1.42	2.47
Nursery stock	0.12	0.17	0.11	–
Total	100.00	100.00	100.00	100.00

TRENDS IN AGRICULTURAL INPUTS

The growth of agricultural output depends on changes in the principal factors of production (land, labor, and capital) and the rate of technical change. This section attempts: 1) to measure the changes in factor inputs over the three-decade period; and 2) to measure the growth rate of both partial and total productivities. As subsequent analysis will make clear, "technological change" or increases in factor productivity account for more of the rise in output than do increases in the quantity of factor inputs. In this chapter, however, no attempt will be made to explain the nature and sources of the rise in factor productivity. That is a subject left to later chapters, but, given the limited knowledge we have of productivity change anywhere in the world, never really satisfactorily "explained."

The indexes of major factors of production and aggregate input are depicted in Figure 5, and growth rates of inputs are presented in Table 18. Of the two primary inputs to agricultural production—land and labor—cultivated land area increased only 15 percent over the thirty-year study period—an average annual compound rate of 0.53 percent.

From 1946 to 1952 the area of total cultivated land remained at almost a constant level due to a negative growth rate of 1.19 percent per year during the Korean War period. Total cultivated land increased at an annual compound rate of about 0.53 percent from 1952 to 1960. From 1960 to 1968 this rate accelerated to 1.68 percent per year with cultivated upland growing much more rapidly than the area of paddy—2.88 percent per year versus 0.81 percent. The rapid expansion of cultivated upland followed the passage of the Land Reclamation Act in 1962 and was motivated by the increased demand for fruits and other cash crops, on the one hand, and by a government subsidy program using U.S. PL 480 grains for reclamation on the other. Since 1968, the area of both cultivated upland and paddy have been declining at rates of 0.98 percent and 0.37 percent per year from 1968 to 1973 respectively. This indicates that

FIGURE 5 Indexes of Inputs to Agriculture
(1970 constant prices, three-year moving averages,
semi-log scale)

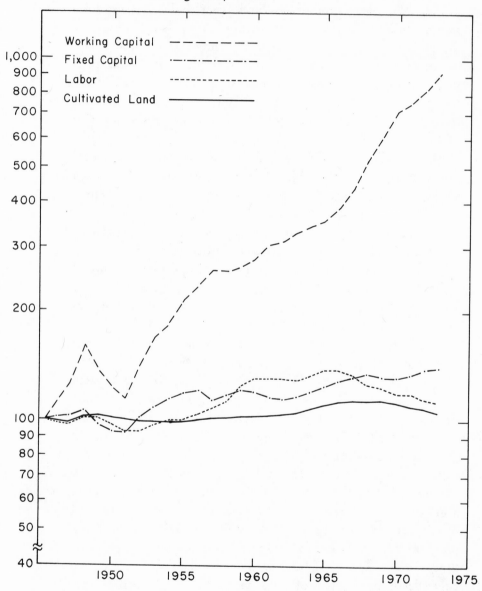

TABLE 18 Growth Rates of Inputs to Agriculture, 1946–1973
(%)

| Period | Inputs | | | | |
	Cultivated Land	Labor Used	Fixed Capital	Current Inputs	Total Input
1946–1952	0.0012	–0.87	0.54	6.92	2.15
1946–1949	1.20	0.49	–0.08	13.08	3.13
1949–1952	–1.19	–2.21	1.18	1.10	1.17
1952–1954	0.52	3.25	7.23	12.46	3.42
1954–1973	0.70	0.42	1.03	8.72	1.72
1954–1965	1.21	3.16	0.95	6.06	2.51
1954–1960	0.53	5.23	0.78	6.89	2.98
1960–1965	2.04	0.74	1.16	5.07	1.95
1965–1973	0.0027	–3.23	1.15	12.48	0.65
1965–1970	0.47	–2.57	0.99	14.46	1.10
1970–1973	–0.77	–4.33	1.40	9.25	–0.10
Whole Period 1946–1973	0.53	0.34	1.37	8.59	1.94

Note: Growth rates are annual compound rates of increase between three-year averages of the data centered within the years shown. For a more detailed breakdown of the components of fixed capital and current inputs, see Appendix B.

additional cultivated land brought about by reclamation was not enough to compensate for that converted for urban development, industrial sites, and highway construction.

Labor input, estimated in terms of man-equivalent units of labor actually used in agricultural production, increased at a rate of 0.49 percent per year between 1946–1949 and declined at a rate of 2.21 percent per year during the period of the Korean War. The agricultural labor force increased markedly after the Korean conflict due to the influx of refugees from North Korea and of workers displaced from devastated urban sectors, in addition to the natural growth of the rural population. The agricultural labor force increased at a rate of 5.23 percent per year during the period 1954 to 1960, but this rate declined to 0.65 percent per year between 1960 and 1966 as the rehabilitation of industries and the expansion of the urban sector began to absorb workers. Since 1966, the agricultural labor input has been consistently declining at an average rate of 3.71 percent per year.

Working capital, or non-farm current input, comprises expenditures for chemical fertilizer, chemicals used to control insects and diseases (that is, insecticides, pesticides, and herbicides), purchased seeds, farm tools, and other minor farming materials. Expenditure for purchased feed, estimated on the basis of imports of feed and by-products of imported food-grains, has also been included.

Working capital increased the most rapidly among all inputs, at an average annual rate of 8.59 percent over the whole study period. After a setback during the Korean War, non-farm current input increased rapidly—at an average annual rate of 9.83—during the 1951–1973 period. The most rapid increase occurred after 1965 with the increased application of chemical fertilizer, pesticides, and minor farming materials.

The various non-farm current inputs are depicted in Figure 6, and the growth rates are presented in Table 19. All the non-farm current inputs showed an increase after World War II and a decrease during the Korean War. After the Korean War, pesticides increased the most rapidly of all non-farm current inputs—at an average annual rate of 26.40 percent during the period between 1954 and 1973. The application of fertilizer increased steadily at slightly less than 8 percent per year. The consumption of non-farm produced feed increased rapidly as well—at 17 percent per year for the 1954–1973 period. This rapid increase reflected the rapid increase in demand for animal proteins in food consumption brought about by rising income levels and population. The demand for minor farming materials has increased very rapidly because of such innovations as the production of fresh vegetables during early and late winter in vinyl houses. This innovative production practice provides opportunity for additional cash income to farmers through the productive use of available

FIGURE 6 Indexes of Non-Farm Current Inputs

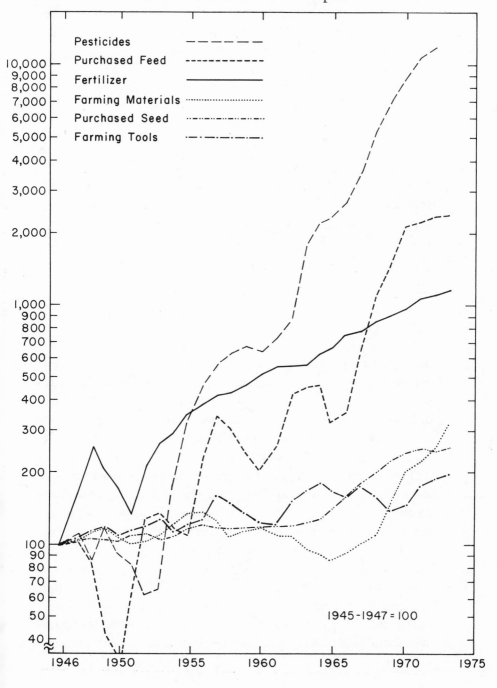

resources—land and labor—otherwise idle during the slack season. It is evident that Korean agriculture is becoming more dependent on purchased inputs as modern inputs are substituted for farm-supplied productive factors.

TABLE 19 Growth Rates of Various Non-Farm Current Inputs, 1946–1973

(%)

Period	Ferti- lizer	Pesti- cides	Minor farm tools	Farming materials	Seeds	Feeds	Total
1946–1952	13.17	–6.07	3.79	0.56	1.37	5.54	6.92
1952–1954	19.26	65.83	–2.96	11.70	2.82	–7.44	12.47
1954–1965	7.62	26.76	3.18	–3.61	2.45	9.60	6.06
1965–1973	8.36	25.90	2.28	17.70	6.66	27.91	12.48
1954–1973	7.93	26.40	2.80	4.85	4.20	19.96	8.71

Fixed capital input is measured as the sum of depreciation charges on farm machinery and equipment, perennial fruit trees, and farm buildings; the service value of draft animals; and irrigation fees. The average annual growth rate over the study period as a whole was 1.37 percent, higher than for either land or labor input. The most rapid increase in fixed capital occurred immediately after the Korean War—an annual growth rate of 7.23 percent from 1952 to 1954.

Since then, fixed capital has grown steadily with an average annual growth rate from 1954 to 1973 of 1.03 percent. The composition of fixed capital has changed in recent years with the rapid increase in farm machinery, such as the power tiller, as substitutes for draft animals and labor.

The growth of non-farm current input since 1970 has been slower than in the period from 1960 to 1970, whereas the opposite is true of fixed capital. This implies that, while land-saving technology has received and is still receiving the main emphasis, labor-saving technology has become increasingly important in recent years, a theme taken up at greater length in Chapter 4.

GROWTH OF PRODUCTIVITIES

AGGREGATE INPUT AND
AGGREGATE PRODUCTIVITY

Inputs have been aggregated by the factor-share approach. However, alternate ways of estimating factor shares are possible, and the aggregate input index (and therefore the aggregate productivity index) will differ depending on the choice of factor-share weights. For .the 1945–1971 period, factor shares estimated in an earlier study were applied with a slight adjustment.[5] For the 1972–1974 period, factor shares were estimated according to the same procedures used for previous years.

The total input index was calculated employing the following chain-linked index formula:

$$I_t = I_{t\text{-}1} \sum_i w_i, t\text{-}1 \frac{q_{it}}{q_i, t\text{-}1} \quad (t = 1,2,3, \ldots) \quad (3.1)$$

where

I_t = index of total input in year t

q_{it} = index of input i in year t

w_{it} = factor share of input in year t.

The trends of total production, total output, total input, and total productivity, both in terms of total production and total output, are compared in Figures 7 and 8. The corresponding growth rates and the relative contribution of increases of input and improvements in productivity are shown in Tables 20 and 21 respectively.

For the whole period, total input grew at an annual rate of 1.94 percent and total productivity at 1.44 percent. Therefore, according to this method of estimation, about 57 percent of total production growth is attributable to the increase of

FIGURE 7 Indexes of Total Production, Total Inputs,
and Total Productivity
(three-year moving averages, semi-log scale)

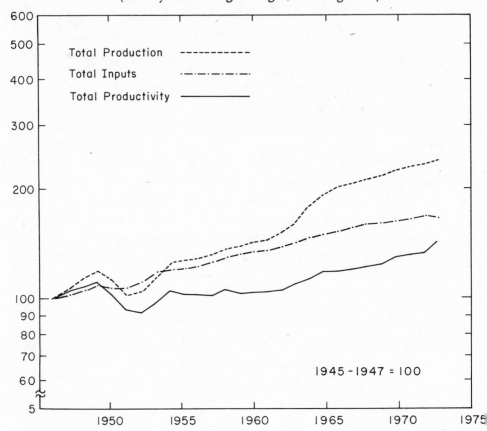

FIGURE 8 Indexes of Total Output, Total Inputs,
and Total Productivity
(three-year moving averages, semi-log scale)

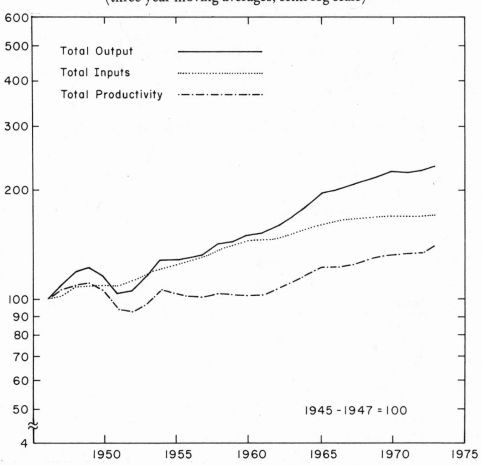

input and the remaining 42 percent to improvement in productivity.

The relative contributions are quite different for different time periods. During 1954–1973, the period after the Korean War, the growth of input accounted for about 50 percent of the growth of production, the remaining 50 percent of total production growth attributable to improvement in productivity. The relative contribution of improvement in productivity had been accelerating over the period. During 1954–1960, total productivity declined at an annual rate of 0.22 percent per year but,

TABLE 20 Growth Rates of Production, Input, and
Productivity
(measured on the total production basis[a])
(%)

	Growth Rate			Relative Contributions	
Period	Production (1)	Input (2)	Productivity (3)	Input (2)/(1)	Productivity (3)/(1)
1946–1952	0.83	2.15	–1.29	259	–155
1946–1949	6.82	3.13	3.57	46	52
1949–1952	–4.83	1.17	–5.92	24	–123
1952–1954	10.87	3.42	7.20	31	66
1954–1965	3.99	2.51	1.44	63	36
1954–1960	2.75	2.98	–0.22	108	–8
1960–1965	5.49	1.95	3.47	36	62
1965–1973	2.79	0.65	2.13	23	76
1965–1970	3.13	1.10	2.01	35	64
1970–1973	2.24	–0.10	2.35	–4	105
1954–1973	3.48	1.72	1.73	49	50
1946–1973	3.41	1.94	1.44	57	42

Note: [a]Total input includes non-farm current input

after 1960, total productivity grew significantly at 3.47 percent per year and accounted for 63 percent of the growth of production from 1960 to 1965. Although the growth rate of productivity declined to 2.13 percent per year from 1965 to 1970, its relative contribution to the growth of production remained at almost the same level—64 percent. From 1970 to 1973, inputs declined at 0.1 percent per year due to the reduction of cultivated land and labor input. In spite of the negative growth in total input, total production grew at an annual rate of 2.24 percent all of which is attributable to productivity gains. This accelerating growth in productivity reflects the productivity impact of

TABLE 21 Growth Rates of Output, Input, and Productivity
(measured on the total output basis[a])
(%)

| | Growth rates | | | Relative Contributions | |
Period	Output (1)	Input (2)	Productivity (3)	Input (2)/(1)	Productivity (3)/(1)
1946–1952	0.88	2.15	–1.24	244.3	140.9
1946–1949	6.97	3.13	3.73	44.9	53.5
1949–1952	–4.87	1.17	–5.96	–24.0	122.4
1952–1954	10.54	3.42	6.88	32.4	65.3
1954–1965	3.89	2.51	1.35	64.5	34.7
1954–1960	2.53	2.98	–0.45	117.8	–17.8
1960–1965	5.55	1.95	3.54	35.1	63.8
1965–1973	2.23	0.65	1.57	29.1	70.4
1965–1970	2.84	1.10	1.72	38.7	60.6
1970–1973	1.21	–0.10	1.33	–8.3	109.9
1954–1973	3.19	1.72	1.44	53.9	45.1
1946–1973	3.21	1.94	1.23	60.4	38.3

Note: [a]Total input includes non-farm current input

the substitution of modern technical inputs for land and labor on an agricultural output whose composition has been shifting away from grain toward livestock, silk cocoon, and market-oriented crops.

TRENDS IN PARTIAL PRODUCTIVITIES

A partial productivity index measures the relationship between output and a single input. Therefore, a partial productivity measure neglects inter-factor substitution. Use of labor productivity as a measure of the progress of technological change is likely to result in upward bias because of the changing input mix (capital-using production process) and output mix (capital-intensive enterprise). Partial productivities, however, are convenient indicators of the efficiency of production. Furthermore, a partial productivity is a good measure of technological progress if technological change is neutral—that is, if the marginal rate of substitution is constant over the study period. Labor productivity also has merit as an indicator of a rise in the standard of living.

The trends of labor and land productivities in Korean agriculture are depicted in Figure 9, and the corresponding growth rates are shown in Table 22. The productivities of both inputs have increased almost at the same rate for the whole period—labor productivity at 2.85 percent and land productivity at 2.65 percent per year. Both productivities have had positive growth rates in most sub-periods except during the Korean War, and also the 1954–1960 period in the case of labor productivity.

The increase in land productivity has been brought about by biological innovation, improvement of land quality—especially expansion of irrigation facilities, increased application of commercial fertilizer and of chemicals to control diseases and insects, and changes of product mix—a shift from crops like the miscellaneous cereals that have a relatively low value of product per unit of land to products of higher value, such as vegetables, fruits, and livestock. As discussed at greater length in Chapter 4, the level of current input per unit of land has increased

FIGURE 9 Indexes of Labor Productivity and Land Productivity
(three-year moving averages)

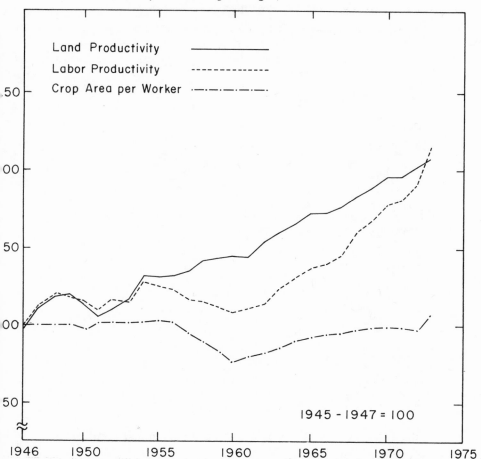

TABLE 22 Growth Rates of Labor and Land Productivities
(measured on the total output basis)
(%)

| Period | Productivity Growth Rates | | Relative Contribution |
	Labor (1)	Land (2)	(2)/(1)
1946–1952	1.77	0.88	50
1946–1949	6.44	5.67	88
1949–1952	–2.70	–3.69	136
1952–1954	7.04	9.99	142
1954–1973	2.76	2.47	89
1954–1965	0.70	2.64	377
1954–1960	–2.55	1.98	–
1960–1965	4.76	3.44	72
1965–1973	5.65	2.23	39
1965–1970	5.57	2.37	42
1970–1973	5.80	2.00	34
Whole period 1946–1973	2.85	2.65	93

substantially, with the consumption of chemical fertilizer and pesticides increasing rapidly.

Labor productivity has been increasing at an accelerating rate since 1954 in contrast to the almost constant growth rate of land productivity. Labor productivity, however, declined at 2.55 percent per year for the period from 1954 to 1960, while land productivity increased at 1.98 percent annually during the same period, mainly due to the deterioration of the man-land ratio brought about by the in-migration of labor force to rural areas after the Korean War. During this period, labor input grew at 5.23 percent per annum compared to 0.53 percent in cultivated land. The index of cultivated land (labor-land ratio) declined from 98 in 1954 to 74 in 1960 (1945–1946 = 100). After 1960, labor productivity grew more rapidly than land productivity, and the growth rate has been accelerating. Labor productivity grew at 4.76 percent per annum compared to 3.44 percent in land productivity from 1960 to 1965 and rose to 5.65 percent for the period from 1965 to 1973 compared to 2.23 percent for land productivity.

Growth in labor productivity depends upon the quantity and productivity of other resources associated with a given quantity of labor and upon technological progress resulting from improvement in the quality of labor force itself. As previously mentioned, the man-land ratio deteriorated up to the early 1960s. Since 1962, however, the man-land ratio has begun to improve, and the index of cultivated land per worker has increased from 74 in 1960 to 103 in 1974 (1945–1946 = 100). Therefore, the growth in labor productivity depended both upon the growth in land productivity and the improvement in the labor-land ratio, with the relative contribution of land productivity to the growth of labor productivity declining over the past 15 years. A more detailed description and analysis of the underlying sources of these trends is undertaken in Chapter 4.

ESTIMATION OF AN AGGREGATE AGRICULTURAL PRODUCTION FUNCTION

It is also possible to attempt to estimate the contribution of various inputs to the rise in agricultural production through the use of an aggregate production function. The Cobb-Douglas form of the production function was chosen and the results from estimating slightly different forms of the basic function are presented in Table 23.

Since all four equations were estimated from time-series data, it is no surprise that the inclusion of a dummy variable for weather improves the overall "fit" of the equation and the

TABLE 23 Aggregate Agricultural Production Function, 1955–1974
(dependent variable is total production)

Equation T-1

$$InY = 2.7462 + 0.6799InLP + 0.0633InNA + 0.3813InFC + 0.0029InWC + 0.0280T - 0.0734D$$

S.E. (3.8916) (0.1300)   (0.0890)   (0.1694)   (0.0819)   (0.00716) (0.01547)

T.V. 0.7057 5.2289**   0.7115   2.2515*   0.0348   3.9083** -4.7465**

R^2=0.992   D.W.=2.1877   S.E. of Reg.=0.0252   F. Statistics (6,12.)=248.047

Equation T-2

$$InY = 5.9444 + 0.5423InLP + 0.07912InNA + 0.2402InFC + 0.0691InWC + 0.02658T$$

S.E. (5.8319) (0.18998)   (0.1384)   (0.2508)   (0.1248)   (0.01112)

T.V. 1.0193 2.8545*   0.57176   0.95785   0.55375   2.3909*

R^2=0.9775   D.W.=2.1952   S.E. of Reg.=0.04064   F. Statistics (5.,13.)=112.8

Equation T-3

$$InY = 5.5790 + 0.7697InLP + 0.1059InNA + 0.1165InWFC + 0.0260T - 0.0705D$$

S.E. (4.6724) (0.1661)   (0.1119)   (0.1561)   (0.0085)   (0.0146)

T.V. 1.1940 4.6328**   0.9462   0.7465   3.040** -4.8399**

R^2=0.9903   D.W.=1.9239   F. Statistics=264.12

TABLE 23 (continued)

Equation T-4

$lnY = 5.5450 + 0.6338lnLP + 0.1339lnNA + 0.2207lnWFC + 0.02268T$
$S.E.$ (5.9491) (0.1790)   (0.1535)   (0.2090)   (0.0113)
$T.V.$ 0.9321 3.5411**   0.8723   1.0558   2.0043

$R^2 = 0.9775$ $D.W. = 2.1311$ $F.$ Statistics $(4.,14.) = 151.794$

where:

Y = Gross agricultural production in 1,000 wŏn at 1970 constant prices
O = Gross agricultural output in 1,000 wŏn at 1970 constant prices
LP = Area of cultivated land in 1,000 hectares
NA = Labor used in agricultural production in 1,000-man-equivalent units
FC = Capital service prices for fixed capital items in 1,000 wŏn at 1970 constant prices
WC = Expenditure on current inputs in 1,000 wŏn at 1970 constant prices
WFC = $WC + FC$
T = Time in years
D = Weather dummy variable: normal year 0, poor rice crop year 1
$D.W.$ = Durbin-Watson statistics
$S.E.$ = Standard error of regression
* = Significant at the 5% level
** = Significant at the 1% level

significance of some of the individual coefficients. In very general terms, there is also consistency between these production function estimates and the previous estimates of the contributions of productivity change to the rise in agricultural production. Specifically, the time trend (T) coefficient is statistically significant and large enough to account for a substantial part of the increase in output. It is also reasonable that the coefficient for land, which is in short supply, should be larger than that of labor, which, at least for much of the period, was in relative abundance.

Unfortunately there are also major problems with these estimates, problems sufficiently serious to reduce if not eliminate their usefulness. In the next chapter, for example, evidence will be marshaled to indicate that the marginal product of labor in agriculture must have been roughly equal to rural wage rates from the mid-1960s on. And yet the marginal product of labor figures that can be derived from these equations suggests that labor surplus conditions (marginal product < wage) prevailed throughout the period.

Even more serious is the statistically insignificant and very low (near zero) coefficient for working capital. Most agricultural technicians who have observed the Korean countryside over the past two decades agree that the rapid increase in availability of such key current inputs as chemical fertilizer and pesticides has had a major impact on farm output, and this theme will be stressed in Chapter 4. And yet the estimates in equations T-1 through T-4 indicate that chemical fertilizer has had little or no impact on agricultural production.

There are various possible explanations for why these estimation procedures lead to implausible results in some cases. The impact of rising amounts of chemical fertilizer, for example, may have been lagged over several years, because it takes farmers time to learn the proper use of increased amounts. Fertilizer must also be used in combination with sufficient and timely supplies of water, so that irrigation (included in fixed capital) and fertilizer (included in working capital) are complements rather than substitutes, a situation that contradicts the basic Cobb-Douglas assumption of a unitary elasticity of substitution between factor inputs.

One could try to make adjustments in the equations presented here in an attempt to take into account these and many other potential sources of bias. But, with only 20 observations for each variable, it is doubtful that the effort would be justified. A priori knowledge, rather than formal tests of significance, would be the only way of judging between alternative formulations of the production function. Under such circumstances it makes more sense to proceed to analysis that will provide one with that a priori knowledge. This is what is done in the next chapter.

FOUR

The Sources of Agricultural Growth

In formal terms, our estimates of agricultural productivity and agricultural production functions provide us with information on what caused the growth described in the previous chapter. It is important, however, to get behind these formal estimates to a more in-depth analysis of the increases in farm inputs and changes in agricultural technology. What follows, therefore, is an item-by-item discussion of the major changes (or lack of changes) that have occurred over the past thirty years in Korean agricultural inputs and technology.

As the discussion develops, it will become clear that it is much easier to go in depth into the contribution of increases in factor inputs to rising output than it is to say anything concrete about the sources of the rise in productivity. In part, rises in productivity are embodied in the steadily improving quality of modern inputs, but many sources of increased productivity derive from such difficult-to-quantify improvements as better-educated farmers and more conscientious and better-trained extension workers. In this chapter, the main focus is on the rise in farm inputs. Changes and improvements in the policies and services that make for increased productivity will be a recurring theme in the chapters that follow, particularly Chapters 5 and 9.

LABOR AND MECHANIZATION

It is impossible to grow anything on a farm without the expenditure of a great deal of human effort, but it is a widely held view that there is more labor in rural East Asia than is needed for agricultural production. In their extreme form, "labor surplus" theories argue that large numbers of workers can be removed from agriculture without affecting production at all. Modified versions of these theories stress that all farm labor is productive, but that the marginal product of the rural work force is below the level needed either to sustain life or to meet some minimal and customary standard of living or rural wage. From the perspective of the analysis in this chapter, the key point is that, if surplus labor conditions exist, an increase in the farm population and work force will not lead to increases in output sufficient to provide that population with a minimal standard of living. Only substantial increases in other inputs (land and capital) or subsidies from the urban sector can then keep farmers from sinking into an abyss of deprivation, and worse.

If any country is a labor-surplus country, Korea would seem to be one. As already indicated, Korea has one of the lowest amounts of arable land per capita anywhere in the world. But the evidence in support of the existence of a labor surplus is weak, to say the least. A surplus may have existed in the late 1940s and the 1950s except, of course, during the Korean War when there was undoubtedly a shortage. But whatever surplus existed seems to have disappeared by the early to mid-1960s.

There are various ways of attempting to determine the existence or lack of a labor surplus. The most direct method is to go out and measure the number of hours an average farm family works in a given year and compare that figure with the number of working hours available to that family. Such a procedure will not tell one much about the marginal product of labor, but it will establish whether there is outright rural unemployment, even if it is disguised by work-sharing arrangements. This method was used by Yong Sam Cho, whose study was based on the 1959 Bank of Korea Farm Household Survey. Although his estimates involved a number of assumptions and the 1959 Farm Household Survey was small in size, the findings are reasonably conclusive. In brief, they are that, in the peak seasons (mainly June, July, and October), there is little or no unemployment or underemployment of family or wage farm workers.[1] In the off seasons, however, particularly in January and February, farmers used less than half of the total amount of work time available to them. Thus, there was a major seasonal unemployment problem in Korea, but the removal of farm workers on a year-round basis would have created (in 1959) a farm labor shortage.

The seasonal pattern of labor utilization in agriculture has changed little since 1959 in the sense that the peak seasons require six or seven times the number of labor hours per farm as the slack seasons.[2] But, since the mid-1960s, there appears to have been a fairly steady and substantial decline in the number of labor hours used annually per farm (not per farm household; see Table 24). This decline has occurred simultaneously with a rise in the acreage per farm worker and an absence of any significant change in acreage per farm household. Thus, in no sense can this decline in number of labor hours per farm be explained in terms of a rising labor surplus as more and more farmers have crowded onto a fixed amount of arable land. Since one can also rule out the possibility of an increasing farmer preference for leisure, it appears, instead, that farm workers are finding more to occupy their time outside of agriculture. We are

TABLE 24 The Changing Land-Labor Ratio, 1962–1975

Year	Farm Household Size (persons)	Workers Per Household (persons)	On-Farm Labor Hours Per Farm Household (hours)	Hired and Exchange Workers on Farm (hours)	Cultivated Land Per Household (hectares)	Acreage Per Farm Worker (hectares)
1933	6.38	—	—	—	1.483	—
1962	6.32	3.39	2,536	722	.892	.263
1963	6.39	3.19	2,662	620	.897	.281
1964	6.44	3.27	2,116	613	.909	.278
1965	6.29	3.15	2,585	722	.959	.305
1966	6.22	3.12	2,557	747	.948	.304
1967	6.12	3.12	2,418	705	.975	.312
1968	6.02	3.00	2,213	589	.999	.333
1969	5.99	2.96	2,181	591	.996	.336
1970	5.92	2.91	2,155	534	.988	.340
1971	5.83	2.92	2,218	497	.995	.341
1972	5.71	2.98	2,075	495	.994	.333
1973	5.72	2.93	2,060	480	.999	.341
1974	5.66	2.86	1,651	396	.967	.338
1975	5.63	2.86	1,708	399	.962	.336

Source: MAF, *Report on the Results of Farm Household Economic Survey, 1974*, pp. 44, 50, 84.

not talking here about migration to the cities, since the data refer to activities of people who have stayed on the farm, but about non-farm activities that can be pursued while retaining residence in the village. These outside activities presumably occur during the peak as well as the slack season, since the decline in work hours per farm has occurred in all months and with all major crops (except vegetables).

Unfortunately, farm household labor data are no longer reported in a way that allows one to make a calculation for the 1970s similar to that of Cho's for 1959.[3] But the above evidence all suggests a growing shortage of labor on the farm during the peak seasons. Another possibility, of course, is that changes in farming practice, notably an increase in mechanization, freed labor for other tasks. But the evidence respecting increases in the level of mechanization does not support this latter interpretation.

Korea, from the 1950s on, appears to have been going through a fairly typical (for Asia) pattern of farm mechanization. The first stage usually involves the mechanization of grain processing. Done by hand, grain processing consumes vast amounts of (mainly female) labor. This first stage began long before 1945 and appears to have been virtually completed by the early to mid-1960s (see Table 25). By that time, a large number of hours had been freed either for other "household" chores or for work on and off the farm. Since the 1960s, improvements in grain-processing machinery have made it possible to do the work required with a reduced number of machines. The second stage of farm mechanization involves the use of power equipment to substitute for human labor in key peak-season activities. Power threshers and power pumps are among the first kinds of equipment to be introduced at this stage. Moving water to the fields with wooden scoops and hand- or animal-driven water wheels uses great amounts of labor and moves relatively small amounts of water. A power pump can move far greater amounts of water to much higher levels at a low cost. Threshing also takes up much time, often (at least

TABLE 25 Agricultural Mechanization, 1951–1975

Year	Grain-Processing Machinery[a] (no.)	Power Pumps (no.)	Power Threshers (no.)	Power Tillers (no.)	Agricultural Motors (1,000 h.p.)
1935 (All Korea)	n.a.	1,341	–	–	33.5
1951	40,111	n.a.	93	–	n.a.
1952	52,236	n.a.	594	–	n.a.
1953	58,164	n.a.	627	–	n.a.
1954	59,816	n.a.	1,703	–	n.a.
1955	62,209	n.a.	850	–	n.a.
1956	67,192	7,463	1,370	–	n.a.
1957	(50,790)	6,618	1,637	–	n.a.
1958	(57,570)	7,097	2,162	–	n.a.
1959	(62,357)	6,460	3,236	– (161)	n.a.
1960	(64,054)	6,911	3,886	12 (153)	n.a.
1961	(66,079)	6,561	4,794	30 (107)	n.a.
1962	91,487	12,292	8,022	93 (148)	n.a.
1963	(68,533)	13,171	9,495	386	n.a.
1964	(84,056)	15,350	14,610	653	376.6
1965	(89,516)	26,029	18,909	1,111	358.3
1966	126,365	29,929	22,338	1,555	521.5
1967	126,941	31,613	25,474	3,819	572.7
1968	126,556	37,796	26,675	6,225	624.6
1969	121,029	49,534	33,878	8,832	670.6
1970	127,909	54,078	41,038	11,884	636.3
1971	120,299	57,896	63,350	16,842	685.3
1972	119,980	60,616	75,532	24,786	698.2
1973	113,072	61,193	85,161	37,660	704.6
1974	111,467	62,863	108,494	60,056	736.0
1975	109,166	65,993	127,105	85,722	679.8

Source: MAF, *Yearbook of Agriculture and Forestry Statistics 1961, 1963, 1968, 1975, 1976.*

Note: [a]excludes barley-cleaning machinery.

where two crops a year are grown) when that time is needed for transplanting the next crop. Korea appears to have begun this second stage of mechanization in a major way in the mid-1960s, and the process was well along but far from completed in the mid-1970s. The increase in threshers, pumps, and similar kinds of equipment (for example, power sprayers) presumably facilitated the decline in labor hours per farm over the past decade. A third stage, the introduction of hand and larger tractors, only began in earnest in Korea in the 1970s. Even with considerable amounts of sharing between families, Korean agriculture could easily make use of several hundred thousand power tillers.

In some countries, mechanization has proceeded quite a long way even though it has been clear that the introduction of machines was putting increasing numbers of people out of work, people who had few alternative occupations. The incentives for such antisocial behavior have ranged from fear of land reform, hence the desire of large landowners to farm as much of their own land as possible, to artificially low prices for farm machinery. In Korea, land reform has long since been completed (see Chapter 10), and there is an upper limit of 3 chŏngbo (3 hectares) on the size of farms. Mechanization appears to be a response to a real shortage of labor at key times, not to the special requirements of a few wealthy landowners.

So far this discussion has dealt mainly with the question of disguised unemployment or the absence thereof. But, as already indicated, of equal or greater interest is whether surplus labor exists in the broader sense where the marginal product of labor is above zero but below an institutionally determined wage. Data in Tables 26 and 27 are presented in an effort to get at this question. Table 26 simply indicates that the average productivity of labor in agricultural activities has been rising. In fact, average productivity would have risen even more than indicated here if we had divided by labor hours spent in farming rather than by the numbers of farm workers. With average productivity rising, it is more than likely that marginal productivity was also rising.[4]

Ideally, one would like to calculate the marginal product of labor directly from a formally estimated production function. Such calculations are possible by using the production functions presented in Chapter 3, but the results are clearly unsatisfactory. The marginal product figures that can be derived from these equations are less than half the rural wage in the 1970s and an even lower percentage of the rural wage in the 1960s. As the discussion that follows will show, such a result is not plausible and merely reflects the biases in the estimated production function itself.

TABLE 26 Agricultural Land and Labor Productivity, 1953–1975

Year	*(1)* *Agricultural Value* *Added per Farm* *Worker* *(1,000 1970 wŏn)*	*(2)* *Cultivated* *Acreage* *per Worker* *(hectares)*	*(3)* *Value Added* *per Hectare* *(1,000 1970 wŏn)*
1953	50.8	.256	198.4
1955	56.9	.265	214.2
1960	56.5	.254	222.4
1965	72.8	.286 (.305)	254.8
1970	93.7	.320 (.340)	292.7
1974	103.9	.317 (.338)	327.0
1975	117.9	–	351.3

Sources and Methodology: (1) Agriculture and forestry (excluding fishery) value added divided by the number of farm workers. The latter figure was obtained by multiplying the number of farm families by the number of farm workers per household as estimated by the Farm Household Survey. The number of workers per household in 1953, 1955, and 1960 was assumed to be the same as in 1962 (3.39) and 1975 was assumed to be the same as 1974 (2.86). (2) Obtained by dividing the number of farm workers into the total cultivated acreage. Figures in parentheses are the estimates of cultivated acreage per worker from the Farm Household Survey.

Studies of urban wages do yield some support for the view that a labor surplus may have existed prior to 1960 but had largely disappeared by the mid-1960s. When an economy has a surplus of labor, the supply of labor to the urban sector is usually

TABLE 27 Rural-Urban Wage Differentials, Males and Females, 1959–1975

Year	(1) Monthly Wage in Manufac-turing	(2) Adult Male Farm Wage Per Month (daily wage X 25)	(3) Monthly Wage in Textile Manufac-turing	(4) Adult Female Farm Wage Per Month (daily X 26)	(5) (1)/(2)	(6) (3)/(4)
1959	2,350	2,440	1,930	1,562	0.96	1.24
1960	2,330	2,408	2,290	1,543	0.97	1.48
1961	2,610	2,653	2,470	1,677	0.98	1.47
1962	2,780	2,875	2,460	1,828	0.97	1.35
1963	3,180	3,575	2,830	2,366	0.89	1.20
1964	3,880	4,975	3,440	3,224	0.78	1.07
1965	4,600	5,525	4,060	3,666	0.83	1.11
1966	5,420	6,400	4,670	4,290	0.85	1.09
1967	6,640	7,675	6,050	5,382	0.87	1.12
1968	8,400	9,525	7,090	6,760	0.88	1.05
1969	11,590	11,575	9,110	8,216	1.00	1.11
1970	14,561	14,475	11,223	10,192	1.01	1.10
1971	17,349	17,375	13,124	12,272	1.00	1.07
1972	20,104	20,075	15,837	14,352	1.00	1.10
1973	22,330	22,150	18,322	16,120	1.01	1.14
1974	30,209	28,525	25,756	20,748	1.06	1.24
1975	38,220	36,675	31,255	27,144	1.04	1.15

Sources: (1) and (3) EPB, *Korea Statistical Yearbooks,* and BOK, *Economic Statistics Yearbooks.* (2) and (4) NACF, *Agricultural Yearbooks* and MAF, *Yearbook of Agriculture and Forestry Statistics.* The average number of days worked per month in manufacturing was about 25, whereas in the textile industry it was about 26 days.

perfectly elastic. Urban wages in real terms, as a result, unless they are propped up by union activities and the like, will tend to stagnate. In societies with relatively perfect labor markets, therefore, urban wages do not rise significantly. In Korea, however, as Roger Sedjo has shown, real urban wages began a marked upward trend in the 1963–1964 period.[5] It is also notable that, from about 1964 on, rural and urban wages have drawn closer together for both males and females.[6] In fact, rural-urban wages are so close that one wonders what happened to the usual rural-urban cost of living differential that one finds in comparisons of this sort.

The key point, however, is that the close correlation of rural and urban wages suggests that workers in both sectors were part of interconnected labor markets. Rising demand for labor in one market required employers of labor in the other market either to raise wages or lose their labor force. If rural labor had been in surplus, farmers hiring outside labor could employ all they required without having to match the wage increases of urban employers (and vice versa).

All evidence, therefore, points in the direction of a rural labor marginal product that was positive throughout the post-Korean War period and equal to farm wages at least from the mid-1960s on. It follows that the rise in the farm labor force up into the mid-1960s made a positive and substantial contribution to farm output. From the mid-1960s through 1975, in contrast, a declining farm labor force was a retarding force on agricultural output.

CULTIVATED ACREAGE

Unlike labor, increases in the acreage under cultivation have not played a significant role in the rise in farm output during the past three decades. Expansion of the acreage under cultivation has been stated to be a major objective in each of Korea's first three five-year plans and yet, toward the end of the third plan,

Korea's total cultivated acreage was only 10 percent above that of 1960 (see Table 28). More significant, total cultivated acreage in 1974 was little if any higher than it was in the 1920s and 1930s (Table 29). The Korean pattern, therefore, is much like that of the rest of East Asia. Centuries of slow but substantial population growth filled up the empty spaces until there was little room for further expansion. As the farm population continued to grow, an effort was made to move up the hillsides and to reclaim tidal lands, but mainly farms got smaller and smaller. The accelerated pace of industrialization and urbanization in the 1960s further exacerbated the problem by removing more and more good land from agricultural uses.

If each of the first three five-year plans called for acreage expansion, it was only in the first plan (1962–1966) that a truly major effort was made. The 260,000-hectare expansion in acreage in that period compares with a 90,000-hectare increase in the previous decade and a net 55,000-hectare *decline* between the end of 1966 and 1974.

The difference between the 1962–1966 period and the years before and after was that large quantities of PL 480 grain were committed to land reclamation in the early 1960s, but not since that time. The Korean government's own commitment to land reclamation, in contrast, has always been rather modest. Private land expansion efforts have existed in all periods but, for reasons that will become apparent, there are severe limits on what can be done without outside subsidies.

It has not proved possible to reconstruct in quantitative terms the full magnitude of U.S. Operations Mission to Korea's (USOM/K's) commitment to land reclamation in the early 1960s, but it was considerable. Of 350,179 tons of wheat flour provided under PL 480-Title II in 1964 through 1966, for example, over 110,000 tons went for upland development and tideland reclamation.[7] PL 480 grain was never the sole source of funding for these projects, of course. The farmers were expected to make a substantial contribution of their own labor and tools equivalent to anywhere from 30 to 60 percent of the total investment in the project according to the accounting conventions used in the reclamation programs at that time. There were also provincial government subsidies in some cases. If a 1967 plan to develop 11,000 hectares of new uplands was typical, 110,000 tons of grain would have made possible the

TABLE 28 Cultivated Acreage, 1952–1975
(1,000 hectares)

Year	Total	Paddy	Upland
1952	1,942.5	1,153.4	789.1
1953	1,939.5	1,152.4	787.1
1954	1,950.4	1,161.0	789.4
1955	1,994.8	1,187.4	807.5
1956	1,991.9	1,188.8	803.1
1957	1,998.8	1,192.9	805.9
1958	2,012.4	1,199.7	812.7
1959	2,016.2	1,202.9	813.3
1960	2,024.8	1,206.3	818.6
1961	2,032.6	1,210.9	821.7
1962	2,062.7	1,223.1	839.6
1963	2,079.7	1,228.1	851.5
1964	2,171.0	1,261.1	909.9
1965	2,256.4	1,286.2	970.2
1966	2,293.1	1,287.1	1,006.0
1967	2,311.9	1,290.5	1,021.4
1968	2,318.8	1,289.3	1,029.5
1969	2,311.2	1,283.0	1,028.1
1970	2,316.7	1,283.6	1,033.1
1971	2,271.3	1,264.8	1,006.5
1972	2,242.3	1,259.4	982.8
1973	2,241.3	1,262.6	978.6
1974	2,238.4	1,268.9	969.5
1975	2,239.7	1,276.6	963.1

Source: MAF, Agricultural Development Corporation, *Yearbook of Land and Water Development Statistics 1975*, pp. 16–17.

TABLE 29 Cultivated Acreage in Six Provinces, 1918–1974
(1,000 hectares)

Year	North Ch'ungch'ŏng	South Ch'ungch'ŏng	Province North Chŏlla	South Chŏlla	North Kyŏngsang	South Kyŏngsang	Total
1918–1929	158	244	235	409	390	279	1,715
1930–1937	159	248	240	420	383	278	1,728
1938–1945	159	250	246	432	383	277	1,747
1952–1959	143	236	239	366	344	354	1,682
1960–1963	150	256	239	380	351	259	1,635
1967	197	295	259	386	393	274	1,804
1974	176	292	249	361	382	264	1,724
1975	176	292	250	352	381	263	1,724

Source: The 1918–1963 data were derived from Jin Hwan Park, *An Economic Analysis of Land Development Activities in Korea* (Department of Agricultural Economics, Seoul National University, 1969), p. 8. The 1967 figures are from MAF, *Yearbook of Agriculture and Forestry Statistics 1968*, p. 66. The 1974 figures are from MAF, Agricultural Development Corporation, *Yearbook of Land and Water Development Statistics 1975*, p. 16.

development of over 140,000 hectares of new upland.[8] The Second Five-Year Plan called for the opening up of 200,000 hectares of new uplands with PL 480 to bear 67 percent of the total cost (315,000 metric tons of grain).[9] The Second Plan, therefore, would have turned land development into even more of an aid-based effort. The first year of the Plan, however, coincided with a USOM/K evaluation report that was critical of the Korean government for failure to develop an adequate land use and development policy and program even after many years of assistance in this area. The report recommended that future assistance in this area be limited to advisory services only.[10] Whether or not this report was the main cause, 1967 appears to mark the end of large PL 480 support for new land development.

The problem was not that it was physically impossible to find land suitable for reclamation, although such land was not plentiful. One estimate of the upland area suitable for conversion into cultivated land as of the early 1970s was 200,000 hectares. The remaining 6.5 million hectares of "forest" land were on slopes too steep (over 24 degrees), or in areas too remote from roads and people, or were retained in forest for conservation purposes. Of the 200,000 hectares available, over one-third (74,000 hectares) were suitable only for pasture.[11] To this potentially developable upland must be added the tideland areas suitable for reclamation, a figure for which a national estimate is unavailable.

It is not difficult to understand why many land reclamation efforts depended so heavily on subsidies, but others would appear to be profitable without subsidies. Large tideland development projects were particularly expensive. One project initiated in 1963 (and not completed) involved investment of 520 wŏn per p'yŏng (approximately 1/3,000 of a hectare) when existing paddy land in that region was selling for only 250–300 wŏn per p'yŏng. Small tideland reclamation projects fared much better. In one sample of nine projects, the average investment cost per p'yŏng (including contributed labor) was 77 wŏn or about half the unit price of paddy land in those areas. And upland development was the most economical of all. Investment costs in the 1960s on these fields were only 24 to 28 wŏn (only 9 wŏn if farmer labor is excluded) against an average price of upland at that time of about 100 wŏn per p'yŏng.[12]

Why then, when the subsidies ended, did all tideland and upland development fall off so sharply? In part the answer is that some of the farmers who owned forest land already farmed as much land (about 2 hectares) as their family labor could handle. To cultivate more land they had to hire labor from outside, and hiring labor for upland was economical only if the new land was planted to cash crops.[13]

But rational economic calculations of this kind do not appear to be the only explanation for why land development depended on subsidies. In one village, for example, corruption and mismanagement sapped the morale of villagers involved in a tideland reclamation project. Even though the project was subsidized with PL 480 grain, once the initial enthusiasm and momentum were lost, the whole effort fell apart, and the tides washed away what had already been accomplished.[14]

In many projects, the question of who benefits and who actually does the work of reclamation is crucial. Given the large numbers of rural poor, including many families with little or no land of their own, it should not be difficult to find enough people who will work simply for the privilege of owning their own farm at the end of the process. All one has to do is identify the right people, lend them enough money (at interest) to tide them over the period of development, and sit back and watch land development proceed. But, of course, it doesn't work this way.

The landless poor don't develop the uplands because they don't own the forest land that could be converted to crops. They can't even rent the land once developed or work on it for wages, as already pointed out, because often the land isn't productive enough to provide a return for both the landlord and the tenant (or farm laborer). Tideland areas may not be privately owned, but the villagers who live beside them look on them as their own. Experiments in Korea to bring in the poor from the outside to develop and settle on the land have met with determined and usually successful resistance from the villagers already there.[15] Those villagers may be too busy or too prosperous to bother themselves with the grueling labor that is involved, but they have nothing to gain from turning over their "rights" to outsiders. An economist might say that the solution was to compensate the villagers for their "loss." But how do you calculate the magnitude of this "loss," and what would it do to the cost of the project?

There is no neat solution to these problems. Where the returns on land development are very high, no doubt solutions have been found, with or without outside subsidies. But most land that remains to be developed cannot promise such high returns. For the latter kind of land, high costs or institutional barriers of one sort or another often make it necessary to resort to subsidies. When the subsidies ran out in the late 1960s, therefore, land development dropped off sharply.

The implications of low returns on newly developed land, of course, go beyond the issue of subsidies. If the yields are generally low on such land, it follows that renewed efforts to expand the cultivated acreage are not the key to the solution of meeting Korea's long-term demand for agricultural products. Tables 30 and 31 present data on how new upland (the main form of newly opened land) has been used and the yields that have been achieved on that land. Only in the case of soybeans were the yields on newly developed land higher than on existing land. In fact, if all new upland had been planted to grain, the 24 percent increase in the upland area in the years 1962–1967 would have led to an increase in total national grain output of no more than 8 to 10 percent.[16] The development and planting to grain, where possible, of all remaining suitable upland would have an even smaller impact on total grain output.

In practice, as already mentioned and as the data in Table 30 make clear, only about half the new upland has been planted to

TABLE 30 Crops Planted on New Upland During
First Three Years After Development
(%)

	First Year	Second Year	Third Year
Wheat and barley	7.9	17.9	18.8
Summer grains (soybeans, corn, sweet potatoes, etc.)	19.0	36.7	35.1
Mulberry trees	7.5	15.1	14.1
Fruit trees	2.9	7.5	10.5
Other	62.7	22.8	21.5
Total	100.0	100.0	100.0

Source: These figures are based on a sample of 925 farms reported in Jin Hwan Park, pp. 90–96.

TABLE 31 Comparison of Yields on Existing and New Upland
(kg. per tanbo)

	Existing Upland	New Upland	
		3 Years After Development	5 Years After Development
Wheat	203.6	164.2	190.0
Barley	290.0	165.3	146.7
Sweet potatoes	1,912.0	716.3	1,097.3
Soybeans	57.0	63.5	69.1
Tobacco	173.6	112.4	122.9

Source: Jin Hwan Park, p. 121.

grain (as compared to 75 percent of existing upland). The other
half of the new upland has been planted to cash crops and mulberry and fruit trees. Because the value per hectare of these cash
crops is higher than in the case of grain, the impact of new land
on the value of agricultural output has been greater than the
above percentages based on a calculation in terms of grain alone
would suggest.

Inclusion of past and potential tideland reclamation areas

would further increase the total impact of new land development on agricultural output, but not in any dramatic way. The 75,000 hectares of new paddy land created in the first half of the 1960s probably increased total grain output by no more than 3 or 4 percent.

To summarize, opening up of new land has had a significant impact on Korean agricultural development in only one period, the first half of the 1960s. Although there have been numerous barriers of one sort or another placed in the way of land development efforts, the main problem has been that there isn't much unused land left that can be put into crops at an acceptable cost. It has even been a problem to prevent a net decline in the cultivated acreage because good farmland is steadily being alienated to industrial and urban uses. Nor has a resort to increasing levels of double-cropping been a solution to the lack of land. The double-cropping index in 1975 was no higher than it was in the early 1960s.

FIXED CAPITAL AND AGRICULTURAL GROWTH CYCLES

In conventional economic analysis, the usual practice is to separate land and capital, and to treat them as inherently different kinds of inputs or factors of production. But it should be apparent from the previous discussion that new land, at least, has many of the same characteristics as capital. The value or price of newly developed tidelands, for example, derives not so much from properties inherent in the soil, but from the labor and other inputs used to convert a part of the sea into a rice paddy. In fact, most agricultural land in Korea and other densely populated East Asian nations should be thought of more as a capital input than a "natural resource." Korean farmers did not migrate to the Korean equivalent of the State of Iowa, sink their plows in the soil, and begin to grow corn. Unlike mid-western American cornfields, rice paddies must be constructed.

The fields must be level (or some of the rice will be under water); there must be ridges of land constructed between the fields, which require frequent maintenance (or the water won't stay in the fields at the proper level); and, ideally, paddy fields have a guaranteed supply of water. This latter "requirement" may involve anything from the construction of a pond with ditches connecting the pond with paddy fields to a large dam and a major network of canals and ditches.

Even in Korea, of course, agricultural land has some inherent or natural resource value. Some notion of this inherent value can be obtained by comparing the prices of fully developed and undeveloped farm land. The average price of paddy in 1974 (according to the Farm Household Survey) was 1,010 wŏn per p'yŏng (3025 p'yŏng = one hectare). Upland was worth 530 wŏn per p'yŏng, but forest land was only 70 wŏn per p'yŏng. Since it is difficult for city people to buy farmland legally, it is likely that these prices are determined in large part by the productivity of the land in agricultural uses, not mainly by the land's recreational, security, or prestige value. The value of forest land, however, undoubtedly understates the value of completely undeveloped potential farmland, because much forest land cannot be so developed. Unfortunately there are no prices for "completely undeveloped potential farmland" in Korea because little or no such land exists. Even much still undeveloped upland is, in a sense, partly developed, since much such land is not far from roads, farm buildings, electricity, and other forms of infrastructure needed to make that upland productive. A precise estimate of that portion of land that is capital and that is a "natural resource," therefore, is not possible without more research and analysis, but there is every reason to believe that the capital component is several times the value of the "natural resource" component.

Confusion over the relationship between land and capital has contributed to various unrealistic notions about the productivity of capital in agriculture. The capital-output ratio, for example, although it is not really a measure of the return to

TABLE 32 Fixed Capital Formation in Agriculture, 1953–1975
national accounts data
(billions of 1970 wŏn)

	Gross Fixed Capital Formation	Consumption of Fixed Capital	Net Fixed Capital Formation	Fixed Capital Stock
1953	6.18	2.45	3.73	90.19
1954	5.34	2.58	2.76	92.95
1955	6.80	2.24	4.56	97.51
1956	7.73	2.92	4.81	102.32
1957	10.17	2.64	7.53	109.85
1958	8.35	2.95	5.40	115.25
1959	9.48	2.86	6.62	121.87
1960	11.10	3.51	7.59	129.46
1961	14.27	3.23	11.04	140.50
1962	11.24	4.60	6.64	147.14
1963	17.44	4.35	13.09	160.23
1964	19.25	7.45	11.08	172.03
1965	23.74	7.27	16.47	188.50
1966	35.08	7.67	27.41	215.91
1967	31.48	9.69	21.79	237.70
1968	35.30	11.03	24.27	261.97
1969	39.91	13.76	26.15	288.12
1970	52.37	15.37	37.00	325.12
1971	55.79	18.45	37.34	362.46
1972	71.97	21.43	50.36	412.82
1973	75.54	25.81	49.73	462.55
1974	98.42	27.71	70.71	533.26
1975	123.22			

Source: These data are those estimated by the BOK (column 1) as adjusted by
Wontack Hong, Factor Supply and Factor Intensity of Trade in Korea. We
have converted Hong's figures from dollars back into 1970 wŏn at 310.6 wŏn
= $1 (Hong's rate).

capital, is nevertheless often used by planners as a proxy for the rate of return. When one excludes land, the capital-output ratio (aggregate or incremental) is invariably very low. One estimate of the capital stock in Korean agriculture in 1968, for example,

TABLE 33 Fixed Capital Formation in Agriculture—
The Missing Elements, 1953–1974
(billions of 1970 wǒn)

	(1) Fixed Capital Stock (net of land)	(2) 80% of Value of Cultivated Acreage	(3) Fixed Capital Stock (incl. land) (1) + (2)
1953	90.19	985.65	1,075.84
1954	92.95	991.19	1,084.14
1955	97.51	1,013.76	1,111.27
1956	102.32	1,012.28	1,114.60
1957	109.85	1,015.79	1,125.64
1958	115.25	1,022.70	1,137.95
1959	121.87	1,024.63	1,146.50
1960	129.46	1,029.00	1,158.46
1961	140.50	1,032.97	1,173.47
1962	147.14	1,048.26	1,195.40
1963	160.23	1,056.90	1,217.13
1964	172.03	1,103.30	1,275.33
1965	188.50	1,146.70	1,335.20
1966	215.91	1,165.35	1,381.26
1967	237.70	1,174.91	1,412.61
1968	261.97	1,178.41	1,440.38
1969	288.12	1,174.55	1,462.67
1970	325.12	1,177.35	1,502.47
1971	362.46	1,154.27	1,516.73
1972	412.82	1,139.54	1,552.36
1973	462.55	1,139.03	1,601.58
1974	533.26	1,137.55	1,670.81

Note: The average price of cultivated land in 1970, according to the Farm Household Survey, was 210 wǒn per p'yǒng or 635, 250 wǒn per hectare.

is 303.9 billion wŏn (in 1968 prices). The resulting capital to gross value of agricultural output and agricultural value added ratios are 0.57 and 0.72 respectively. Similarly, if one uses national accounts data to estimate the incremental increases in the capital stock (see Table 32), the resulting incremental ratios are also very low. The impression left by these ratios is that major increases in agricultural investment will lead to large gains in farm output.

If one includes land in the capital stock, the capital-output ratios change dramatically. Household data in Table 34, for example, suggest a ratio of anywhere from 4:1 to 7:1. The rate of increase in the national capital stock also falls dramatically if one includes land. Without land, the annual rate of increase between 1953 and 1974 was 8.8 percent but, with the land included, the rate falls to 2.1 percent (see Table 33). The incremental capital-output ratio remains low, since the *increase* in land has been small and investment in *new* land is included in the additions to capital, even though existing land is not included in the capital stock. But does this low incremental ratio reflect high returns to capital, or is most of the output side of this ratio explained mainly by increases in inputs other than capital?

For an individual contemplating an investment in a Korean farm, the largest part of that investment goes for the purchase of land. If land is not a reproducible asset but a "national resource," then what is capital to the individual is not capital to society taken as a whole. But if, as suggested above, most of the value of farmland comes from the investment in that land, then capital for the individual is also capital for society.

Which, then, is correct? Is the rate of return on capital in agriculture high as suggested by low capital-output ratios or is it much lower in conformance with the experience of individual investors? A precise answer is not possible, but what evidence exists suggests that the return is generally low. Specific projects, to be sure, can have a high rate of return. A feeder road into a village that previously depended on walking paths, for example,

TABLE 34 Asset-Income Ratios in Agriculture, 1962–1975

	(1) Fixed Assets of Farm Households	(2) Agricultural Receipts (excl. inventory apprec.)	(3) (1)/(2)	(4) Farm Agricultural Income (excl. asset apprec.)	(5) (1)/(4)
1962	217,891	64,817	3.36		
1963	327,223	90,545	3.61	66,162	4.95
1964	323,493	110,293	2.93	85,966	3.76
1965	396,224	113,629	3.49	86,450	4.58
1966	440,993	119,580	3.69	89,603	4.92
1967	496,285	138,693	3.58	104,057	4.77
1968	547,405	149,082	3.67	108,935	5.03
1969	616,643	187,534	3.29	140,045	4.40
1970	743,032	222,667	3.34	168,640	4.41
1971	858,270	284,542	3.02	219,884	3.90
1972	1,204,824	346,898	3.47	272,285	4.42
1973	1,783,382	398,407	4.48	308,464	5.78
1974	2,737,872	501,911	5.45	379,402	7.22
1975	3,963,203				

Source: MAF, Report on the Result of Farm Household Economic Surveys.

may have a very high return, but that is because all the other forms of capital needed to make that village productive (buildings, irrigation systems, and so on) were already in place and only the road was needed for the village to use the full potential of its resources.

The road example brings out the essential complementarity of many kinds of agricultural fixed capital. Nowhere is this clearer than in the relationship between land and water. To grow crops one must have land and water; either alone is obviously insufficient. For many crops, timely rainfall is an adequate source of water, but rice paddies require large amounts of water at just the right time. Rainfall can and does supply some of this need, but it is an inadequate and unreliable source. Irrigation systems, therefore, play a key role. Investment in irrigation, however, is subject to diminishing returns. The construction of an irrigation system where previously there was none may make

FIGURE 10 Diagram of Land and Water as Complementary Inputs

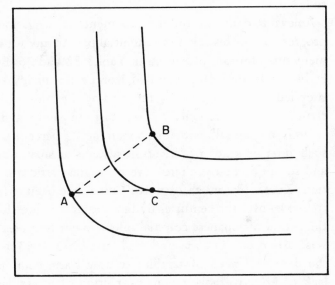

Investment in Irrigation

it possible to convert the acreage from dry-land crops to paddy. The return on such an investment is likely to be high. But further investment in that system will make more marginal improvements, increasing the efficiency with which the water reaches the fields or making the system resistant to abnormally long drought, for example. The ability to overcome abnormally long drought may affect productivity in only one year in five or one in ten. The return on investment in such improvements, as a result, is likely to be low.

This essential relationship is illustrated by Figure 10. If land is fixed, investment in irrigation will move one along a line from A to C. By the time C has been reached, the return on further investment in irrigation will have fallen to zero. If the cultivated acreage is expanding together with continued investment in irrigation (movement from A to B), in contrast, the return to these inputs jointly will remain high. Korea, like the rest of East Asia, is moving along a line more like A to C than A to B. Cultivated land is nearly fixed and has been so for a long time, while investment in irrigation has been proceeding for a long time.

Relevant data on the development of irrigation and the progress of reforestation (an essential part of any water management program) are presented in Tables 35 and 36. Even by the early 1950s, over 40 percent of Korea's rice paddy acreage was supplied by an irrigation system and by the mid-1970s, this had risen to 70 percent. Only 7 percent of the paddy land depended entirely on rainfall, with the remaining 23 percent supplied by such inferior kinds of irrigation systems as small ponds, wells, and so on. Similarly, there have been great strides in reforestation. Korea, like much of north China, had been stripped of its tree cover over the centuries by families in search of fuel. During the period of Japanese colonial rule, a major reforestation effort was carried out, but was undone in the 1945–1953 period when the desperate need of families to stay warm combined with a lack of effective government protection of forest land worked once again to strip the hillsides bare. Except in more remote

TABLE 35 The Expansion of Irrigation of Paddy Fields
(1,000 hectares)

| | | Irrigated | | Non-Irrigated | |
	Total Paddy Acreage	Benefited by Irrigation Association	Completely Irrigated Area	Partially Irrigated	Rain Field
1952	1,226.3	183.7	351.3	278.8	412.6
1956	1,093.2	197.8	340.8	279.4	275.2
1960	1,202.9	236.4	392.7	285.5	288.3
1965	1,198.9	281.2	421.1	298.7	197.9
1970	1,183.5	304.1	543.7	223.2	112.6
1974	1,268.9	309.1	583.5	288.6	87.8
1975	1,276.6				

Sources: NACF, *Agricultural Yearbook 1961, 1968, 1975*; MAF, *Yearbook of Agriculture and Forestry Statistics 1952, 1961, 1976.*

TABLE 36 Reforestation

(1,000 hectares)

	Denuded Forest Land	Wooded Area	Forest-type Land
1950	3,544	2,925	6,469
1955	3,318	3,353	6,671
1960	2,788	3,913	6,701
1965	1,244	5,369	6,614
1970	860	5,701	6,611
1974	730	5,930	6,641

Sources: MAF, *Yearbook of Agriculture and Forestry Statistics 1952*, pp. 232–233. This yearbook gives a figure for total forested land and the area "required to be reforested." "Wooded Area was derived as a residual." The 1956–1973 data are from MAF, *Yearbook of Agriculture and Forestry Statistics 1968, 1975*, and *1976*. The minor discrepancies between the totals in column 3 and the addition of columns 1 and 2 are due to "unenumerated forest land."

areas, Korea in the early 1950s was more brown than green. By the 1970s, however, reforestation had been largely completed. One can quarrel with the government's choice of trees (much of the planting was in varieties of little commercial value), but at

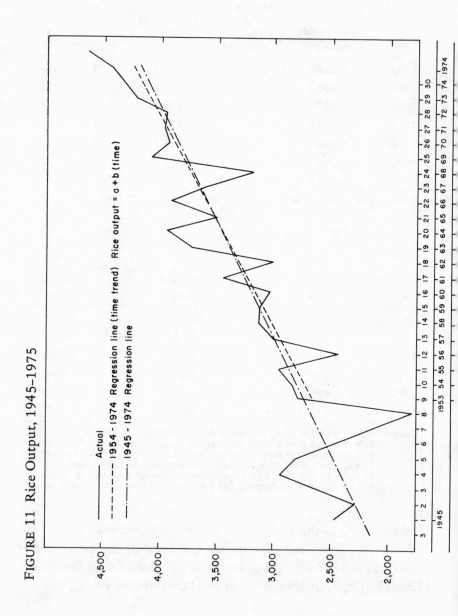

FIGURE 11 Rice Output, 1945–1975

least the worst problems of erosion and rapid runoff of water were brought under greater control.

Improved control over the supply of water has not only raised yields in good years; it also appears to have reduced some of the fluctuation in farm output from year to year. Firm conclusions in this respect may be premature, but data in Figures 11 and 12 suggest that the sharp fluctuations that characterized rice output and, to a lesser degree, farm output as a whole have been considerably reduced in the 1970s. Better weather may be part of the explanation, but the fact that the amount of rain-fed paddy land in 1972–1974 was only half the total of 1967–1968, the last major period of drought, must also be part of the explanation.

Korea, on the basis of a strong foundation from the period of Japanese colonial rule, therefore, has continued to make major advances in freeing agriculture from the vagaries of the weather. But the very magnitude of that progress when combined with a more or less fixed amount of arable land suggests that the return to further investments of this kind is low and falling. It does not follow that all investments in this area will have a low rate of return or that those with a low rate of return should not be built. It does suggest that the number of projects with a high rate of return will not be large and that others must be justified on grounds other than economic efficiency (the protection of farm household income from debilitating fluctuations, for example).

The discussion here has concentrated on water because, together with buildings and the development of arable land itself, water is the main form of rural capital or infrastructure. Roads are another form of rural capital and one where major changes have been of much more recent origin than in the case of water. Roads, however, can be more effectively discussed in the chapter on regional development, and hence are only mentioned here. Similarly, farm machinery, another form of fixed capital, was discussed above in connection with its role as a substitute for labor. The essential point, however, is that

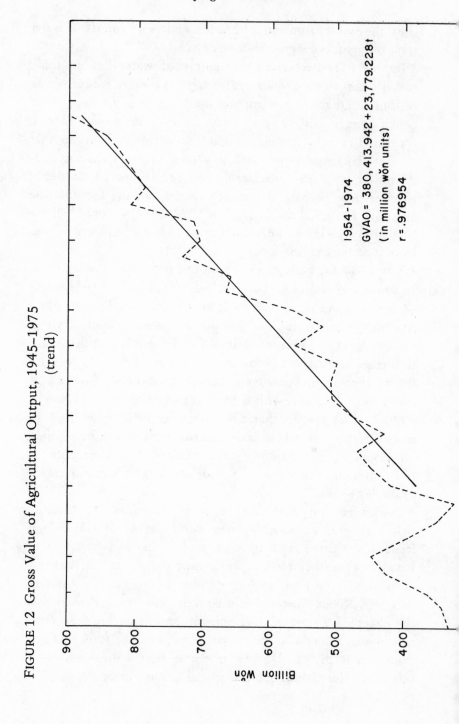

FIGURE 12 Gross Value of Agricultural Output, 1945–1975
(trend)

1954-1974
GVAO = 380, 413.942 + 23,779.228 t
(in million wŏn units)
r = .976954

Billion Wŏn

increases in fixed capital in rural Korea (including new land development) have not been a major source of rises in farm output. The problem has been less a question of lack of investment in fixed capital in this sector than of the limited opportunities for highly productive investment of this type.

CURRENT INPUTS

Despite the inconclusive results from the formal estimates in the previous chapter, it was the rapid rise in current inputs, not labor and fixed capital, that accounted for most of the rise in farm output, or at least that part of the rise that could be attributed to increased inputs at all. Many different kinds of current inputs were involved, ranging from imported feed for animals to a variety of chemical pesticides and fungicides. There was also a large increase in the use of electricity in rural areas from negligible levels in the late 1940s to 25.2 million kwh in 1962 and 56.6 million kwh in 1973.

The two current inputs of greatest importance, however, have been improved plant varieties (seeds) and chemical fertilizers, and on these we shall concentrate here. Improved plant varieties and fertilizer are basically complementary inputs; that is, without large increases in the availability of chemical fertilizer, little of the potential rise in productivity from the new varieties could be realized. Similarly, without the introduction of new varieties, the increased application of chemical fertilizer would face sharply diminishing returns.

Korean farmers, unlike those in much of the rest of the developing world, have had half a century of experience with modern methods of seed selection. The physical proximity of Japan made it comparatively easy to introduce plant varieties developed there into Korea, and the Japanese Government General through its extension service made vigorous efforts in this regard. Since independence was regained, there has been a gradual evolution in the rice and other crop plant varieties in use in Korea. The introduction of the *t'ongil* and *yusin* varieties of rice in the 1970s was thus a continuation of a process begun long before. The *t'ongil* and *yusin* varieties, however, differed from earlier varieties not so much in their potential impact on productivity (they made possible a 30 percent rise in yields) as in their origin. These newest rice seeds were based on varieties developed originally for tropical areas at the International Rice Research Institute in the Philippines.

Korean experience with chemical fertilizers like that with improved plant varieties also goes back long before 1945. As part of the effort to turn Korea into Japan's rice bowl, the Japanese Government General promoted the use of chemical fertilizers and the increased application of organic fertilizers. By the late 1930s, Korean farmers were using on the average over 20 kilograms of chemical fertilizer (mostly nitrogen) per hectare (see Table 37), not large by present-day standards, but enough to have a considerable impact on yields.

TABLE 37 Chemical Fertilizer Consumption 1915–1938 (all Korea)

| | Gross Weight (1,000 tons) | | Nutrient Content (N+P+K) | |
	Ammonium Sulphate	Other	Per 1,000 Tons	Per Hectare (kg.)
1915	0.06	0.47	0.1	0.02
1925	12.5	8.8	4.7	1.0
1932	164.6	67.8	59.8	12.4
1938	311.5	201.1	105.2	21.7

Source: Chōsen Sōtokufu, *Chōsen no nōgyō* (1936), pp. 148–150; and (1940), pp. 196–198. Conversions to nutrient content were made by the author. 1 ton=2000 pounds=910 kilograms.

The division of the country in 1945 left all the fertilizer plants north of the 38th parallel and also disrupted imports; but recovery of imports up to and beyond the Japanese colonial period levels was rapid. Even before the outbreak of hostilities in 1950, chemical fertilizer consumption per hectare was four times and more the level of the 1930s, and the Korean War itself proved to be only a brief interruption in the pattern (see Table 38). During the latter half of the 1950s, per hectare consumption held steadily at about 110 kilograms but, beginning in 1960, fertilizer consumption per hectare began a fairly steady rise that averaged 8 percent a year through 1974.

Throughout the 1950s and into the 1960s, U.S. economic assistance played a dominant role in these imports (see Table 39). Without U.S. aid and given the heavy demands on non-aid sources of foreign exchange, it can be safely assumed that the supplies available to Korean farmers would have been a small fraction of actually achieved levels. The impact on farm output would have been considerable. If the amount of nitrogen had been cut from 80 to 40 kilograms per hectare, for example, grain output may have fallen by well over half a million tons (10 percent or more of total grain output).[17]

But beginning in the early 1960s, the role of U.S. aid changed dramatically. No longer were aid resources the primary supplier of farm demand for nitrogen. Instead, aid funds went mainly for the import of nutrients (phosphates and potash), whose need in Korea was recognized by soil scientists but not by the farmers. Nitrogen was supplied first with Korean-earned foreign exchange and increasingly over time by newly built domestic fertilizer plants (see Table 40). By the early 1970s, Korean fertilizer output had reached a level that doubled the supply of nutrients to the farmers when compared with the early 1960s. Imports of nitrogen were virtually eliminated, with only potash continuing to be supplied mainly from abroad. For a brief period (mainly 1969–1972), Korea even became an exporter, albeit somewhat prematurely, of significant amounts of nitrogenous fertilizers.

Long-term farmer experience with chemical fertilizers, large commitments of aid funds, and in the 1960s the rapid pace of Korean industrialization, therefore, all combined to turn Korea within three decades into one of the world's major consumers

TABLE 38 Chemical Fertilizer Consumption, 1945–1975

	Total (1,000 metric tons)				Per Hectare of Arable Land (kg./hectare)	
	N	P	K	Total	Total	N
1945	1.242	0.749	0.346	2.337	1.1	0.6
1946	31.388	8.819	7.631	47.838	23.3	15.3
1947	84.723	21.574	1.090	107.387	52.3	41.3
1948	67.985	48.885	9.951	126.821	61.8	33.1
1949	132.692	39.518	28.090	200.300	97.6	64.6
1950	14.598	1.217	–	15.815	8.1	7.5
1951	50.404	19.523	–	69.927	36.0	26.0
1952	125.276	1.194	6.974	133.444	68.7	64.5
1953	90.647	19.341	2.659	112.647	58.1	46.7
1954	115.468	50.438	1.920	167.826	86.0	59.2
1955	146.476	28.218	8.847	183.541	92.0	73.4
1956	158.699	53.781	8.116	220.596	110.7	79.7
1957	143.939	68.520	6.547	219.006	109.6	72.0
1958	171.685	66.758	5.019	243.462	121.0	85.3
1959	161.786	57.241	6.017	225.044	111.6	80.2

TABLE 38 (continued)

| | Total | | | | Per Hectare of Arable Land | |
| | N | P | K | Total | Total | N |
	(1,000 metric tons)				(kg./hectare)	
1960	217.128	55.206	7.090	279.424	138.0	107.2
1961	210.867	80.788	16.839	308.494	151.8	103.8
1962	201.298	87.580	18.917	307.795	149.2	97.6
1963	191.729	94.371	20.995	307.095	175.1	92.2
1964	173.152	153.571	37.422	364.145	167.7	79.8
1965	217.925	123.489	51.684	393.098	174.2	96.6
1966	239.693	124.796	58.782	423.271	184.6	104.5
1967	277.556	132.722	76.213	486.491	210.4	120.1
1968	285.919	121.361	71.180	478.460	206.4	123.3
1969	320.223	130.749	83.717	534.689	231.4	138.5
1970	355.550	124.354	82.998	562.902	243.0	153.5
1971	347.318	165.030	92.789	605.137	266.4	152.9
1972	372.585	170.945	104.172	647.702	288.9	166.2
1973	411.236	232.176	149.796	793.208	353.9	183.5
1974	449.383	231.877	155.399	836.659	373.8	200.8
1975	481.524	237.637	167.047	886.208	395.7	215.0

Source: MAF, *Yearbook of Agriculture and Forestry Statistics*, various years.

TABLE 39 Imports and Exports of Chemical Fertilizer,
1951–1975

	Imports (1,000 metric tons)						Exports
	AID Financed			Korea Foreign Exchange			(1,000 metric tons
	N	P	K	N	P	K	of all nutrients)
1951	43.9	19.3	0	5.4	0	0	0
1952	91.1	0	7.0	33.2	0.1	0	0
1953	89.9	18.6	2.7	0.6	0.4	0	0
1954	96.4	50.2	1.9	18.7	0	0	0
1955	127.7	28.2	8.8	18.7	0	0	0
1956	157.3	53.8	8.1	50.9	4.2	0.4	0
1957	143.9	68.5	6.5	36.6	11.6	0	0
1958	171.7	66.8	5.0	75.5	15.1	0.7	0
1959	102.4	55.5	6.0	76.5	11.6	2.0	0
1960	199.7	55.2	7.3	n.a.	n.a.	n.a.	0
1961	138.7	74.4	16.8	41.4	6.4	0	0
1962	1.5	15.8	0	11.4	24.1	0	0
1963	.4	113.0	29.1	161.3	5.1	0	0
1964	0.4	131.5	32.7	84.9	7.2	2.8	0
1965	27.7	161.8	82.1	145.8	3.9	3.7	0
1966	18.8	164.2	136.5	156.2	5.2	5.6	0
1967	5.6	179.1	110.9	127.2	6.1	6.4	9.2
1968	1.9	70.5	68.4	111.1	3.0	10.1	11.5
1969	0.8	46.5	21.6	35.1	2.2	58.1	45.7
1970	0	0	0	2.4	2.4	1.9	55.8
1971	0	0	0	16.2	12.2	63.2	82.4
1972	0	0	0	11.3	10.1	93.4	88.0
1973	0	0	0	9.8	13.2	106.3	27.2
1974	0	0	0	2.8	26.1	122.3	0
1975	0	0	0	9.9	35.0	196.0	0.5

Sources: MAF, *Yearbook of Agriculture and Forestry Statistics,* various years.
USOM/K, *Rural Development Program Evaluation Report, Korea, 1967,*
p. 82.
Korean Traders Association, *Statistical Yearbook of Foreign Trade* (1969
through 1975 editions). All phosphate fertilizers imported were converted
to nutrient (P) at a rate of 20%. Compound fertilizer, both exports and
imports, was assumed to be 22-22-11.

TABLE 40 Domestic Production of Chemical Fertilizer,
1945–1975

(unit: M/T)

	N	P	K	Total
1945	424	215	111	750
1946	480	520	124	1,124
1947	141	1,386	172	1,699
1948	–	–	–	–
1949	–	–	–	–
1950	–	–	–	–
1951	–	200	–	200
1952	–	1,143	–	1,143
1953	162	285	–	447
1954	–	146	–	146
1955	–	442	–	442
1956	–	–	–	–
1957	–	–	–	–
1958	–	–	–	–
1959	–	–	–	–
1960	6,145	–	–	6,145
1961	29,814	–	–	29,814
1962	37,382	–	–	37,382
1963	44,895	–	–	44,895
1964	64,916	–	–	64,916
1965	75,271	–	–	75,271
1966	80,863	1,691	–	82,554
1967	155,694	21,231	9,513	186,438
1968	321,557	121,205	42,138	484,900
1969	366,517	145,918	48,589	561,024
1970	400,553	139,543	49,745	589,841
1971	408,001	144,686	46,785	599,472
1972	418,193	162,569	54,506	635,268
1973	447,255	159,292	65,172	671,719
1974	514,061	166,195	69,750	750,006
1975	582,740	195,475	81,509	859,724

Source: Various issues of MAF, *Yearbook of Agriculture and Forestry Statistics.*

of chemical fertilizers. By 1972 Korea was actually using more nitrogen per hectare than Japan (Table 41), although, if phosphates and potash are included, Japanese fertilizer consumption continued to surpass Korean levels, but not by a large margin.

Throughout most of the post-1945 period, Korea also

TABLE 41 Fertilizer Consumption—International
Comparisons, 1972
(kg. per hectare of arable land)

	N	P	K
Japan	138	135	113
United States	40	24	21
Mexico	19	6	1
India	11	4	2
Korea (1972)	166	76	46
(1975)	215	106	75

Source: These data were derived from data compiled by the FAO (except for the Korean data) and published in the 1975 and 1976 editions of the MAF *Yearbook of Agriculture and Forestry Statistics.*

TABLE 42 Organic Fertilizer Consumption, 1926–1962
(kg. per tanbo; average on total cultivated acreage)

	Farm Supplied Fertilizers (total)	Barnyard Manure	Night Soil
(all Korea)			
1926	476	308	116
1938	812	567	145
(South Korea only)			
1955	1,315	970	213
1960	1,362	1,039	196
1962	1,334	992	186

Sources: For 1926 and 1928, derived from Chōsen Sōtokufu, *Chōsen no nōgyō* (1936), pp. 151–152; (1940), pp. 200–201. Data in original table in tons which were converted to metric tons (0.909 metric tons = 1 ton). 1955–1962 figures are from the 1956, 1961, and 1963 editions of the MAF, *Yearbook of Agriculture and Forestry Statistics.*

remained a heavy user of organic fertilizers, particularly barn-yard manure and night soil. Statistics in this area, of course, are subject to a considerable margin of potential error, but it appears that Japanese efforts to push greater application of organic fertilizers in the 1920s and 1930s met with a considerable measure of success (Table 42). After 1945, use of organic materials continued at a comparable or higher level into the early 1960s. By the mid-1960s, however, the use of organic fertilizers began to decline (Table 43). The already mentioned reduction in the number of farm workers, together with the increasing availability of chemical substitutes, probably accounts for much of this decline. Aside from their aesthetic shortcomings, manure, night soil, and other forms of compost require large

TABLE 43 Organic Fertilizer Consumption, 1963–1970
(kg. per tanbo)

	Chemical Fertilizers	Urea	Farm Fertilizers	Barnyard Manure	Night Soil
	Barley				
1963	35.6	5.0	1,046	672	303
1966	30.7	9.7	1,175	763	343
1967	30.1	13.8	973	584	333
1968	33.1	16.0	821.1	503.8	263.7
1969	31.3	15.8	895.7	535.5	252.9
1970	30.3	15.4	875	575.1	242.8
	Rice				
1963	31.1	3.5	385	301	8
1966	34.8	8.8	334	260	7
1967	35.2	15.9	316.1	265.5	5.3
1968	37.2	11.0	325.8	272.2	2.7
1969	37.0	14.1	312.7	266.2	2.7
1970	34.5	14.1	338.7	289.0	4.3

Source: 1963–1971 MAF, *Report on the Results of Farm Household Economic Surveys.*

amounts of labor for preparation and for transport to the fields, particularly when compared with urea. Unfortunately, data on organic fertilizer use are unavailable for years after 1970, but it is likely that the decline in use has continued.

Precise estimates of the impact of increases in chemical fertilizer consumption on agricultural production are impossible. To be effective, fertilizer must be combined with other inputs, notably water and improved seed varieties, and the impact of fertilizer alone cannot easily be separated out when time series data are used. It is also necessary to know how much fertilizer was used on each kind of crop. Various attempts have been made, however, to estimate yield response functions for fertilizer, and these can be used to get some notion of the overall impact of increased supplies of these chemicals. On two major varieties of rice, for example, the difference in yields between maximum usable level of chemical fertilizer and no chemical fertilizer was a 20 to 30 percent difference in yields. On barley, the difference surpassed 50 percent.[18] With the introduction of the *t'ongil* varieties of rice in the early 1970s, the role of chemical fertilizers was further enhanced. Chemical fertilizers have also made an important contribution to the rapidly expanding cash crop sector, notably vegetables and fruits. Without the rapid increase in chemical fertilizer consumption, it can safely be argued, Korean agriculture would have grown at a rate far below that actually achieved.

The Korean experience with chemical fertilizers, however, has not been without its problems. Problems connected with setting appropriate fertilizer prices will be discussed in a subsequent section. A major shortcoming in the 1950s and well into the 1960s was the way the Korean government handled the distribution of fertilizer. The greater portion of chemical fertilizer distribution in this period was in the hands of the government, which in theory charged farmers a price well below what would have prevailed on an open market. In practice, government management was so chaotic that farmers seldom received fertilizer at the proper time or the proper place. To supplement government supplies, farmers had frequently to resort to private suppliers and black markets where prices reached double and triple those charged by the government. In the early 1960s Agency for International Development (AID) (then the International Cooperation Administration, ICA) encouraged an expansion of private as against government supplies, but to little avail.[19] Into the 1970s, fertilizer distribution was entirely in the hands of the agricultural cooperatives which, in effect, meant the government. Legal private sales had been totally eliminated. More effective and timely distribution policies, however, have reduced the black market to a more modest level.

The problem of distribution has not solely been a question of administrative difficulties and the elimination thereof. Farmers have been reluctant to use phosphates and potash at any price. Stories are even heard of farmers throwing away potash in order to use the bags. It is unclear why farmers have responded with such lack of enthusiasm for potash, but it may be because the correlation between use and yield is less obvious and less immediate than in the case of nitrogen. The government has attempted to counter these tendencies by selling fertilizers in compound or mixed form, a measure that involves a degree of coercion. The main burden of selling the farmers on phosphates and potash, however, has remained with the extension service.

INCREASES IN PRODUCTIVITY

Like most discussion of the sources of increases in agricultural production, this chapter has concentrated for the most part on the rise in measurable inputs. But in Chapter 3 it was shown that roughly half of the rise in farm output could not be directly attributed to increases in inputs. Can one say anything meaningful about the sources of this rise in productivity?

To begin with, one of the major sources of increased productivity in Korean agriculture has already been dealt with in this chapter, namely, improvements in plant varieties. The methods used in the previous chapter to estimate the contribution of seed to farm output do not begin to capture the full impact of the new varieties. Nor is it likely that the standard indexes for other inputs, such as chemical fertilizer or even water, fully capture improvements in quality as contrasted to increases in quantity. Urea, for example, is a more efficient and easier-to-use source of nitrogen than many of the fertilizer varieties available in the 1940s and 1950s.

Second, as already mentioned in Chapter 3, another part of the rise in productivity can be attributed to the shift of farmers away from grain to cash crops, livestock, and the like. In effect, these items provide the farmer with more output value for a given level of input. Another way of stating this is to say that the burgeoning urban market, together with the development of transport and other marketing facilities, made it possible for the farmer to use his own inputs more efficiently. Because it is impossible to measure that part of the improvements in transport and marketing that affects agriculture, this item is relegated to the residual. Similarly, improvements in the marketing of inputs to farmers would appear there as well. The gradual elimination of the extreme forms of inefficiency in the distribution of chemical fertilizer, for example, must have had a significant impact on productivity.

In the chapter that follows, an attempt is made to capture these sources of productivity rise through the concept of "isolation." As Albert Keidel shows, a significant proportion of the output difference among regions can be "explained" by a region's degree of isolation. The relevance of this finding to the discussion here is that improvements in transport and marketing, by reducing this isolation, have made an important contribution to the rise in agricultural productivity. If the cross-section measurements of the contribution of reduced isolation can be applied to changes over time, then one has a rough method of getting at the quantitative significance of these sources of increased productivity.

Finally, mention must be made of the improving quality of

farmers and farm management. The rapidly rising education levels of Korean farmers are discussed in Chapter 11 and at length in a separate study in this series. Following standard practices in human capital theory, one could attempt to measure the contribution of education to agricultural development. But only a true believer in such reasoning could use urban wage differentials for various levels of education as a guide to measure the relative productivities of, say, a high school and a primary school graduate in the agricultural sector of a less developed country such as Korea. It is not just the quality of the farmer that is improving. There has also been a marked improvement in the quality of the extension services providing those more educated farmers with information on new methods and techniques. The nature and quality of government services and other government policies that affect agriculture will be taken up again in Chapters 6 through 9.

In formal terms, improvements in government services and policies lead to rises in productivity by increasing either the quantity of agricultural inputs or by improving the efficiency with which inputs are selected. Thus, the quantitative contribution of these services and policies shows up in formal sources of growth analysis in both the residual (that is, as a contribution to the rise in productivity) and in the contribution of specific inputs (for example, the irrigation systems built with government funds or the fertilizer bought with government credit).[20]

FIVE

Regional Agricultural Production and Income

Farm output and real farm income in Korea have increased considerably since the end of World War II. Yet, in spite of the relative homogeneity of all of rural South Korea, the regional impact of these increases in output and income has not been even. In fact, shifts over thirty years have left once relatively productive and wealthy rice-specializing provinces lagging behind areas where farmers have been able to profit from growing markets for the produce of drier "upland" fields.

This reversal in relative positions has taken place in spite of price changes favoring the major food-grains produced in Korea's traditional "rice bowl." It is interesting that the lagging areas have also shown both a more concentrated application of pesticides and chemical fertilizers and a greater incidence of double-cropping. These factors, however, have all been over-shadowed by the growing profitability of upland cash crops.

Once relatively underutilized land resources in the hillier provinces combined with access to urban and export markets have given those areas striking increases in income from vegetables, fruits, livestock, and certain special crops. Parallel increases have not been realized in the more isolated wet paddy regions. In addition, regions closer to urban centers have received electric power sooner than the predominantly rice-growing regions, where opportunities for non-farm wage income are also fewer. These developments are additional factors in the emerging pattern of regional production and income differentials.

This chapter will examine the evolution of such regional differences in some detail. Two types of information are used. Over thirty years of administrative data for each of South Korea's nine provinces have been combined with seventeen years of crop price and farm consumer index data. Analysis from these sources gives a general picture of the evolution of regional patterns since 1945. A more detailed study of regional differences is provided by 1970 census and administrative data for each of South Korea's 167 mainland counties. Study of this second set of information confirms and highlights the geographically less detailed provincial results.

THE REGIONS OF SOUTH KOREA

Although this chapter will dwell on the differences among the various regions of South Korea, it is important to say at the outset that rural Korea is remarkably homogeneous when compared with countries like Brazil, Malaysia, Indonesia, and India, where racial, linguistic, and cultural differences inject external factors hindering economic analysis and where extreme crop specialization in different areas accounts for much of the variety in regional production and income. Of almost complete racial homogeneity, Koreans everywhere speak essentially the same dialect of their Ural-Altaic-related language with only some regional differences in pronunciation and word usage. Rice is by

far the predominant crop throughout the South Korean peninsula, and significant cultivation of a common set of secondary
crops is found in all provinces. Because of this relative
homogeneity, it is particularly rewarding to analyze the economic factors that have influenced the degrees of inequality
that do exist among regions.

The poorest region in South Korea through the whole period
of this study has been the very large northeastern area
represented by Kangwŏn province. Korea is an extremely
mountainous country, with temperatures cool for its latitude
and a monsoon pattern that brings as much as half of the
year's rainfall up from the south in the critical agricultural
months of June, July, and August. Given its northern location
and rough terrain, Kangwŏn province, sometimes known as
"potato country," has the least desirable extremes of these
national characteristics. The growing season is short and the
rains late for the necessary early rice planting; arable land is
scarce, and the rugged mountains leave much of the province
isolated from other parts of the country.

Contrasting with the northeast is the traditional rice belt
represented by the three provinces which contain Korea's
southwestern plains and valleys. South Ch'ungch'ŏng province
and North and South Chŏlla provinces enjoy less frost, early
rainfall, and generous endowments of arable land, most of which
is suitable for wet-paddy rice cultivation. In addition, warm
temperatures allow farmers in the southern provinces to double-
crop winter barleys on more than two-thirds of their paddy
land. These same southern provinces, however, have also suffered
from poor transportation and communication contacts with the
urban and industrial areas to the north and southeast. It is the
two Chŏlla provinces that have been eclipsed most conspicuously
in terms of per capita production and income by provinces
recently profiting from cash crops.

The central region occupied by North Ch'ungch'ŏng and
North Kyŏngsang provinces has a higher share of its cultivated
land in the form of dry upland fields and hence less of a tradition

of relying on rice and barley. Since the mid-1960s, these two provinces have shown a dramatic increase in output and income from vegetables, fruits, and special crops.

The hybrid regions of Kyŏnggi and South Kyŏngsang provinces are at opposite ends of the country. The wealthiest of the mainland farm regions for over thirty years, Kyŏnggi province in the north surrounds the huge and rapidly growing Seoul metropolis, which had an urban population of over 7 million in 1975. The province's striking feature, however, is its much lower density of farm population per hectare of cultivated land when compared with other areas. It is a hybrid region because, while it has taken good advantage of the profitability of vegetables, fruits, and livestock, the importance of rice in its total output and income approaches levels found in the rice bowl to the south.

South Kyŏngsang province, embracing the port of Pusan, is a hybrid of a different sort. It contrasts sharply with Kyŏnggi province in that it has by far the highest density of farm population per cultivated hectare in all of South Korea. Like Kyŏnggi province, South Kyŏngsang has shown a rapid increase in the importance of vegetables and relies heavily on rice production, but because is is more mountainous it has many sections without access to Pusan or other markets, and considerable shares of its total output are in barleys and potatoes, a characteristic of the most isolated regions.

The last of the nine provinces, Cheju Island, is well to the south of the Korean peninsula, and with its very different climate and volcanic origins it shares neither the emphasis on rice cultivation found throughout the mainland nor the severe winters and shorter growing seasons of the other regions. Very poor for much of the post-World War II period, Cheju Island began rapid development of its orange industry in the latter 1960s, and in very recent years this single crop has given the island the highest average per capita agricultural output in all of South Korea.

Cheju Island is a striking example of a phenomenon that has significantly affected other parts of the country also. The rapid development of its all-important orange industry has depended on access to outside markets and a learned awareness of the opportunities those markets might represent. The emergence of modern industrial urban areas throughout much of South Korea has presented rural areas with similar opportunities to adjust their cropping and marketing patterns and to take advantage of both the new composition and the new volume of demand for farm products. Areas closer to the burgeoning cities have had better access to these new opportunities, while isolated regions have not. Roads, railroad lines, and harbors provide the channels of access, and the growing network of these transportation facilities, in particular paved highways, when added to the geographical distribution of industrial centers, primarily in Seoul, Pusan, and Taegu, superimposes on the traditional climatic and geographical factors a second set of determinants of regional crop output and per capita income differentials in rural Korea.

RURAL LAND

A brief description of the differences in provincial land endowments explains a great deal about regional variation in Korean agriculture. Mountains are the most distinctive feature of South Korea's geography and, as Table 44 shows, most provinces have more than two-thirds of their total area in mountainous terrain. The first two columns show that, although Kangwŏn province is large in size, it is also extremely mountainous. The three least mountainous provinces are the ones described above as comprising the rice belt; it is, furthermore, interesting to note that Kyŏnggi province is also well below the national average. The dry-field provinces, however, are much more rugged, and these basic characteristics of provincial terrain, when reflected in the shortage of paddy, are fundamental factors underlying regional crop output patterns.

The distribution of Korea's mountainous regions is seen vividly in Figure 13, "Mountainous Counties of South Korea."

TABLE 44 Land and Cultivated Land

Provinces	Total Area (km²)	Mountains as a Share of Total Area (percent)	Area of Cultivated Land (km²)				
			1949	1955	1965	1975
Kyŏnggi	11,071	67.0	3,994[a]	2,870	3,162	3,072
Kangwŏn	16,827	89.5	1,126[a]	1,281	1,570	1,549
North Ch'ungch'ŏng	7,436	73.8	1,387	1,443	1,655	1,764
South Ch'ungch'ŏng	8,764	61.0	2,323	2,469	2,872	2,919
North Chŏlla	8,058	64.3	2,261	2,385	2,524	2,503
South Chŏlla	12,084	65.4	3,075	3,277	2,774	3,617
North Kyŏngsang	19,805	77.8	3,419	3,465	3,792	3,817
South Kyŏngsang	11,966	74.1	2,484	2,513	2,735	2,660
Cheju Island	1,820	68.2	366	382	478	495
South Korea	97,831	73.5	20,535[a]	20,085	22,562	22,396

Sources: Total area and private non-cultivated land use: EPB, *Korea Statistical Yearbook, 1976*, pp. 17, 22. Cultivated land: MAF, *Yearbook of Agriculture and Forestry Statistics*, various years.

Note: [a]Because pre-Korean War boundaries for Kyŏnggi and Kangwŏn were changed by the Armistice, these data cannot be compared directly with those of later years.

FIGURE 13 Mountainous Counties of South Korea

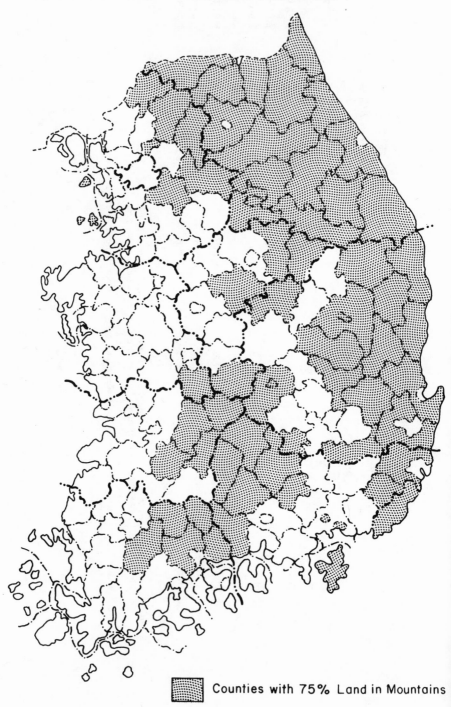

Counties with 75% Land in Mountains

The northeast and east are heavily mountainous, as is much of South Kyŏngsang province as the result of a branch of the Taebaek Mountains which extends to the south and west as far as South Chŏlla province.

The share of rice paddy land in total cultivated land has remained relatively stable since the Korean War but, as Table 45 shows, those regions with more mountains also have less paddy, with the exception of hybrid South Kyŏngsang province. With this exception, the highest concentrations of paddy land are in the rice-bowl provinces.

TABLE 45 Paddy Land, 1955, 1965, 1975

Provinces	Paddy as a Fraction of Total Cultivated Land %			Double-Cropped Paddy as a Share of Total Paddy %		
	1955	1965	1975	1955	1965	1975
Kyŏnggi	61.3	60.5	60.8	3.4	9.1	25.7
Kangwŏn	35.4	35.7	37.3	2.5	4.2	6.7
North Ch'ungch'ŏng	48.3	46.4	45.0	31.1	35.0	45.6
South Ch'ungch'ŏng	66.7	62.0	61.7	32.2	32.6	56.6
North Chŏlla	73.9	67.1	68.0	48.8	50.3	76.3
South Chŏlla	62.7	58.9	59.4	50.0	58.8	79.8
North Kyŏngsang	56.0	54.9	54.9	45.2	62.3	61.7
South Kyŏngsang	67.0	66.5	66.7	53.8	71.1	81.9
Cheju Island	2.1	2.1	2.1	25.8	40.7	52.4
South Korea	59.8	57.0	57.0	37.4	45.1	60.0

Sources: MAF, Yearbook of Agriculture and Forestry Statistics, relevant years.

Finally, as a result of both differences in land endowment and differences in climate, the incidence of "double-cropped" paddy land is much greater in the southern and rice-bowl regions, although the increase in such land has been very rapid everywhere since the Korean War. As can also be seen from Table 45, the southern regions have always had an advantage over the other areas in this respect, although Kyŏnggi province increased its double-cropped share of paddy remarkably in the ten years following 1965.

It is not clear to what extent the superior natural endowments of the southern and rice-belt region have been evened out by man's artificial improvements. Certainly this is not the case for the consumption of chemical fertilizers; that has been concentrated in the south at least since the 1950s. Table 46 demonstrates this, although data for the 1950s are inconsistent with those for later years and cannot be depended on for an accurate comparison over time.

Equally as important as chemical fertilizer is paddy irrigation; here again the southern provinces are very well irrigated. But Table 46 shows that the mountainous provinces of Kangwŏn, North Kyŏngsang, and North Ch'ungch'ŏng are also well irrigated, reflecting, perhaps, the ready availability of water from mountain streams and springs. Significant also is the respectable degree of irrigation in Kyŏnggi, which is less mountainous and which has a larger share of its land in paddy.

Flat land has, it is clear, added to the advantage given to the rice-belt provinces by their more temperate climate. It is not clear whether more northerly dry-field regions have compensated much for their disadvantage by artificially improving the soil. Even if they have to some extent, the available data confirm the superiority of the overall agricultural environment of the south and southwestern regions.

REGIONAL AGRICULTURAL OUTPUT

Over the thirty years since 1945, as previous chapters have indicated, the agricultural sector has undergone a substantial structural change which has seen the significance of rice production in total output greatly reduced. Rice remains, of course, extremely important, and the regions that supply the urban areas are vital to the Korean economy. But regions that have also been able to respond to the growing urban demand for a

TABLE 46 Fertilizer and Irrigation, 1955, 1965, 1975

Provinces	Chemical Nutrient Consumption (kilograms/hectare)			Securely Irrigated Paddy as a Fraction of Total Paddy %		
	1955	1965	1975	1955	1965	1975
Kyŏnggi	56	132	321	33.9	32.6	43.6
Kangwŏn	43	133	330	48.4	53.1	57.5
North Ch'ungch'ŏng	62	160	320	29.6	34.5	47.7
South Ch'ungch'ŏng	58	163	351	29.3	31.7	40.1
North Chŏlla	62	191	438	22.0	27.3	32.1
South Chŏlla	64	199	418	30.9	37.7	49.6
North Kyŏngsang	68	167	363	33.7	38.8	54.5
South Kyŏngsang	73	228	405	30.5	35.9	48.2
Cheju Island	56	171	438	39.2	39.8	71.1
South Korea	62	174	374	30.8	35.1	45.9

Sources: Calculated from chemical nutrient (nitrogen, phosphorous, and potassium) consumption data, irrigated paddy data, and total cultivated land from MAF *Statistical Yearbook of Agriculture and Forestry,* relevant years.

varied diet have shown more rapid growth in total output than those regions that have not.

Structural changes, already significant when viewed at the aggregated national level, become much more so when broken down into provincial statistics. Declines in major crops of one region have, in many instances, been compensated for in the national average by increases in emerging major crops in other regions. As a result, the structural changes in agricultural output experienced by the individual regions have, in many cases, been much more severe than those shown by the figures for South Korea as a whole.

To illustrate the variation in structural change from region to region, Tables 47 and 48 give the shares in total agricultural output of major product categories for five selected provinces in five years. The years were chosen because they had few harvests that were either extremely bad because of drought or blight or extremely good because of extraordinary rainfall, temperature, and sunshine conditions. The five provinces were chosen as representatives of major regions. Kyŏnggi province is the richest and most versatile region; Kangwŏn province is the poorest. North Ch'ungch'ŏng province represents the "upland" region, while South Chŏlla typifies the isolated rice- and barley-producing areas. Finally, South Kyŏngsang province is a southern hybrid region with its own urban and isolated sections and the full range of Korean agricultural products.

Table 47 gives some idea both of those regions that traditionally concentrate most heavily in food-grain production and those regions which have most dramatically reduced their dependence on such crops. South Chŏlla province, representing the rice bowl, along with hybrid South Kyŏngsang province, shows both high shares of output in food-grains and a slower rate of reduction in that share. It is worthwhile noting that neither of these provinces is the extreme in this respect, since data from the same sources show that food-grain production as a share of North Chŏlla's total farm output fell from an extremely high 80.3 percent in 1938 to only 62.1 percent in

1974. By contrast, regions with poor endowments in paddy land have experienced a much more precipitous decline. Kangwŏn and North Ch'ungch'ŏng provinces (with roughly a third and a half of cultivated land in paddy, respectively, throughout the period) experienced drops of as much as half for the former and a third for the latter.

Table 47 also shows very clearly how these same upland regions shifted emphasis in total output towards vegetables. (The contribution of fruits to this category is not great, a

TABLE 47 Shares of Rice, Barleys, Vegetables, and Fruits
in Total Agricultural Output, by Region, 1938–1974
(%)

	1938	1947	1960	1966	1974
	Rice and Barleys				
Kyŏnggi	60.7	62.8	64.3	58.5	48.7
Kangwŏn	61.3	49.8	50.8	36.6	33.3
North Ch'ungch'ŏng	61.7	71.2	64.9	50.2	40.9
South Chŏlla	72.1	77.8	67.3	62.2	59.9
South Kyŏngsang	73.6	73.2	67.6	63.3	56.9
South Korea	70.3	72.2	66.0	58.4	52.1
	Vegetables and Fruits				
Kyŏnggi	26.5	22.9	16.2	22.7	23.2
Kangwŏn	13.0	15.3	13.2	24.7	23.1
North Ch'ungch'ŏng	15.8	9.8	11.0	18.9	31.4
South Chŏlla	7.7	6.2	9.1	11.8	13.0
South Kyŏngsang	10.5	8.6	11.0	12.5	20.1
South Korea	13.4	10.1	11.9	16.6	21.4

Source: Calculated by combining data on individual output levels for all product categories with individual 1970 prices for those products. The value output levels thus computed were then used to calculate the shares of categories in total output. The 1951–1975 annual provincial data for individual products were obtained from individual years of the MAF *Yearbook of Agriculture and Forestry Statistics.* Similar data for the years 1936–1950 were obtained from a special edition of the same yearbook published in Pusan in 1952. Crop prices were obtained from the NACF *Summary of Rural Prices,* various years. For details, see Albert Keidel, "South Korean Regional Farm Products and Income, 1910–1975," PhD dissertation, Harvard University, 1978.

maximum increase of 3 percentage points for North Ch'ung-ch'ong province and less for the other provinces shown.) The decline in the share of vegetables everywhere from 1938 to 1947 and the lack of significant increase through 1960 reflect two different phenomena. From a very good year in 1938, yields of many major vegetables declined precipitously through the war years and suddenly increased again in 1948 and subsequent years. Shortage of chemical fertilizers seems to have been responsible for this pattern. For the period 1947 to 1960, the insignificant change in the vegetable share in some northern areas points up an interesting phenomenon in Korea's increase in vegetable output. From the Korean War to the early 1960s,

TABLE 48 Shares of Livestock and Special Crops in Total
Agricultural Output, by Region, 1938–1974
(%)

	1938	1947	1960	1966	1974
	Special Crops and Silk Cocoons				
Kyŏnggi	4.5	1.3	1.0	1.7	2.0
Kangwŏn	5.5	1.4	3.2	5.7	7.9
North Ch'ungch'ŏng	8.9	2.7	2.1	10.8	12.6
South Chŏlla	12.4	4.7	2.8	3.0	5.0
South Kyŏngsang	10.7	2.9	1.6	2.8	3.4
South Korea	8.0	2.8	1.9	4.7	6.0
	Livestock				
Kyŏnggi	n.a.	6.9	12.3	9.6	21.0
Kangwŏn	n.a.	11.4	18.8	16.1	15.2
North Ch'ungch'ŏng	n.a.	7.5	13.7	9.7	7.5
South Chŏlla	n.a.	7.0	15.6	11.4	10.9
South Kyŏngsang	n.a.	9.2	15.6	11.4	13.0
South Korea	n.a.	7.6	14.1	10.4	12.1

Source: See note to Table 47.

Note: Regional shares for crop categories shown in this and the preceding table do not sum to 100 because of the omission of other crop categories: pulses, potatoes, special crops, miscellaneous grains, and mulberry.

rice-bowl provinces increased output in vegetables much faster than the northern and upland regions, but in the following fifteen years exactly the opposite was true, and to a greater degree.

Table 47, therefore, shows how the decline in rice and the increase in vegetable output changed the structure of agricultural production in the various regions of South Korea. Special crops and livestock production have played a similar though less important role in the diversification of Korean agriculture. Table 48 shows the change in the shares of total output in these categories for the same selected provinces. Livestock output since 1945 has increased in two different phases. The first phase lasted for most of the first twenty years of the period and was characterized by increases in all types of live-stock, but most notably pigs, chickens, and rabbits. Further-more, the increases in this period were distributed rather evenly throughout all parts of South Korea. In the second phase, from the mid-1960s to the mid-1970s, these more traditional forms of animal husbandry ceased to grow significantly, but output of dairy cattle and hence dairy products showed a spectacular increase averaging over 35 percent a year, with two-thirds of the output in Kyŏnggi province and Seoul. It is this second phase of expansion in livestock production, with its emphasis on dairy products for the urban market, that has caused the major regional impact.

Changes in output of special crops and cocoons have also played a significant role, because of increases in some regions and declines in others. The Japanese had encouraged the produc-tion of cotton and other fiber products as raw materials for light manufacturing industries, and output of these products, concentrated in southern and rice-bowl provinces, showed a steady decline after the end of World War II. Since the 1950s, the most significant increases have occurred in tobacco and sesame, the latter extremely important in Korea as a flavoring and as a source of oil. Although output of sesame favors southern provinces such as South Chŏlla, the output of tobacco is much more significant in value terms, and its production is concentrated very heavily in North Ch'ungch'ŏng and North Kyŏngsang provinces, which together form the principal upland dry-field region. As a result, shifts in production of special crops since the Japanese period have added to the overall pattern of change in Korean agriculture favoring the drier and once less productive regions at the expense of traditional rice-growing provinces.

The shifts in agricultural composition since World War II and the differential regional impact of those shifts which became pronounced in the fifteen years after 1960 have caused an adjustment in the contributions of the various provinces to South Korea's total agricultural output. Output in provinces with vegetables, fruits, livestock, and special crops grew faster and increased their shares in national farm production. Table 49 shows the degree to which this has been true for all the provinces of South Korea. In the period before 1960, the traditional rice areas grew more rapidly than the dry-field and hybrid areas, but nowhere was the average annual growth rate above 4 percent. After 1960, however, North Ch'ungch'ŏng grew at an average rate of 6.2 percent, and all of the upland provinces outperformed those in the rice-bowl region. As a result, between 1947 and 1975, the upland and northern provinces increased their percentage shares of national farm output, while the shares of southern and rice-bowl provinces declined.

In summary, as the structure of the Korean economy has shifted away from agriculture, so has the structure of agriculture itself shifted away from the cultivation of rice. Increased emphasis on vegetables and other crops produced primarily for the market has favored northern and dry-field regions, while provinces in the traditional food-grain regions and in the isolated mountainous northeast have tended to lag. After the discussions of prices and regional farm population which follow, it will be seen that these shifts in output have had a significant impact on regional differences in per capita gross farm income.

TABLE 49 Provincial Agricultural Output by Share
and Growth Rate, 1938–1974
(%)

	Share in Total Agricultural Output			Average Annual Growth Rates		
	1938	1947	1974	1938–1947	1947–1960	1960–1974
Kyŏnggi	14.8	12.7	14.0	–3.8	2.6	4.7
Kangwŏn	2.0	4.7	4.8		2.0	4.7
North Ch'ungch'ŏng	7.7	6.4	8.1	–4.1	2.1	6.2
South Ch'ungch'ŏng	14.2	11.5	12.4	–4.5	3.2	4.0
North Chŏlla	13.4	12.5	10.8	–3.0	3.8	1.7
South Chŏlla	15.0	16.9	15.8	–0.9	2.6	3.4
North Kyŏngsang	18.7	18.3	18.0	–2.4	1.6	4.7
South Kyŏngsang	14.2	15.7	13.7	–1.1	1.9	3.6
Cheju Island	–	1.3	2.5	–	1.5	10.4
South Korea	100.0	100.0	100.0	–2.2	2.6	4.0

Sources: The output was valued in 1970 constant prices. See the footnote to Table 47.

REGIONAL TERMS OF TRADE

Generally speaking, as previous chapters have shown, prices Korean farmers received increased faster from the 1960s into the 1970s than did the prices they paid; thus their terms of trade improved. But this was more the case for prices of basic food-grains than it was for other crops, and thus the trend of improving terms of trade has been more pronounced in the southern rice- and barley-growing regions than in the dry-field upland provinces. This section briefly analyzes crop prices combined in a weighted index which uses regional weights instead of average national ones.

The figures used in this section are only an approximation of actual terms of trade and were computed by taking the ratio of output calculated in current prices and deflated by the farm consumer index to output calculated in constant 1970 prices.

Technically, therefore, the weights used are not theoretically valid,[1] but the approximation is close enough to a true terms-of-trade calculation to give some idea of the different regional impacts of farm product price changes.

The sharpest contrast in regional terms of trade is between those for North Ch'ungch'ŏng province, representing the dry-field region, and those for South Chŏlla province, representing the southwestern food-grain region. As Table 50 demonstrates, after a long period of relatively similar changes for the two provinces in the 1960s, the terms of trade for South Chŏlla improved dramatically after 1970. At the same time, those for the upland region represented by North Ch'ungch'ŏng improved, but at a slower rate. When these figures are compared with the average national ratio terms of trade, it becomes clear that farmers in the three northern provinces shown here all had slower increases than farmers in the south in the purchasing power of their output.

Ignoring the influence of the bad rice and barley harvests in the early 1960s and the failure of the extremely important red pepper crop in 1969, Table 51 shows clearly how prices for Korea's staple food-grains have risen much faster than the farm consumer index, while those for other products, and vegetables in particular, have not. Tables 47 and 48 have shown how the southern and rice-bowl regions concentrate in food-grain production, while the upland dry-field regions to the north depend more on vegetables, fruits, and special crops. As a result, the southern provinces have been able to profit more from the increase in food-grain prices, while the value of output further to the north has been based more heavily on prices that have not increased significantly in relation to the rural cost of living.

There seem to have been two major factors influencing the changes in prices outlined above. On the one hand, exceptionally good and exceptionally bad harvests have caused shifts in supply which have dramatically affected prices. On the other hand, government purchase of farm products and government price

TABLE 50 Regional Purchasing-Power Ratios, 1959–1975
(1970 = 100)

	Kyŏnggi	Kangwŏn	North Ch'ungch'ŏng	South Chŏlla	South Kyŏngsang
1959	72.5	72.2	72.5	72.1	71.9
1960	80.3	80.3	80.8	80.4	81.9
1961	85.8	85.2	89.4	87.6	88.7
1962	85.5	86.3	92.7	91.3	91.8
1963	118.9	116.2	123.0	119.4	118.4
1964	120.7	118.5	129.6	126.3	130.6
1965	110.1	103.9	105.7	103.2	106.9
1966	97.5	100.4	101.8	96.1	97.7
1967	94.2	101.8	93.7	96.3	97.6
1968	86.2	90.2	80.7	91.7	91.7
1969	93.1	92.9	89.0	96.5	94.2
1970	100.0	100.0	100.0	100.0	100.0
1971	100.0	101.5	101.6	108.0	105.7
1972	104.4	107.9	108.3	117.6	111.1
1973	105.2	112.4	105.3	116.1	110.7
1974	108.5	110.9	106.8	117.0	113.8
1975	112.7	112.0	109.0	121.9	116.4

Sources: The ratios are obtained by the division of provincial output valued in current prices and deflated by a farm consumer index on the one hand, by provincial output valued in constant 1970 prices on the other. See the text and the note to Table 47 for price and output sources.

supports have been responsible for some of the important trends in the late 1960s and early 1970s.

Table 52 documents the importance of these two factors for rice and barley, the two major food-grain crops. For each crop, the table gives three series: a purchasing-power index reflecting changes in the crop's price; an index of harvest output by year; and government purchase of the crop by year as a percentage of the total harvest. The influence of crop failure can be seen by looking at the early 1960s. Drought in 1962

TABLE 51 Purchasing Power Ratios for Major
Crop Categories, 1959–1975
(1970 = 100)

	Rice (I)	Barleys (II)	Vegetables (III)
1959	72.5	81.6	59.8
1960	80.7	97.2	70.7
1961	92.7	121.1	49.6
1962	88.8	122.4	56.3
1963	127.5	189.2	103.2
1964	125.0	194.9	77.1
1965	101.4	123.8	84.7
1966	95.4	104.5	92.0
1967	92.8	109.2	70.4
1968	91.2	98.7	57.3
1969	102.5	104.8	64.8
1970	100.0	100.0	100.0
1971	109.8	120.0	81.7
1972	122.2	136.6	73.0
1973	116.7	128.9	67.5
1974	125.4	121.7	70.3
1975	128.5	137.4	78.2

Sources: See text and note to Table 50.

caused a poor rice crop, and the drop in output, following the bumper year in 1961, is credited with causing the severe rise in prices experienced in the following year. Rice is harvested late in autumn, and much of the crop is actually sold in the following year, when the price effects of the crop's size become most apparent. This lag in price changes for rice complicates regional terms-of-trade analysis, and will be discussed below. A red rust blight was responsible for the disastrous barley harvest in 1963 and, since barley is harvested in the spring, the sudden increase in the barley price for that year is attributable directly to the crop failure.

TABLE 52 Prices, Output, and Government Purchase of Rice and Barley, 1959–1975

	(I) Rice Purchasing-Power Ratio (1970=100)	(II) Rice Output Index (1970=100)	(III) Gov't Collection as Share of Rice Crop (%)	(IV) Barley Purchasing-Power Ratio (1970=100)	(V) Barley Output Index (1970=100)	(VI) Gov't Collection as Share of Barley Crop (%)
1959	72.5	83.2	6.1	81.6	63.0	3.8
1960	80.7	79.9	4.5	92.2	63.2	3.0
1961	92.7	94.7	8.3	121.1	68.2	3.8
1962	88.8	79.8	8.9	122.4	77.7	1.2
1963	127.5	95.1	8.2	189.2	22.8	0.0
1964	125.0	100.4	6.0	194.9	83.2	0.1
1965	101.4	88.9	8.5	123.8	95.0	3.6
1966	95.4	99.5	9.0	104.5	104.2	6.5
1967	92.8	91.5	7.9	109.2	99.0	3.9
1968	91.2	81.1	4.8	98.7	104.4	4.7
1969	102.5	103.8	7.9	104.8	105.8	8.3
1970	100.0	100.0	9.3	100.0	100.0	8.8
1971	109.8	101.5	12.6	120.0	93.2	7.9
1972	122.2	100.5	12.8	136.6	95.5	16.6
1973	116.7	106.9	11.6	128.9	84.7	19.1
1974	125.4	112.8	16.8	121.7	81.4	22.4
1975	128.5	118.6	17.2	137.4	99.2	28.8

Sources: For columns (I) and (IV), see the note to Table 50; for other columns, see Albert Keidel, cited above.

The year 1969 was the first for the new government two-price
subsidy system for rice and barley, whereby the government paid
a higher price to farmers than it received when the grain was
sold to consumers, after adjusting for handling costs (see Chapter
8). As columns III and VI in Table 52 show, the fraction of the
total harvest for each crop purchased by the government began
a trend of steady increase from 1969, and at the same time the
purchasing power of rice and barley output increased significantly and remained high through the rest of the period for
which data have become available. Government purchase of
food-grain crops does not vary significantly from province to
province but, given the larger share of these crops in the total output of southern and southwestern provinces, it is clear that the
two-price government subsidy program for rice and barley has
been of greater assistance to those provinces than to more
northern upland regions.

Because the weights used in constructing the terms-of-trade
indexes used in this section reflect current value weights to a
certain extent, the presence of the one-year lag in rice price
changes exaggerates the degree to which terms of trade improve
as the result of a bad harvest. The large rice component of the
end-of-year farm products inventory, when valued at prices for
the year just ended, will have a lower value than it would if it
were valued at the higher price it will actually be bringing during
the following year. The low output and the low price combine to
produce a weight for the bad year which is lower than the
actual effective weight. By contrast, the weight for the following
year will be too large, because it will combine the larger harvest
with the high prices which should have been applied to the
poorer year. This same exaggeration in terms-of-trade results
occurs when a blight or drought affects one region rather than
another, if one region's share of total output in the affected
crop is significantly different from that for the other region.
This was true of the bad rice harvest in 1968 which was much
more serious for South Chŏlla and the two Kyŏngsang provinces
than for the other areas.

The different terms of trade faced by farmers in different regions introduced above resulted from assuming that the actual prices they received were the same from region to region, and that only the shares of different crops in total output influenced regional variation. It is also interesting to see whether the actual prices themselves are different from one region to the next. Data on regional differences in actual prices received by farmers are difficult to obtain, but Table 53 shows that, if we take the least isolated of rural markets for which data are available, there does exist some difference in the actual price of rice received by farmers. The price received by farmers in Kyŏnggi province,

TABLE 53   1970 Rice Prices by Rural Markets
(wŏn/100 liters of middle-grade rice)

Province	Market Location	Price
Kyŏnggi	Hwasŏng county	6,394
North Ch'ungch'ŏng	Ch'ŏngwŏn county	5,750
South Ch'ungch'ŏng	Nonsan county	6,067
North Chŏlla	Okku county	5,710
South Chŏlla	Kwangsan county	5,882
North Kyŏngsang	Yŏngch'ŏn county	5,855
South Kyŏngsang	Ch'angwŏn county	5,703

Source: NACF, *Summary of Rural Prices 1959–1970* (Seoul, 1971), p. 102.

which is close to Seoul, is almost 5 percent higher than that received in most other provinces. This is not surprising, given the tremendous importance of Seoul in the Korean rice market. But it is interesting to note that North Ch'ungch'ŏng province does not seem to have benefited from rice prices higher than those of the rice-bowl provinces, South Ch'ungch'ŏng and North and South Chŏlla. We can therefore conclude that the terms-of-trade disadvantage of the upland dry-field provinces induced by their greater dependence on non-food-grain crops is not mitigated by regional variation in actual prices received for rice, which is still the most important crop in their output structure.

This section has shown that, while the purchasing power of Korean farm output has improved during the 1960s and early 1970s, the improvement has been more pronounced for southern and rice-belt provinces than for the regions with a smaller percentage of cultivated land in paddy. Thus, while the shift in output structure of Korean agriculture has favored these latter dry-field regions, price movements have tended to offset the impact of that structural change on regional farm income. The net effect of these changes will be discussed in detail in the section on agriculture income below.

PROVINCIAL FARM POPULATION

For determining gross farm income for different regions in Korea, total output and prices are, of course, very important. If other things are equal and one region is more densely populated than another, that region's farmers will be poorer. The share of farming population, however, has fallen rapidly in all provinces (see Table 54). In the thirty years since 1945, the provinces surrounding the growing industrial centers of South Korea have shown the most dramatic reduction in the share of total population working in agriculture.

Changes in provincial farm population since 1945, although significant for other reasons, are particularly important because of their effect on relative farm population density per area of cultivated land. Table 55 gives the actual farm population per square kilometer of farm land, and it is clear not only that the density increased rapidly up to 1965 (allowing for the dislocations of the Korea War) and fell off rapidly thereafter, but also that Kyŏnggi province had by far the lowest density and South Kyŏngsang the highest of all the South Korean provinces throughout the period shown. This relative pattern of farm population density is one of the major reasons for the long-term

TABLE 54 Farm Population by Province, 1949–1975

(1,000s)

	1949	1955	1966	1975	Annual Average Growth Rate (%) 1966–1975
Kyŏnggi[a]	1,408	1,375	1,604	1,345	-1.9
Kangwŏn[a]	719	720	934	728	-2.7
North Ch'ungch'ŏng	923	930	1,111	887	-2.5
South Ch'ungch'ŏng	1,692	1,690	2,026	1,583	-2.7
North Chŏlla	1,648	1,501	1,773	1,455	-2.2
South Chŏlla	2,475	2,368	2,884	2,156	-3.2
North Kyŏngsang	2,466	2,346	2,762	2,167	-2.7
South Kyŏngsang	2,235	1,995	2,206	1,640	-3.2
Cheju Island	214	213	255	231	-1.1
South Korea	14,416	13,300	15,781	12,263	-2.8

Sources: Respective editions of the MAF, *Yearbook of Agriculture and Forestry Statistics.*

Note: [a]Boundaries for Kyŏnggi and Kangwŏn Provinces were different before the Korean War from what they were after as a result of the changed Armistice line. Farm population for 1949 for these provinces has been multiplied by the ratio of post-1950 to pre-1950 total cultivated land (0.7182 for Kyŏnggi and 1.0492 for Kangwŏn), so the usefulness of these data is tentative at best.

TABLE 55 Farm Population Density, 1933–1975
(persons/km² of cultivated land)

	1933	1949	1955	1965	1975	Growth Rate 1965–1975 (%/yr)
Kyŏnggi	333	509	499	550	449	-2.0
Kangwŏn	385	559	563	590	470	-2.2
North Ch'ungch'ŏng	492	665	646	672	504	-2.8
South Ch'ungch'ŏng	471	728	684	707	542	-2.6
North Chŏlla	537	729	631	699	582	-1.8
South Chŏlla	498	805	722	758	596	-2.4
North Kyŏngsang	515	721	678	732	567	-2.5
South Kyŏngsang	565	919	837	857	632	-3.0
Cheju Island	–	585	561	544	468	-1.5
South Korea	469	702	662	701	547	-2.5

Sources: Data for 1933 are computed from statistics in Chōsen Sōtokufu, Shōwa 8 nen nōgyō tōkei hyō (1933 agricultural statistical tables, Seoul, 1937), pp. 7, 9. Other years are calculated from statistics in various years of the MAF Yearbook of Agriculture and Forestry Statistics.

discrepancies observed in per capita farm income in the thirty years since World War II. The pattern has its origins in the traditionally lower productivity of northern land, a relationship fundamentally altered during the Japanese colonial period.[2]

The ranking of provinces by farm population density has not changed since 1945, but the disparity has and, while the density of Kyŏnggi's farmland was little more than half that of South Kyŏngsang in 1949, by 1975 it was more than 70 percent as dense as the densest province. There has been, therefore, a steady trend of reduction in the difference between provincial farm population and densities.

REGIONAL FARM INCOME

Differences in the per capita gross farm receipts of farmers in various provinces give a good indication of the differences in per capita farm disposable income if costs of production and tax burdens as a share of output do not vary significantly from region to region. Studies of regional costs and examination of regional taxation patterns show that, although costs and taxes are not equal, the differences are not great enough to affect the usefulness of gross receipts as an index of net income. This section analyzes regional per capita gross farm receipts since the Japanese colonial period and, in addition, provides data on rural electrification, an important public good made available to the Korean farm village.

Table 56 combines the output and population information presented above and shows the degree to which Kyŏnggi province has, since before World War II, enjoyed a considerably higher per capita product than the other regions. The data show that in 1938 Kyŏnggi had much wealthier farmers than any of the other provinces, but that after 1945 this was no longer so.

TABLE 56 Total per Capita Output by Region, 1938–1975
(1,000 wŏn/capita, 1970 prices)

Province	1938	1947	1960	1966	1970	1975
Kyŏnggi	52.1	29.5	44.3	44.9	61.1	85.5
Kangwŏn	14.0	22.3	28.2	32.1	38.6	59.1
North Ch'ungch'ŏng	32.6	24.7	31.8	42.5	50.5	76.7
South Chŏlla	32.2	26.6	32.4	36.0	43.1	62.0
South Kyŏngsang	32.1	24.6	31.8	43.7	51.1	63.8
South Korea	32.9	27.9	39.0	41.2	48.8	69.5

Sources: Total population figures for 1949–1966 are from Tai-hwan Kwon, "Population Change and Its Components in Korea, 1925–66," PhD dissertation, Australia National University, 1972. Total population figures for 1970 and 1975 are from EPB, *Korea Statistical Yearbook 1973* and *1976.* The 1970 data are based on the 1970 Population Census. Farm population data for 1949–1966 and 1975 are taken from the MAF *Yearbook of Agriculture and Forestry Statistics*, various years. The farm population data were taken from the 1970 Agricultural Census.

Study of per capita rice and vegetable production clarifies the trends behind the regional differences shown in Table 56. The low density of farm population is largely responsible for the high per capita output of rice in Kyŏnggi province shown in Table 57, but it is interesting to note that only in the late 1960s and early 1970s have per capita rice output levels reached the pre-war level, as represented by 1938. What is more interesting is the fact that, in the 1970s, the upland region of North Ch'ungch'ŏng province increased its per capita rice production above that of South Chŏlla province, a traditional rice-bowl area.

TABLE 57 Per Capita Output from Rice and Vegetables
by Region, 1938–1975
(1,000 wŏn/capita, 1970 prices)

Province	1938	1947	1960	1966	1970	1975
				Rice		
Kyŏnggi	31.2	26.9	25.4	22.8	29.2	35.3
Kangwŏn	8.6	9.9	12.0	9.8	12.3	13.2
North Ch'ungch'ŏng	15.7	13.3	15.1	14.4	16.3	24.9
South Chŏlla	18.8	16.6	17.0	15.7	18.5	21.1
South Kyŏngsang	18.4	14.4	16.5	18.3	18.8	24.9
South Korea	19.1	16.8	18.0	17.7	20.2	25.3
				Vegetables		
Kyŏnggi	12.3	9.2	6.7	9.1	14.4	16.1
Kangwŏn	1.4	3.4	3.8	6.9	7.5	9.8
North Ch'ungch'ŏng	4.4	2.3	3.1	8.0	12.7	19.6
South Chŏlla	1.8	1.5	3.1	4.0	4.4	6.1
South Kyŏngsang	2.3	1.8	3.3	5.1	9.8	10.0
South Korea	3.6	2.6		5.8	7.8	10.1

Sources: See footnotes to Tables 47 and 56. For similar data on an annual basis for all provinces, see Keidel.

This is due mainly to the heavy use of new high-yield seed varieties in the upland region, where the soil is said to be particularly suitable to them. It is not surprising, then, that total per capita production in the upland area has risen so dramatically above that of the southwestern region.

This rise is further clarified by the regional per capita output of vegetables, also shown in Table 57. Before World War II, Kyŏnggi province was heavily engaged in the cultivation of vegetables, and since the mid-1960s the northern hybrid region has again profited from its ready access to Seoul. But the per capita output of vegetables in North Ch'ungch'ŏng, with its higher percentage of otherwise less well-used upland fields, has surpassed the level of output in Kyŏnggi. The other regions, and in particular the rice-belt area, lag far behind in this respect.

Using constant prices from just one year to value total output in every year has several drawbacks as a way of accurately measuring income available to farmers in different years and in different regions. Constant prices ignore fluctuations over time in the price of one crop relative to the price of another and, if the farm cost of living rises faster in some years than in others, the impact of such movements will not show in constant-price calculations.

A much truer measure of farm income is obtained by valuing crops at their actual current prices and then adjusting for inflation by dividing through with an index of the farm cost of living. Table 58 shows per capita farm income in the major regions by valuing crops with current prices deflated by the farm consumer price index. Since 1959 when farm price index data became available, the regional trends in purchasing power thus obtained are similar to those shown by the constant-price calculations, but the effects of the improved terms of trade for rice and barley since 1970 and the deteriorating ones for vegetables have modified the degree of the differences. Kyŏnggi province is still the wealthiest by far, and the surge in income in North Ch'ungch'ŏng province is still remarkable. The income calculations show, however, that Kangwŏn province in the northeast, with its lesser emphasis on rice, has not done as well as the constant-price per capita output calculations show. Similarly, while the constant-price data indicate that the southern hybrid region of Kyŏngsang province out-produced the rice-belt province of South Chŏlla, actual income per farmer in the southwest increased more rapidly than and actually surpassed that of South Kyŏngsang, when measured by deflated current prices.

In sum, while per capita output calculated in constant prices less than doubled in the period between 1960 and 1975, actual farm income more than tripled, as shown by the purchasing-power-per-farmer data in Table 58. These data give the best

TABLE 58 Regional Per Capita Gross Farm Income, 1959–1975
(1,000 wŏn in current prices deflated to 1970 prices by farm consumer index)

	Kyŏnggi	Kangwŏn	North Ch'ungch'ŏng	South Chŏlla	South Kyŏngsang	South Korea
1959	30.6	21.2	22.5	23.2	21.8	21.5
1960	32.8	22.3	25.4	24.0	23.1	25.9
1961	37.3	25.5	32.3	32.1	31.4	32.9
1962	34.9	30.3	29.4	29.7	28.3	30.6
1963	50.5	32.1	38.9	34.9	34.2	39.5
1964	57.9	37.5	51.6	47.0	49.1	50.4
1965	38.9	29.2	38.9	36.3	46.7	40.0
1966	46.0	32.2	45.8	40.0	43.3	43.4
1967	46.4	36.8	42.8	30.1	42.1	39.7
1968	41.5	32.0	38.2	30.7	39.1	37.4
1969	55.2	35.1	41.9	41.3	48.9	45.7
1970	60.9	37.1	50.8	41.5	49.9	48.3
1971	63.4	42.5	55.6	49.2	55.2	52.9
1972	61.0	46.0	59.2	55.8	57.2	57.3
1973	70.9	51.9	64.3	55.3	54.2	58.9
1974	83.4	50.1	76.7	59.4	66.8	67.7
1975	95.2	66.3	83.7	75.6	74.3	79.6

Sources: The data were calculated by combining current prices with annual provincial output figures for individual products and deflating the resulting value output with a farm consumer price index. The resulting provincial figures were then divided by annual provincial farm population levels. For data sources, see footnotes to Tables 47 and 56.

TABLE 59 Upland and Rice-Belt Per Capita Income from Selected Crops, 1959–1975 (1,000 wŏn in current prices deflated to 1970 values by farm consumer index)

	Barley		Vegetables		Rice (current prices)		Rice (leading prices)	
	North Ch'ungch'ŏng (I)	South Chŏlla (II)	North Ch'ungch'ŏng (III)	South Chŏlla (IV)	North Ch'ungch'ŏng (V)	South Chŏlla (VI)	North Ch'ungch'ŏng (VII)	South Chŏlla (VIII)
1959	4.1	3.6	1.5	1.5	10.6	12.2	12.1	13.4
1960	4.6	4.0	2.0	1.7	12.0	12.4	13.9	14.0
1961	6.5	6.2	1.7	2.1	14.7	17.5	15.1	17.5
1962	5.3	6.4	2.7	1.4	10.0	14.8	13.4	18.5
1963	3.8	2.1	4.7	1.6	18.0	22.2	19.0	24.0
1964	10.3	8.9	3.8	2.9	19.5	24.5	18.1	21.8
1965	6.0	6.8	4.7	3.4	13.8	16.4	13.3	15.8
1966	8.2	6.5	7.2	4.3	14.8	17.9	14.5	17.6
1967	7.4	7.1	6.3	2.7	13.1	11.3	13.0	11.2
1968	5.4	8.0	6.8	2.9	11.8	9.0	12.7	9.7
1969	6.4	8.1	6.5	2.9	14.9	19.1	14.7	18.8
1970	5.8	8.6	12.4	3.9	15.7	17.0	16.7	18.1
1971	6.1	10.8	11.4	4.2	20.5	21.8	22.0	23.4
1972	6.7	14.4	10.5	3.8	20.3	24.1	19.8	23.5
1973	5.9	12.2	10.3	3.7	23.5	23.9	24.7	25.1
1974	4.7	9.8	12.7	4.4	31.0	26.2	31.7	26.9
1975	5.6	14.1	13.7	6.0	38.0	35.0	36.9	34.0

Sources: See footnote to Tables 47, 50, and 58.

picture of regional farm income differences and changes over time and reflect a combination of output, price, and farm population factors.

The most interesting contrast in the recent development of regional farm income differentials is that between the traditional rice-growing areas and the hillier dry-field regions to the north and east. Table 59 shows the changes in the purchasing power of the per capita income from three important crop categories for these two regions. Rice is extremely important in both regions, as it is throughout Korea, and South Chŏlla province historically had somewhat higher yields per hectare and higher per capita output of rice than the upland region. Columns (V) and (VI) of Table 59 use deflated current prices to show per capita income from rice in these two different regions. The relative significance of barley and vegetables can be seen in columns (I) through (IV). The gap between the two regions due to income from barley is not as large as the reverse gap due to vegetable income. Government price supports and the southern province's ability to double-crop have helped South Chŏlla province's farm income in the early 1970s, but access to markets and a soil suitable for vegetables have given dry-field farmers in the upland region an even greater advantage.

The use of current prices in the analysis above gives us a more accurate picture of regional farm income differentials, but it does not remove all the complications and distortions. As mentioned above in the section on terms of trade, using the current year's rice price to value the same year's output introduces some distortion into the comparison of incomes over time and between regions, especially when extremely good harvests are compared with other relatively poor ones. This is because as much as 60 percent of a year's rice harvest is actually sold by the farmer in the following year, when prices have changed. The last two columns in Table 59, when compared with the previous two columns in the same table, show how this is true. Columns (VII) and (VIII) show per capita income from rice when output is valued by a weighted combination of current and following-year rice prices, with the weights determined by the average shares of the rice harvest sold in the current year and in the following year. This method is itself a simplification, since it uses average yearly prices rather than monthly prices, but it is a great improvement over using only current-year prices.

The significance of this more accurate pricing can be seen by looking at three years: 1962, a bad harvest in all regions; 1967, a bad year in the south but not in the north; and 1972, a good year everywhere, particularly in the south. As a result of the poor harvest in 1962, rice prices rose markedly in 1963 and, when 60 percent of output is valued at the 1963 average price it actually brought, South Chŏlla province's farmers received more income from rice when compared to 1961 rather than less, as the data using current prices would suggest. The percentage rise in the effective rice price was greater than the percentage fall in South Chŏlla province's rice output. In North Ch'ung-ch'ŏng province the crop failure was more serious, so there was no rise in income, but the drop in 1962 rice income was much less severe than the current-year price calculations would lead us to believe. The 1967 failure of the southern rice crops because of drought reveals a different phenomenon. Because the poor harvest was more localized, it did not have as great an effect on total national output nor on rice prices in the following year. In fact, the terms of trade for rice deteriorated in 1968, and the regional impact of the bad weather was made even more severe. The effective adverse movement in the purchasing power of rice affected the northern regions as well, of course, but such movements were easier to bear, since the harvest was respectable. Both 1962 and 1967 show that, by including in our analysis the actual price movements caused by poor harvest, the regional impact of the crop failures is made more pronounced if the failure is localized and less pronounced if the failure is wide-spread enough to increase prices by a large enough percentage to compensate for the percentage fall in output.

The reverse impact, that of a good year, on actual farm income can be seen by analyzing the 1972 bumper rice harvest. The use of current prices to value the 1972 crop does not show the influence of the drop in prices the following year and, as columns (VII) and (VIII) show, the benefit accruing to the south because of its high yields was deflated to the point where actual rice income per capita was lower when compared to other years rather than higher, as current price calculations would indicate.

The analysis in this section has shown the increase in rural income from farm production for different regions of South Korea. But farm production is not the only source of income for agricultural households, and two other classes of income need to be analyzed. Non-farm income includes a variety of elements, such as transfer payments from the government and from urban relatives, both farm and non-farm wage income, and income from side businesses such as fishing and forestry. In addition, public goods, such as schools, roads, and civil order, can make valuable contributions to farm income, both directly and indirectly. Road construction is discussed in some detail in the next section on rural isolation, but electrification of rural villages is a public good which has brought illumination, communication (television), and an alternative source of power for farm machinery.

Regional data on non-farm income are scarce and deal only with traditional forms of supplementary income from forestry and fishing. As data from the Ministry of Agriculture's *Farm Household Economic Survey* indicate, however, other forms of wage income have not changed very much in the ten years preceding 1975. On the other hand, it is safe to assume that the regional distribution of non-farm wage income favors areas closer to urban centers, thereby increasing the relative income of farm communities close to Seoul, Pusan, and Taegu.

Since the beginning of the rural electrification program in 1965, the share of farm households in South Korea with electric power has increased from 12 percent to over 60 percent, but regionally the rice-belt provinces have been somewhat slower to receive benefits from the government-sponsored program and, by the mid-1970s, the provinces with the large urban areas have also shown the highest rate of rural electrification. The data in

Table 60 show the pattern of rural electrification and lead us to conclude that, to a certain degree, they complement the regional pattern of farm income from crop sources.

TABLE 60   Electrified Rural Households as a Share of Farm
Population by Province, 1964–1974
(%)

Province	1964	1967	1971	1974
Kyŏnggi	29.1	33.1	56.5	83.8
Kangwŏn	7.5	17.5	37.2	59.1
North Ch'ungch'ŏng	13.8	16.6	31.1	62.1
South Ch'ungch'ŏng	12.3	14.8	26.5	49.8
North Chŏlla	12.8	17.0	29.6	58.5
South Chŏlla	8.0	12.7	27.2	55.1
North Kyŏngsang	10.6	14.5	30.2	63.1
South Kyŏngsang	13.0	18.1	33.8	62.9
Cheju Island	1.3	13.8	33.4	51.4
South Korea	12.6	17.4	33.0	61.1

Sources: Electrified household data are from Yun-sik Chang, *Han'guk chŏllyŏk suyo mit kagyŏk ŭi punsŏk*, and from The Korean Electric Association, *The Electric Yearbook, 1974*, p. 149. Total farm households are from Keidel.

Finally, a much more dramatic picture of regional per capita farm income and the degree to which it is influenced by urban centers is given by data for the 167 counties of South Korea for 1970. In an effort to document the recovery from drought in the late 1960s, the Ministry of the Interior used national income accounting methods to calculate county income from agriculture, forestry, and fishing. When the agricultural household income is converted to per capita figures, the range falls between incomes of below 20,000 wŏn per year for the poorest counties of Kangwŏn and South Chŏlla provinces and incomes of over 80,000 wŏn per year for farm communities close to Seoul. The geographical distribution of rich and poor counties can be seen by looking at Figure 14. It is easy to see how the counties with per capita income above 48,000 wŏn per year are clustered around Seoul and Pusan, with other pockets of well-to-do

FIGURE 14 Per Capita Farm Income by County, 1970

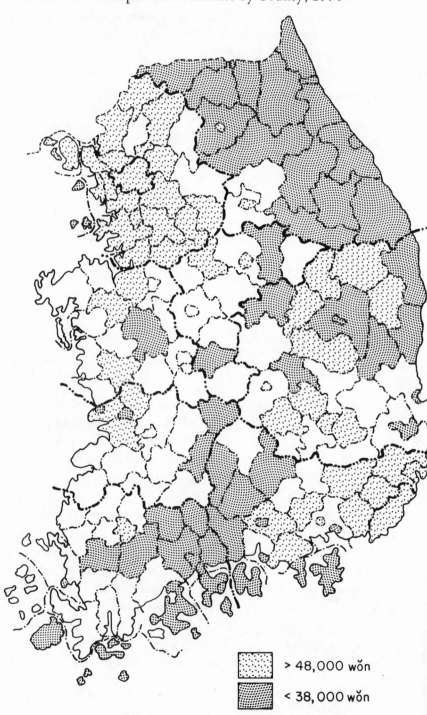

> 48,000 wŏn

< 38,000 wŏn

farmers in the vicinity of provincial capitals and industrial centers. By contrast, the counties with per capita income of less than 38,000 wŏn per year are concentrated in the remote and mountainous sections of the nation. How well this regional distribution of farm income coincides with a more sophisticated analysis of rural isolation will be seen in the next section.

In general, both provincial and county data confirm that, where farmers have been able to grow vegetables and obtain access to urban markets, their incomes have risen above those of the traditionally good grain regions, in spite of movements in prices which give an advantage to rice and barley cultivation.

RURAL ISOLATION

Land, climate, and population have by themselves determined the basic patterns of Korea's regional farm production and income for much of the twentieth century but, since the early 1960s, urbanization and industrialization have increased very rapidly, and the degree to which rural areas have been in contact with these modern developments has significantly affected their farm economies. Paved roads and expressways have been the major catalyst in reducing rural isolation.

Although total domestic freight more than tripled in the dozen years (1963–1975) for which there are data, freight moved by highway increased by a factor of more than 4.5. The share of total freight transported by road rose from 45.3 percent to over 60 percent, while the share of freight moved by rail declined. Just as significantly, the average distance hauled for a highway cargo almost doubled.

Highway cost-benefit studies in Korea confirm that highway traffic moves much more rapidly over paved road than it does over gravel, and that the costs in terms of vehicle depreciation are lower. The paving of national highways and the construction of limited-access freeways bear much of the responsibility for the increase in road freight traffic mentioned above. From before World War II until the early 1960s, only a small proportion of Korea's highways were paved, and much of the overland traffic required jeeps and other versatile vehicles. Beginning with the First Five-Year Plan in 1963, however, increasing attention was paid to paving non-urban roads and, as Table 61 shows, the share of national inter-city roads paved had reached 44 percent by 1975. The construction of expressways to link major cities and regions of Korea was begun in 1969, and the sudden ability to move freight and passengers at high speeds from many parts of the country to urban centers revolutionized cash-crop farming for the areas most directly affected.

Early impetus to road pavement came through United States aid in the form of earth-moving vehicles, paving equipment, and technical advice. The major consideration was military, but the repair of roads damaged in the Korean War and the paving of some vital sections provided immediate benefits in the form of improved transportation; the aid has had lasting influence through the transfer of technology and the encouragement of an expanded motor vehicle transportation system. Table 62 gives the actual levels of highway aid for the ten years following the Armistice and also shows the share of total highway investment supported by the aid. During the ten-year period, U.S. aid provided 18.9 percent of the funds invested in highway construction, with the bulk of the aid going to paving and a lesser share to the building of bridges.

As helpful as this early aid may have been for the development of a Korean highway system, the progress in the 1960s and early 1970s completely overshadows the scale of the work in the 1950s. The scope and rapidity of the highway paving program can be seen in Figures 15, 16, 17, and 18, which give the paved national highways and expressways for 1963, 1967, 1971 and 1975 respectively. The most dramatic expansion in the paved system has been in the 1970s, and the recent completion of the expressway through South Chŏlla province will perhaps alter the pattern of that area's retrogression in comparison to more northern provinces. In looking at the map for 1971, it is

TABLE 61 Paved Roads, 1936–1975

	National and Provincial Paved Highway (km)	Share of Total Roads Paved (%)	Limited Access Expressway (km)	National Paved Roads (km)	Share of National Roads Paved (%)
1936	339	2.1	–	293	4.8
1944	827	4.9	–	746	12.2
1947	809	5.3	–	737	14.3
1950	403	2.6	–	364	7.0
1953	329	2.1	–	294	5.1
1956	369	2.3	–	332	5.8
1959	612	3.8	–	554	9.7
1961	779	4.8	–	721	12.6
1963	919	5.4	–	865	14.9
1965	1,113	6.1	–	1,042	17.7
1967	1,494	7.9	–	1,442[a]	17.6[a]
1969	1,772	9.4	472	1,652	20.4
1971	2,542	13.4	655	2,302	28.3
1973	3,358	17.5	1,013	2,869	34.6
1975	4,325	22.6	1,142	3,620	44.0

Sources: Unpublished data sheet from the Ministry of Construction (MOC), compiled from their *Statistical Yearbook 1960* (for 1936–1959 data), individual years of the same yearbook (1961–1966), and individual provincial road reports (1967–1975).

Note: [a]There was a reclassification of roads in 1966, with over 2,000 km of provincial roads reclassified as national roads. As a result, the data are not consistent between 1965 and 1967.

FIGURE 15 Paved Highway in South Korea, 1963

FIGURE 16 Paved National Highway in South Korea, 1967

FIGURE 17 Expressways and Paved National
Highway in South Korea, 1971

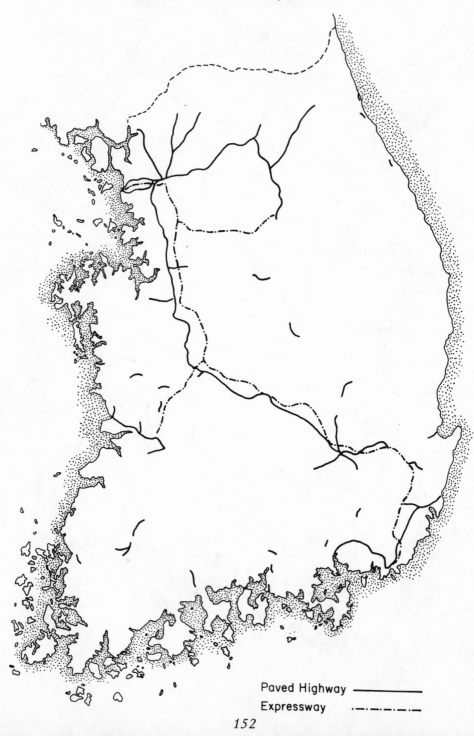

Paved Highway ————
Expressway —·—·—·—

FIGURE 18 Expressways and Paved National
Highway in South Korea, 1975

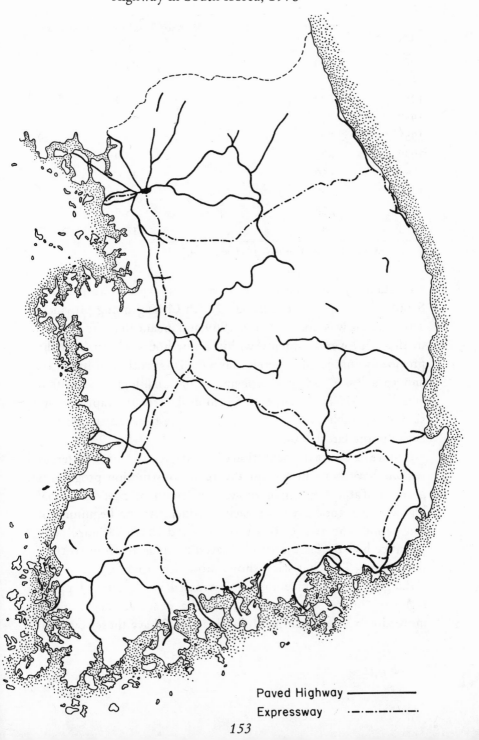

Paved Highway ⸻
Expressway ⸱–⸱–⸱–⸱–⸱

TABLE 62 U.S. Assistance to Korean Road Investment,
1954–1963

	Amount ($U.S. 1,000)	Share of Total Road Investment (%)
1954	2,280	41.5
1955	2,271	17.4
1956	670	21.7
1957	937	11.2
1958	3,902	28.5
1959	828	7.9
1960	950	11.4
1961	1,209	17.3
1962	2,028	25.5
1963	583	10.5

Sources: MOC, Statistical Yearbook 1963 and 1964.

important to note that the portion of the expressway between Seoul and Taejŏn runs through North Ch'ungch'ŏng province. This section was completed in 1969 and meant that the farmers in that area were suddenly within a few hours by truck of the previously inaccessible Seoul markets. This section of highway, and to a lesser extent the paving of national highways in that area by 1967, are in large part responsible for the rapid expansion of vegetable production and of income from vegetables that began in the late 1960s.

A more statistical presentation of the degree to which different regions have benefited from the road construction program is given in Table 63, which shows the length of paved national highway per total area for each province. At the beginning of the period, the two Chŏlla provinces and North Ch'ungch'ŏng province had by far the fewest paved roads in relation to their total area. The statistics show, however, that, while North Ch'ungch'ŏng province rapidly increased its density of paved highways in the late 1960s, the Chŏlla provinces began much more slowly. For example, only since 1973 have these southern rice-belt provinces been able to match the scale of paving in the more mountainous South Kyŏngsang region. It is also significant that the poorest region, Kangwŏn province, has shown both a low level of paving and a low rate of growth in its paved highway density.

TABLE 63 National Paved Road Density
(km of paved road/1,000 km² of area)

Province	1964	1967	1969	1971	1973	1975
Kyŏnggi	18	26	30	32	38	44
Kangwŏn	9	13	15	16	17	23
North Ch'ungch'ŏng	3	10	12	19	27	41
South Ch'ungch'ŏng	9	13	16	25	40	45
North Chŏlla	2	4	8	12	23	28
South Chŏlla	3	5	6	16	19	31
North Kyŏngsang	9	11	15	21	24	34
South Kyŏngsang	7	10	11	17	25	29
Cheju Island	20	34	63	130	142	142
South Korea[a]	8	12	15	22	27	35

Sources: Calculated from paved national road data in individual years of the MOC, *Statistical Yearbook* and total province area from EPB, *Korea Statistical Yearbook.*

Note: [a]Does not include the special cities of Seoul and Pusan.

These statistics, together with the maps, give a good idea of how road paving and access to urban areas have correlated with some of the production and income patterns presented earlier in the chapter. A more detailed analysis of the same relationship is given below, using county data for the single year 1970.

The degree to which a certain region or county is isolated from the modern industrial sectors of the national economy depends both on how far it is from various urban industrial centers and on the size of those centers. A county close to a small manufacturing center will not benefit as much as another county somewhat further away from a center which is nevertheless much larger. Figure 19 shows the results of considering each

FIGURE 19 Least and Most Isolated Counties in South Korea, 1970

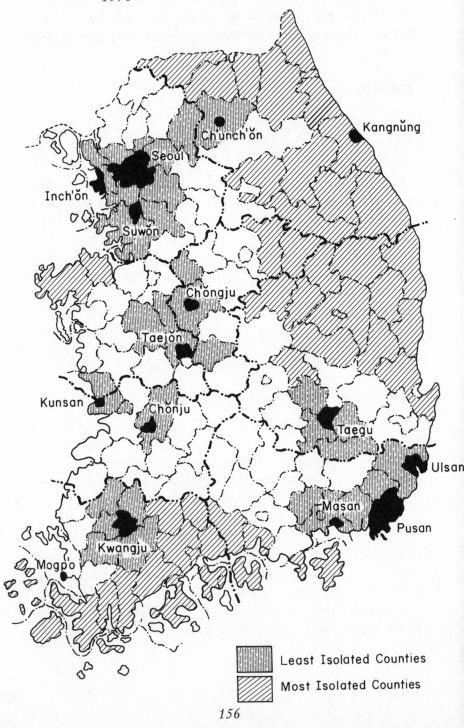

Least Isolated Counties

Most Isolated Counties

county's distance from Korea's fourteen largest centers of industrial employment in 1970 as well as the relative size of the corresponding centers. By the method used, a county at a certain distance from, say, five small urban areas might be isolated to the same degree as another county the same distance from a single larger city. A "gravity model" of the form $I = a(U^b R^c / D^d)$ was used, where the isolation, I, of a county from a city is calculated from the size of the rural county's population, R, the size of the urban area's industrial employment, U, and the distance by national highway between the county and the city, D. The parameters a, b, c, and d were estimated using separate data on frequency of bus transportation between the same industrial centers and outlying rural areas. The value of d turned out to be 1.500.[3] The individual indexes of isolation for a county from all the urban areas were summed to give the county's overall isolation index.

The maps, of course, show that counties close to the largest industrial centers are least isolated. But the maps also give a more graphic analysis of the degree of relative isolation for more distant counties. For some of the western counties in North Ch'ungch'ŏng province, proximity to Taejŏn, to the provincial capital of Ch'ŏngju, and to Seoul combine to make them relatively accessible to the modern sector. In South Chŏlla province, only the counties around the city of Kwangju turn out to have access to industrial population centers. The calculation of the isolation indexes did not consider the degree to which the national roads linking the counties with cities were paved. From the information in the highway maps above, we can be certain that the advantage given to North Ch'ungch'ŏng by its physical location is greatly amplified by the paved road connections to Seoul which existed in 1970.

The same maps of isolation differences show that the mountainous northeast and certain sections of South Chŏlla and South Kyŏngsang provinces have the least access to industrial centers of all the regions of South Korea. It is interesting to compare the map of regional isolation in 1970 (p. 156) with the map of regional farm per capita income (p. 146) in the previous section. Although the correlation is far from exact, it is clear that, in general, the more isolated farms are also the poorer farms. In fact, more sophisticated statistical analysis shows that a county's isolation is very significant in explaining its per capita farm income in relation to that of others.[4] Given the importance of harbors and of transportation and communication corridors within the Korean peninsula, it is safe to assume that the location of the urban industrial centers was not determined by the location of productive agricultural regions, although that may have had some bearing on the historical development of Seoul as the dynastic capital and largest city. It can be safely concluded, then, that the correlation between isolation and farm income proves the importance of urban markets and knowledge of urban markets in modifying the traditional productivity and profitability of South Korea's agricultural regions.

CONCLUSION

This chapter has shown that, although Korea's agriculture is by many standards homogeneous from one region to another, differences in production and income patterns do exist. Furthermore, the relationships between the various regions have been changing as the influence of modernization and urban demand for vegetables and other crops have begun to favor regions that historically were not able to support as dense a population as the traditionally more prosperous areas.

The single most important factor in explaining long-term per capita farm production and income differences is probably the great range in levels of farm population density per area of cultivated land. This pattern of differences has its origins in historical patterns of productivity, but by World War II those patterns had already altered to the point where farmers in Kyŏnggi province, with larger farms, also had a level of productivity which, though lower than that of southern areas, was high enough to give them a better standard of living.

In the thirty years since 1945, the traditional pattern has been altered further as a result of fruit and vegetable cultivation in upland provinces and in areas with direct access to cities. With the increasingly rapid construction of paved highways and expressways, this trend towards the production of non-grain crops is likely to increase and become regionally more widespread.

PART THREE

Government Policies

SIX

Government Investment Policy and Plans

THE SITUATION BEFORE 1945

Government agricultural investment policy before 1945 was part of overall Japanese colonial economic policy carried out in Japan's own interests—the major objectives being increased production of food and raw materials to satisfy Japan's own growing demand. The first rice production plan was formulated upon completion of the cadastral land surveys to cover the period 1918–1926 and set a rice production target of 1.3 million metric tons (M/T). Measures were taken to improve and expand irrigation facilities and to convert irrigable upland into paddy land. Wasteland and tideland conversion projects were implemented, and improved crop varieties and chemical fertilizer were distributed. A total of 63 million yen was provided as direct subsidy and 75 million yen as a credit loan to implement various projects.

As a result, many advanced agricultural practices were adopted by farmers, and the level of agricultural production was increased, although the compulsory measures used by the Japanese authorities created much resentment on the part of Korean farmers. But the production target was not met, so formulation of a second rice production plan was necessary.

The second rice production plan was launched in 1927 with a target of a 1.2 million M/T increase in rice production. A total of 325 million yen was invested to expand and improve irrigation facilities on 350,000 hectares of paddy land, and a nationwide quasi-governmental body called the Korean Agricultural Association was organized (in 1926) to undertake various agricultural programs throughout the country. But the second rice production plan was suspended midway in 1930 because of a sharp drop in rice prices in Japan due to the Great Depression; this reduced Japan's interest in raising Korean rice production. The third rice production increase plan was formulated in order to meet the growing demand for rice arising out of Japan's military expansion and subsequent involvement in World War II; this plan was suspended with Japan's defeat in the war.

Along with the efforts to increase food production, the Japanese government encouraged farmers in the sourthern part of Korea to cultivate cotton under the "cotton production increase plan" in order to supply the cotton textile industry in Japan. With Japan's occupation of the producing area in northern China, however, cotton production increase efforts in Korea received less attention. Sericulture and livestock propagation, particularly propagation of Korean native cattle, also received emphasis during the Japanese colonial period.

Although Japan's interest in developing Korean agriculture was based on a desire to provide food and industrial raw materials to meet Japan's own needs, various measures, including the introduction of improved crop varieties, improved cultivation methods, dissemination of chemical fertilizers, and the expansion of irrigation facilities, did contribute to raising the productivity of Korean agriculture, as indicated in Chapter 4.

THE YEARS FROM 1945 TO 1960

The U.S. Military Government replaced the Japanese Colonial Government in August 1945 and took over the colonial administrative system without modification. Amidst social confusion after Liberation from Japanese rule, the U.S. Military Government was obliged to concentrate all its efforts on maintaining social order, and the major emphasis in economic policy, if any, was directed toward reducing the post-war inflationary spiral. There were no policies or programs designed to expand farm production. Policy instead was oriented toward securing an adequate flow of food-grain for urban consumers either out of current domestic farm output or through importation of foreign grain. Government investments and loans to agriculture were severely limited by the small size of the overall budget, and maintaining existing irrigation facilities and importation of fertilizer were about all the budget was capable of financing.

With the establishment of the Government of the Republic of Korea in 1948, considerable efforts were made to shift to a new agricultural policy. The most notable effort was the land tenure reform carried out under the Land Reform Act of June 1949, an effort that will be discussed at length in Chapter 10. In 1950, the Grain Management Law was passed to replace the compulsory rice collection system with a government purchase system at government-determined prices.

Public investment and loans in agriculture started in 1949 when the government formulated an economic rehabilitation budget based on counterpart funds generated by U.S. aid delivered by the Economic Cooperation Administration (ECA). But the investment of these counterpart funds was withheld, and a major part had to be spent on combating runaway inflation. Under the economic stabilization program formulated in 1950, counterpart funds were again to be used as a source of government investment and loan projects, but the whole stabilization program was suspended when the Korean War broke out in June.

Following the outbreak of the Korean War, general budget expenses almost tripled over the level of the previous year to a total of 235.6 million wǒn. Government finance was shifted to a wartime system, and defense outlays totaled 132.4 million wǒn or 56.2 percent of the total budget expenditures. The agricultural sector received only 6 million wǒn or 2.5 percent, but the special account for grain management was set at 113.1 million wǒn, an increase from the previous year. Securing of military grain and the distribution of relief grain to refugees emerged as one of the priority issues at the time with the annual quantity of rice needed for government distribution estimated at about 600,000–700,000 M/T. Because of enormous wartime budgetary requirements, resulting in monetary inflation and an upward spiraling of grain prices, the government was obliged to discontinue the direct purchases of rice from farmers. Instead, an attempt was made to secure rice by means of farmland reimbursement for land distributed to tenants at the time of land reform. Thus, two special accounts mentioned above increased continuously during the 1950–1953 period.

With the Armistice agreement signed in July 1953, policy emphasis shifted from wartime efforts to recovering from war damage, building new plant facilities, and alleviating the rapid inflation in general prices. Beginning in 1954, programs were actively launched with U.S. aid, and the emphasis in U.S. aid policy changed from relief to economic development with the financing of imports of raw materials and capital equipment. Major sources of public investment and loans were grant aids from the Foreign Operations Administration (FOA), counterpart funds from U.S. grants usable under the Korea-U.S. agreement of the Combined Economic Board (CEB) for Economic Rehabilitation and Financial Stabilization Program concluded in 1953, and the special account for economic rehabilitation established in 1954.

Overall food policy was further altered in a major way with the signing of the U.S. Farm Surplus Importation Agreement in 1955. American farm products imported under this agreement were wheat, barley, raw cotton, corn, sorghum, and tallow, with wheat and barley accounting for about 50 to 60 percent of the total value of imports. The sale proceeds of these imported commodities in the domestic market provided "counterpart funds," which enabled the Agriculture Bank to establish an agricultural production loan fund amounting to 39 million wŏn and a rice lien program.

In 1958 the Agriculture Bank was established as the sole credit institution in rural areas. The Office of Agricultural Extension was organized to assume the responsibilities for the development of improved farm technology, and a five-year food production plan was formulated for the period 1958 through 1962. Subsidies and loans were provided to farmers for increased food production, and the water fee was reduced. With the economic confusion surrounding the Student Revolution of April 1960, implementation of the five-year food production plan had to be suspended, but was resumed under the government of the Chang Myon regime in late 1960 with a major emphasis on expansion of irrigation facilities. Despite the various efforts described above, farm policies during the fifteen-year period from 1945 to 1960 were subordinated to the government's concern with industry and general economic stability, and, as a result, the impact of government efforts on the agricultural sector was very limited.

THE FIRST FIVE-YEAR ECONOMIC DEVELOPMENT PLAN PERIOD (1962-1966)

Economic policy under the First Five-Year Plan (FFYP), formulated shortly after the Park Military Revolution in May

1961, shifted emphasis from rehabilitation and stability to expansion; this had a direct effect upon agricultural development.

The basic goal of the plan was to build a foundation on which to attain self-sustaining economic growth. The agricultural policies geared to this economic goal included: 1) development of viable farm units through regional farm development; 2) expansion of arable land and pastures; 3) maintenance of the prices of farm products at "reasonable" levels; 4) promotion of the livestock industry; 5) development of forestry resources; and 6) fostering of farm organizations at various levels.

These policies, however, were not fully implemented. The creation of optimum farm units was stalled because such a comprehensive program could not be undertaken with an average farm holding of less than one hectare. The idea of abolishing the ownership ceiling of three hectares was intermittently suggested with a view to expanding farming scale, but was never realized. The policy of supporting the prices of farm products at "reasonable" levels could not be implemented because of funding difficulties, and emphasis was shifted instead to maintaining farm prices, particularly rice prices, at a low level in favor of urban consumers. Specific policy action was not taken to develop large-scale commercial livestock farms, due to limited feed resources and lack of transportation, processing, and marketing facilities.

Various institutional reforms, however, were carried out. The most notable was a program to liquidate the usurious debts of rural areas in order to relieve farmers and fishermen of their heavy debt burden. The Law on Settlement of Usurious Debts of Farm and Fishing Communities was enacted in 1961 and required that all loans with interest rates exceeding 20 percent per annum be reported to the authorities. A total of 4.8 billion wǒn was reported in loans by the end of 1961, of which 2.5 billion were declared usurious. Credit debentures were issued to creditors by agricultural cooperatives for their loans receivable, and debtors were required to repay their debts to the agricultural cooperatives on an installment basis. Other institutional reforms included a new Agricultural Cooperatives Law promulgated to merge the Agriculture Bank into the National Agricultural Cooperative Federation and thus to integrate cooperative marketing and agricultural credit services. The Office of Agricultural Extension and the Bureau of Community Development were also merged to create the Office of Rural Development under the Ministry of Agriculture and Forestry.

In addition to these reforms, the government launched various programs to boost agricultural production, including expansion of irrigation facilities, slopeland reclamation by bench-terracing methods, increased supply of fertilizer and pesticides, and the strengthening of agricultural research. A total of 10 billion wŏn or 22.2 percent of total agricultural investment was allocated for expanding and improving irrigation facilities, and 53,000 hectares of paddy land were brought under irrigation during the plan period. Beginning in 1964, paddy consolidation projects also assumed increasing importance. These projects involved consolidation of small irregularly shaped paddies into single, larger units of uniform shape. Included in this effort were improvements of irrigation and drainage ditches, installation of on-farm water control structures, and construction of feeder roads to provide better access to the fields. The first trial projects were carried out in North Kyŏngsang province on approximately 6,000 hectares of the existing paddy land under local government financing. Having received a favorable reaction from the farmers, the government decided to expand the project scale through direct subsidies and, in 1965, the government invested 85 million wŏn to consolidate 18,000 hectares of the existing paddy land and increased this investment outlay in 1966 to 188 million wŏn for 19,500 hectares of paddy land. Tideland development and slopeland development projects were also intensively carried out during 1962–1966, as indicated in a previous chapter. A total of 2.6 billion wŏn or roughly 8 percent of the total investment in the agricultural sector was allocated to this effort.

Among the most noteworthy efforts during this period were the accomplishments of the Office of Rural Development in the field of agricultural research and extension. The government spent about 2 billion wŏn or 4.4 percent of total agricultural investment outlays in this sector during 1962–1966. A total of 125,000 M/T of improved seeds of various crops, such as rice, barley, soybean, and corn, were distributed to local farmers; 100,500 demonstration plots by 14,600 efficient farmers were established to disseminate improved fertilizer application and improved farming techniques; and emphasis was also placed on training local volunteer leaders and rural youth through 4-H Club activities.

In the farm mechanization program, the major attention of the government was directed to distributing water pumps and powered sprayers and dusters together with such hand-driven equipment as deep plows and hand carts. Total investment and loans for this program amounted to 757 million wŏn during 1962–1966, of which about 310 million wŏn, or 40 percent, was spent to subsidize the price of water pumps. Successive crop failures in 1962 and 1963 due to drought and plant diseases were major reasons for this government concentration on the distribution of water pumps and powered sprayers and dusters.

As a measure to attain self-sufficiency in fertilizers, the government stepped up building fertilizer plants during the First Plan period. Following the building of the Ch'ungju fertilizer plant with an annual capacity of 80,000 M/T of urea, Honam and a second complex fertilizer plant were built. Moreover, the government made fertilizer available for farmers at subsidized prices, lower than those on the international market, and extended credit for the purchase of fertilizer at relatively low interest rates.

One of the main improvements in farming in this period was in the area of blight control. Until the beginning of the 1960s, agricultural chemicals had been used almost entirely for cultivation of fruits and vegetables. The Law on Prevention of Epidemics in Agriculture, enacted in December 1961, marked a turning point for insect and disease control in Korean agriculture, and annual consumption of agricultural chemicals of various kinds increased from 7,400 M/T in 1962 to 12,600 M/T in 1966, approximately a 70 percent increase during the five-year period.

Data on government investment in this period are presented in Tables 64, 65, and 66.

TABLE 64 Total Investment and Loan, 1962–1966
(million wŏn)

	Amount	Ratio
Total Government Investment in all Sectors	170,347	
In Agriculture, Forestry & Fisheries	45,185	100.0
In Agriculture	33,637	74.4
Improvement of Production Basis	14,600	32.3
Irrigation Facilities	10,035	22.2
Land Consolidation	273	0.6
Slopeland Reclamation	1,726	3.8
Tideland Reclamation	2,566	5.7
Agricultural Production Increase	7,616	16.9
Cash Crops	609	1.3
Sericulture	1,404	3.1
Livestock	1,971	4.4
Agricultural Experiment & Extension	1,970	4.4
Agricultural Mechanization	765	1.7
Others	5,467	12.1
In Forestry	4,633	10.3
In Fisheries	6,555	14.5

Source: MOF, *Summary of Financial Implementation* for FY 1962–1966.

TABLE 65 Total Investment and Loan, 1964–1966
(million wŏn)

	Investment		Loan		Total	
	Amount	Ratio	Amount	Ratio	Amount	Ratio
Total Government Investment in all Sectors	95,872		19,982	100.0	115,864	100.0
In Agriculture, Forestry & Fisheries	23,011	100.0	7,708	100.0	30,719	100.0
In Agriculture	16,895	73.4	5,978	77.6	22,873	74.4
Improvement of Production Basis	8,961	38.9	1,699	22.0	10,660	34.7
Irrigation Facilities	5,936	25.8	774	10.0	6,710	21.8
Land Consolidation	273	1.2			273	0.9
Slopeland Reclamation	329	1.4	925	12.0	1,254	4.0
Tideland Reclamation	2,423	10.5			2,423	7.9
Agricultural Production Increase	1,933	8.4	1,736	22.5	3,669	11.9
Cash Crops	236	1.0	92	1.2	328	1.1
Sericulture	911	4.0	203	2.6	1,114	3.6
Livestock	367	1.6	270	3.5	637	2.0

TABLE 65 (continued)

	Investment		Loan		Total	
	Amount	Ratio	Amount	Ratio	Amount	Ratio
Agricultural Experiment & Extension	1,356	5.9			1,356	4.4
Agricultural Mechanization	757	3.3	8	0.1	756	2.5
Others	2,374	10.3	1,970	25.6	4,344	14.1
In Forestry	2,752	12.0			2,752	9.0
In Fisheries	3,364	14.6	1,370	17.8	4,734	15.4

Source: MOF, Summary of Financial Implementation for FY 1964–1966.

TABLE 66 Farm Machinery Supply, 1962–1966

(1,000 wŏn)

Kind of Machine	Supply	Investment and Loan Central Govt.	Local Govt.	Loan	Farmer	Total
Power Tiller	1,360	88,835		5,850	174,996	269,681
Plow	28,644	14,915		1,887	22,785	37,700
Power Sprayer and Duster	7,158	73,783			156,523	232,193
Lever-type Hand Sprayer	1,503	2,430			3,404	5,834
Back Bearing Hand Sprayer	30,038	21,591			58,230	79,821
Hand Duster	12,080	5,284			16,555	21,839
Pump	8,675	310,032	310,032			620,064
Engine	12,324	84,487			21,124	105,611
Thresher	47	2,870	170		139	3,179
Seed Cleaner	66	13,368	4,032			17,400
Potato-grinding Machine	878	3,492			1,496	4,988
Rear Car	31,654	51,239			127,103	178,342
Soil-moving Car	99	37,630			18,543	56,173
Perforator	147	5,953			5,880	11,833
Others	21,890	41,100			20,476	61,576
Total	156,563	757,009	314,234	7,737	627,254	1706,234

Source: Ministry of Agriculture and Forestry.

THE SECOND FIVE-YEAR ECONOMIC
DEVELOPMENT PLAN PERIOD (1967-1971)

The basic goals enunciated in the Second Five-Year Plan were: 1) to modernize the industrial structure; and 2) to accelerate economic growth so as to attain a self-sustaining economy. Agricultural production was planned to grow at an annual rate of 5 percent over the Plan period, and the major objectives in the agricultural sector were an adequate supply of food and enhancement of the social and economic position of farmers.

Included in the development measures were: 1) increased foodgrain production and attainment of self-sufficiency in staple food by 1971; 2) development of intensive farming areas under the principle of suitable-crop-for-suitable-area; 3) farm price supports to provide greater incentives for increased production; 4) promotion of the livestock industry through a stable supply of feed; and 5) increased agricultural exports through promotion of exportable farm products.

In order to strengthen the legal foundation for the implementation of various farm policies and programs, the government enacted the Basic Agricultural Law in 1967. In addition, the investment and loans budgets for the agricultural sector were substantially expanded. In the agricultural sector, the highest priority was given to the development of land and water resource projects. Construction and improvement of small irrigation systems was by far the largest program, accounting for about 70-80 percent of the total investment and loans in land and water development programs in most years during the Plan period. Large-scale development projects did not become important until 1971 and 1972 when large investments were initiated for the P'yŏngt'aek-Kŭm River irrigation projects. Following severe droughts in 1967 and 1968, the government embarked on tube well construction on a massive scale in 1969 and 1970.

Government financial support was also given to paddy consolidation, upland reclamation, and tideland development projects. Original targets were to perform consolidation work on 200,000 hectares of tideland into paddy land, and to bring 126,000 hectares of hillside land under cultivation by 1971. But actual performance on these projects lagged far behind the original target in the plan. A total of 95,000 hectares of paddy land was rearranged, 42,000 hectares of hillside land reclaimed, and 1,200 hectares of tideland developed into paddy land during the Plan period.

Efforts were also directed toward introduction and diffusion of improved varieties of major crops, such as rice, barley, and soybean. A seed-improvement program was expanded and carried out intensively during the Plan period. Agricultural research and extension services were continuously expanded to improve farming technology, particularly to improve the efficient application of fertilizer and pesticides. The increase in fertilizer use was due in part to the government policy of making sufficient supplies available to farmers at subsidized prices. The government made lime available to farmers, for example, at full subsidy to correct soil acidity, and use of agricultural lime has increased from 175,000 M/T in 1966 to almost 500,000 M/T in 1971.

The Second Five-Year Plan gave added impetus to the farm mechanization process. As indicated in the previous chapters, rapid industrial and urban growth during the 1960s stimulated off-farm migration at a pace that resulted in a decline in both farm population and labor, and motivated the government to promote greater mechanization. In addition to providing financial and technical assistance, the government established a unified machinery inspection program in 1966 to test manufactured and imported machinery for compliance with government specification. The government also provided direct subsidies of grants and credit to farmers for machine purchase. A total of 12 billion wŏn was spent for farm machinery during the Plan period, of which approximately 9 billion were spent by the government in the form of subsidies and credit loans. Of the total spending on farm machines, the government subsidized about 39.5 percent, which compared with 44.4 percent during the First Five-Year Plan period.

To boost agricultural production and to upgrade farm income, the government also initiated a program of developing "specialized production areas" beginning in 1968. The program was launched by selecting 43 project areas for promotion of specialized production of sericulture, livestock, fruits, vegetables, and marine products. To implement this program, the government allocated 34 billion wŏn during the 1968–1971 period, of which subsidies accounted for 8 billion wŏn and loans for 26 billion wŏn. This program was not pushed forward as originally planned, however, because marketing, processing, and storing facilities were totally inadequate to the task of absorbing the increased farm products from the specialized areas. Moreover, severe fluctuation in the prices of fruits, vegetables, and other cash crops proved to be a major impediment to individual farmers considering specialization. Sericultural production, however, was a success story owing to the continuous increase in the export demand. Supporting data on investment in the Second Five-Year Plan period are presented in Tables 67–75.

THE THIRD FIVE-YEAR ECONOMIC DEVELOPMENT PLAN PERIOD (1972–1976)

During the 1960s, although the agricultural sector grew at an annual average rate of 4.5 percent, supply failed to keep pace with the rising demand, and imports of food-grains rose. In addition to this growing grain trade deficit, the government also became concerned with the growing income disparity between rural and urban households. Growing food deficits and a greater income disparity, therefore, were the two major developments which caused government to put much more emphasis on agricultural development in the Third Five-Year Plan of 1972–1976.

Agricultural development strategies were laid down within

TABLE 67 Total Investment and Loan, 1967–1971

(million wŏn)

					Actual			
			Investment		Loan		Total	
	Amount	Composition Ratio	Amount	Composition Ratio	Amount	Composition Ratio	Amount	Composition Ratio
Total	401,090	100.0	627,076	100.0	142,727	100.0	769,803	100.0
Agriculture, Forestry & Fisheries	93,838	23.4	138,825	22.1	59,924	42.0	198,749	25.8
Mining & Manufacturing	19,868	5.0	75,156	12.0	25,573	17.9	100,729	13.1
Social Overhead Capital Formation & Other Services	287,384	71.6	413,096	65.9	57,230	40.1	470,326	61.1

Source: MOF, *Summary of Financial Implementation for FY 1967–1971.*

TABLE 68 Government Investment and Loan, 1967–1971
(million wŏn)

	Investment		Loan		Total	
	Amount	*Ratio*	*Amount*	*Ratio*	*Amount*	*Ratio*
Total Investment	627,076	100.0	142,727	100.0	769,803	100.0
Agriculture, Forestry & Fisheries	138,824	100.0	59,944	100.0	198,768	100.0
Agriculture	106,893	80.0	45,708	76.3	152,601	76.8
Improvement of Production Basis	34,711	25.0	11,292	18.8	46,003	23.1
Irrigation Facilities	28,807	20.8	8,169	13.6	36,976	18.6
Land Consolidation	2,382	1.7		–	2,382	1.2
Slopeland Reclamation	130	0.1	3,073	5.1	3,203	1.6
Tideland Reclamation	3,552	2.6	50		3,602	1.8
Agricultural Products	13,328	9.6	14,847	24.8	28,175	14.2
Agricultural Experiment & Extension Service	6,937	5.0	465	0.8	7,402	3.7
Cash Crops	932	0.7	2,475	4.1	3,407	1.7
Sericulture	3,260	2.3	2,900	4.8	6,160	3.1
Livestock	3,425	2.5	5,063	8.4	8,488	4.3

(continued)

TABLE 68 (continued)

	Investment		Loan		Total	
	Amount	Ratio	Amount	Ratio	Amount	Ratio
Agricultural Mechanization	4,708	3.4	4,107	6.9	8,815	4.4
Others	39,592	28.5	4,559	7.6	44,151	22.2
Forestry	15,472	11.1	892	1.5	16,364	8.2
Fisheries	16,459	11.9	13,344	22.3	29,803	15.0

Source: MOF, *Summary of Financial Implementation for FY 1967–1971*

TABLE 69 Special Program for Farm Income Increase, 1968–1971

	Outline of Program		
	Specialized Areas	Farm Households Involved	Total Investment
	90	410	49.7
		(1,000s)	(billion wŏn)

	Investment and Loan				
	Planned (A)	%	Realized (B)	%	B/A %
	(billion wŏn)		(billion wŏn)		
Sericulture	9.3	18.6	9.2		98.9
Cash Crops	11.3	22.8	9.7		85.4
Livestock	19.4	39.1	18.1		93.3
Fisheries	3.1	6.3	3.0		96.8
Forestry	1.6	3.2	1.5		93.8
Others	2.3	4.6	2.3		100.0
Interest Subsidy	2.7	5.4	1.4		51.9
Total	49.7	100.0	45.2		90.8

Source: Ministry of Agriculture and Forestry.

TABLE 70 Source of Investment, 1968–1971

	Planned (A)	Realized (B)	Ratio (B/A)
	(billion wŏn)	(billion wŏn)	(%)
Total	49.7	45.2	90.8
Subsidy	7.6	7.0	91
Central Government	2.3	2.2	96
Local Government	3.9	3.6	91
Grain	1.4	1.2	86
Loan	28.1	25.3	90
Financial Fund S.A.	9.5	8.5	89
Bank Loan	14.9	14.4	97
Others	3.7	2.4	65
Farmers Burden	11.3	11.6	103
Interest Subsidy	2.7	1.3	48

Source: Ministry of Agriculture and Forestry.

TABLE 71 Major Achievements of Special Program for Farm Income Increase

	Unit	(A) Planned	(B) Realized	B/A %
Mulberry Plant	Million Plants	189	187	99
Mushroom	1,000 p'yŏng	223	214	96
High-Quality Vegetable	1,000 p'yŏng	417	416	100
Orange	1,000 Plants	3,638	3,638	100
Korean Cattle	1,000 Heads	193	184	95
Milk Cow	Head	4,371	4,285	98
Oyster	Each	8,220	8,194	100
Clam	Hectare	3,644	3,644	100
Chestnut	1,000 Plants	4,765	4,765	100

Source: Ministry of Agriculture and Forestry

TABLE 72 Comparison of Farm Incomes, 1968–1971
(1,000 wŏn)

	1968	1969	1970	1971
General Farm Household (A)	179	218	256	286
Farm Household in Special Program (B)	195	242	288	333
B/A	108.9	111.0	112.5	116.4

Source: Ministry of Agriculture and Forestry.

Note: The farm household income figures used here could not be adjusted in accordance with the procedures explained in Appendix C without destroying the comparability between the two types of households.

TABLE 73 Improvement of Production Basis, 1967–1971
(hectares)

Year	Irrigation Achieved	Land Consolidation		Slopeland Reclamation		Tideland Reclamation	
		Planned	Achieved	Planned	Achieved	Planned	Achieved
1967	15,000	23,197	23,246	10,942	16,785		
1968	1,481	15,159	15,972	40,000	13,500		
1969	228,392	10,000	13,527	20,000	7,690		
1970	83,444	68,700	15,380	27,400	2,953		342
1971	14,000	83,000	26,907	27,400	1,137		846
Total	342,317	200,056	95,032	125,742	42,065	20,000	1,188

Source: Ministry of Agriculture and Forestry.

TABLE 74 Irrigation Project, 1968–1971

(area in hectares)
(cost in million wŏn)

	1968		1969		1970		1971		Total	
	Area	Cost	Area	Cost	Area	Cost	Area	Cost	Area	Cost
New Development										
Tube Well	1,411	99	97,744	3,403	27,522	981	–	–	126,677	4,483
Infiltration Gallery	13	2	58,205	4,253	24,111	2,130	1,855	392	84,184	6,777
Pumping Station	–	–	18,149	2,579	13,911	2,181	4,399	1,623	26,459	6,383
Feed Canal	–	–	8,438	422	1,355	102	1,458	178	11,251	702
Weir	–	–	7,515	502	3,107	494	1,420	361	12,042	1,357
Reservoir	37	6	15,098	3,057	10,333	4,125	6,203	3,497	31,671	10,685
Sub-Total	1,461	107	205,149	14,216	70,339	10,013	15,335	6,051	292,284	30,387
Supplementary Irrigation & Drainage (Reclaimed Tideland)	20	2	23,243	2,347	13,105	1,338	4,865	2,062	41,233	5,749
Total	1,481	109	228,392	16,563	83,444	11,351	20,200	8,113	33,517	36,136

TABLE 75 Farm Machinery Supply, 1967–1971

(1,000 wŏn)

Kind of Machine	Supply	Investment and Loan				
		Central Govt.	Local Govt.	Loan	Farmer	Total
Farm Tractor	60	33,982		25,307	62,572	121,861
Power Tiller	16,112	1,107,904		1,990,251	2,254,563	5,352,718
Plow						
Power Sprayer & Duster	64,024	706,009		1,592,835	420,424	2,719,268
Power Sprayer	3,700	179,082		261,252	79,213	519,547
Lever-type Hand Sprayer	403,131	14,572		210,000	90,000	314,572
Hand Duster						
Pump	22,285	2,527,898				2,527,898
Thresher	408	17,482	563		11,581	29,626
Dryer	39	1,928	1,686		242	3,856
Seed Cleaner	39	5,769	4,486		1,332	11,537
Cutter	142	14,873		13,795	5,518	34,186
Sowing Machine	9	3,351	39,315			42,666
Rear Car	25,424	26,818			113,851	140,669
Trencher	16	56,675				56,675
Feed Mixer	133	11,522		13,794	5,518	30,834
Others						
Total		4,707,865	46,050	4,107,234	3,044,814	11,905,913

Source: Ministry of Agriculture and Forestry.

the framework of the broad objective of balanced development between agriculture and industry and included: 1) the achievement of a more equitable income distribution and improved rural infrastructure through rural electrification, farm road construction, and housing improvement; 2) accelerated expansion of food-grain production and the achievement of self-sufficiency in rice; 3) development of land and water resources in order to achieve all-weather farming; 4) accelerated farm modernization through increased mechanization; and 5) improved marketing, storage, and processing facilities to provide adequate services required by commercialized agriculture.

The original plan called for investments in the agricultural sector nearly four times as large as during the Second Plan (1967–1971) and eight times as large as during the First Plan (1962–1966). During the 1972–1975 period, a total of 340 billion was invested in agriculture, fisheries, and forestry from the public sector, accounting for 23.2 percent of the total government investment and loans. Of the total government investment and loans in the agricultural sector, land and water development projects received the largest share, about 94 billion wŏn or 27.5 percent. During the Second Plan, the government emphasized small-scale irrigation projects, such as construction of weirs, pumping stations, and tube well irrigation, but emphasis was shifted to larger-scale development projects for 1972–1976.

The integrated plan for Four Major River Basins development was set out in 1971 and included the construction of 13 dams and power plants. In addition, afforestation and erosion control projects on watershed areas of the Han, Kŭm, Naktong, and Yŏngsan Rivers were also planned. The government planned to invest 560 billion wŏn for these projects during 1971–1981, of which 165 billion wŏn were earmarked for agricultural projects on areas of 215,000 hectares. Projects for developing multi-purpose large-scale farming areas were undertaken in the Kŭm River and P'yŏngt'aek areas, where reservoirs, tidal dikes, water pumping and draining plants, and water canals were under construction at a cost of 48 billion wŏn by 1976. The two tidal dikes completed at Asan and Namyang Bays in 1974 are the biggest projects in Korean history in the area of water resource development. Completion of these projects brought 11,000 hectares of sea-land under cultivation and irrigated 16,000 hectares of existing farmlands. Paddy consolidation projects continued to be promoted to facilitate farm mechanization and increase farming efficiency. The total area of paddy fields rearranged under this program amounted to 251,000 hectares by 1976.

In order to minimize the transfer of arable land for purposes other than farming, the government enacted the Farmland Expansion Promotion Law in 1974. All farmland was classified into two categories—absolute and relative farmland. The law strictly prohibited the use of farmland classified in the absolute category for non-agricultural purposes.

The most notable effort in the field of rural development during 1972–1975 was the New Community Movement (Saemaul Undong), which became a national movement in 1973. Emphasis in this movement was shifted from basic projects for improving the living environment to income-generating projects based on cooperative work among villagers. Farmers were encouraged to carry out productive projects, including cultivation of cash crops, raising livestock, repairing river dikes, and so on. The government supplied technical and financial assistance to rural communities for improving facilities for education, health, housing, roads, electrification, communication, and various other projects designed to improve rural living conditions.

During 1972–1975, government investment and loans designed to improve the rural living environment increased to 75 billion wŏn, constituting approximately 22 percent of total agricultural investment. Of the total amount, 51 billion wŏn, or 68 percent, was expended as government subsidy and 24 billion, or 32 percent, as credit loan. Through the implementation of these projects, 41,500 kilometers of farm roads were constructed; the number of rural households benefiting from electrification increased to 1,643,000 in 1975, or about 65 percent of the total farm households in Korea. In addition, the number of ri (villages) and tong (administrative units comprising on the average 3–4 natural villages) installed with communication networks reached more than 10,000, and 1,595,000 rural households, or 77 percent of the total in Korea in 1975, had improved roofs.

Farm mechanization received increased attention in the Third Five-Year Plan. Rising rural wage rates and labor shortages during the peak-demand season were what motivated the government. In spite of significant increases in the use of powered farm equipment during the Second Plan period (1967–1971), the absolute level of mechanization was still low. There were only 16,842 power tillers in use in 1971, an average of one tiller for 147 farm households. Although the use of power-driven threshers, water pumps, sprayers, and dusters was relatively widespread, there was still only one thresher for 39 farm households, one water pump for 43 farm households, and one power-driven sprayer for 36 farm households.

The Third Plan called for an increase of 90,000 power tillers during 1972–1976. The target increase for power threshers was 60,000 and that for power sprayers 53,500. By 1975, power tillers in use increased to 85,722, more than five times the level in 1971, or an average of one power tiller per 29 farm households. There were 65,993 water pumps, an average of one pump per 38 farm households; 137,698 power-driven sprayers and dusters or one per 18 farm households; and 127,105 power threshers or one per 20 farm households.

Total government outlays for farm mechanization during 1972–1975 amounted to 3.8 billion wŏn, of which 1.5 billion were provided as direct subsidy and 2.3 billion as loans for purchase of farm equipment. Farmers could apply for loans up to 70 percent of the purchase price of power tillers and 50 percent of the cost of other equipment. The subsidy component in farm equipment purchases, which averaged 30 to 50 percent of the purchase price during most of the 1960s, was reduced to 10 percent or even less during the Third Plan period except for water pumps, which continued to receive a higher subsidy.

Increased attention in research was devoted to improving crop varieties, with highest priority placed on rice. One of the most noteworthy achievements was the development of a new rice variety IR 667 (*t'ongil*), which yielded on the average 25–30 percent more than traditional varieties when grown on well-irrigated paddy and accompanied by improved cultural practices. In spite of such weaknesses as poor cold resistance and unsatisfactory milling qualities, the acreage planted to this variety rapidly increased from 200,000 hectares in 1972 to 306,000 hectares in 1975. Research focus was also placed on barley, wheat, and soybean as part of an effort to encourage double-cropping.

Improvement of agricultural marketing services also received a high priority in the Third Plan. There was considerable public and private investment in marketing facilities, but inadequacies were found in facilities for storage, processing, and distribution. An investment of 42 billion wŏn was allocated during the 1972–1976 period, with most of the proposed investment going to the agricultural cooperatives for the government's grain supply management and for the buffer-stock operation of selected farm products. During 1974–1975, a total of 980 warehouses were built by private enterprises or individuals under government financial support, each with an average capacity of storing 1,000 M/T of grains. Total costs for building these warehouses were estimated at 21,560 million wŏn, of which 15,092 million were government loans and the remaining 6,468 million were borne by individuals. Data on the Third Five-Year Plan period are presented in Tables 76–80.

TABLE 76 Government Investment and Loan, 1972–1975
(million wŏn)

	Investment		Loan		Total	
	Amount	Ratio	Amount	Ratio	Amount	Ratio
Total Government Investment in all Sectors	1,334,813	(100.0)	128,095	(100.0)	1,462,908	(100.0)
Agriculture, Forestry & Fisheries	280,442	100.0 (21.0)	59,686	100.0 (46.6)	340,128	100.0 (23.2)
(Agriculture)	244,277	87.1	54,097	90.6	298,374	87.7
Agricultural Production	26,428	9.4	–	–	26,428	7.8
Cash Crops	2,782	1.0	2,424	4.1	5,206	1.5
Sericulture	2,832	1.0	673	1.1	3,505	1.0
Livestock	4,889	1.7	879	1.5	5,768	1.7
Improvement of Production Basis	83,251	29.7	10,376	17.4	93,627	27.5
Agriculture Mechanization	44	0.0	14,982	25.1	15,026	4.4
Agricultural Experiment & Extension Service	6,943	2.5	–	–	6,943	2.0
Improvement of Rural Environment	51,083	18.2	23,553	39.5	74,636	21.9
Others	66,025	23.5	1,210	2.0	67,235	19.8

TABLE 76 (continued)

	Investment		Loan		Total	
	Amount	Ratio	Amount	Ratio	Amount	Ratio
Forestry	23,020	8.2	450	0.8	34,470	10.1
Fisheries	13,145	4.7	5,139	8.6	18,284	5.4

Source: MOF, *Summary of Financial Implementation for FY 1972–1975*

TABLE 77 Large-Scale Agricultural Comprehensive Development Project, 1967–1975
(million wŏn)

Year	Domestic	Foreign Loan	Total
1969	10	–	10
1970	326	321(1,018)	647
1971	1,484	1,237(3,236)	2,721
1972	5,031	3,395(8,729)	8,426
1973	8,103	7,121(17,798)	15,224
1974	10,982	6,836(16,320)	17,818
1975	13,255	8,044(16,586)	21,299
Total	39,191	26,954(63,687)	66,145

Source: Ministry of Agriculture and Forestry (Fisheries).

Note: Figures in parentheses are U.S. $1,000.

TABLE 78 Large-Scale Agricultural Comprehensive Development Plan

States of Proceeding	Sources of Loan	Number of Regions	Area Under Development (hectares)	Projected Costs			Others
				Total	Domestic Source	Foreign Loan	
				(billion wŏn)			
Total		19	339,941	899.5	506.2	393.3	
Under Construction	IBRD	4	67,173	132.8	90.1	42.7	Kŭm River, P'yŏngt'aek, Yŏngsang River (I), Kyŏngju areas.
Conclusion of an Agreement	OECF	3	29,802	93.7	62.2	31.5	Sapkyo Stream Kyehwa Island Ch'ŏngnyong
	ADB	1	8,166	23.2	14.0	9.2	Imjin area
Under Survey	IBRD	4	101,500	226.1	120.4	105.7	Yŏngsan River Ok Stream
	ADB	1	26,600	51.3	26.6	24.7	Miho Stream Nam River
Standing Regions for Development Plan		6	106,700	372.4	192.9	179.5	

Source: Ministry of Agriculture and Forestry (Fisheries).

TABLE 79 Farm Machinery Supply, 1972–1974
(number of machines)

Kind of Machine	Supply
Power Tiller	39,039
Power Sprayer	10,088
Power Sprayer & Duster	46,642
Pump	8,985
Thresher	6,871
Cutter	186
Sowing Machine	3
Tractor	111
Other (Dryer)	117
Total	112,042

Source: Ministry of Agriculture and Forestry (Fisheries).

TABLE 80 Farm Machinery Supply Plan, 1975
(number of machines)

Kind of Machine	Supply
Power Tiller	29,130
Power Sprayer	6,000
Power Sprayer & Duster	17,859
Pump	4,000
Thresher	5,000
Tractor	200
Sowing Machine	1,500
Cutter	200
Total	63,889

Source: Ministry of Agriculture and Fisheries.

SEVEN

Agricultural Credit Policy

AGRICULTURAL CREDIT BEFORE LIBERATION

Before the introduction of a modern rural credit system into the country, Korea's agricultural credit system was centered around capital advanced by landowners and credit extended by pre-modern commercial capital. In addition, farmers made use of *kye* (financing clubs), a kind of mutual credit system prevalent in the non-agricultural sectors as well.

With Japan progressively making economic inroads into Korea, however, the Japanese began to set up banks—modern financial institutions—and invest their capital in them. The business of these banks, however, was mostly confined to large cities and did not involve agricultural credit at all.

Finally in 1906, the Nonggong (agro-industry) Bank was founded to serve as a financial institution specializing in agricultural credit. In 1907 a "Financial Association" was established with the professed goal of providing modern cooperative financing under the Local Financial Associations Decree. In 1908, the Oriental Development Company was founded with the aim of providing long-term production promotion loans and carrying out development projects.

The Nonggong Bank was engaged in providing loans through mortgages on real estate, using both private capital and government funds. The bank's loans were for land reclamation and irrigation projects, construction and improvement of buildings for agro-industry, and purchases of machinery and equipment for agro-industry. The Financial Associations made short-term agricultural production loans in the main through a cooperative financing system, and their clients were limited to farmers in and above the small-farmer class. The Oriental Development Company set up a financing department through which loans were provided to agricultural organizations involved primarily in reclamation projects, settlers, and agricultural financing institutions. Although these three organizations appeared to be engaged in different types of business, they had, in fact, the common purpose of supporting the Japanese colonial policy through the provision of Japanese capital.

The Nonggong Bank was reorganized in 1917 into the Korean Industrial Bank, which played the role of a central bank for Financial Associations. The reorganized bank was engaged only in making loans secured by real estate, with the Financial Associations becoming exclusively involved in providing agricultural credit to small- and medium-scale farmers (see Table 81).

When an agricultural crisis broke out in 1930, a campaign to boycott Korean rice was waged in Japan, thereby inflicting a heavy blow on Korea's rural economy. Such developments compelled the Financial Associations to reorganize and reinforce their financing operations. Thus, in 1933, the Decree Establishing the Korean Federation of Financial Associations was promulgated and, with the establishment of the Federation, a strong nationwide network of rural credit institutions came into

TABLE 81 Agricultural Loans by Type of Financial Institution,
1907–1936

(million yen)

Year	Financial Associations	Industrial Bank	Tōyō Takushoku Kaisha	Other Bank	Total
1907	0.02	0.14	–	–	0.16
1912	1.70	1.70	–	0.20	3.60
1917	3.80	4.30	4.00	0.20	12.30
1922	32.40	42.60	32.30	1.20	108.50
1927	–	106.60	26.40	2.70	135.70
1932	88.10	178.00	34.20	12.00	312.30
1936	150.60	219.70	35.80	21.80	427.90

Source: NACF, *Agricultural Yearbook 1965.*

being. The Federation was authorized to float Korean Financial
Debentures to raise lendable funds, and thus it became unneces-
sary for it to borrow money from other banks. This resulted in a
decline of the business of the Korean Industrial Bank and the
Oriental Development Company, both of which were engaged
primarily in providing agricultural credit to big landowners, and
these organizations were thus unable to carry out their intended
functions fully.

Toward the end of the 1930s, the Japanese economy was
gradually shifted onto a wartime economic basis, and the
financial resources of the country began to be increasingly
diverted to defense and basic industries; the availability of loans
for the agricultural sector inevitably decreased in relative
terms (Table 82). Strong emphasis was laid on increased savings
by the public in an effort to meet war expenses. As part of the
savings-encouragement program, the Japanese authorities in
1943 instituted a tax depositing system, tax reductions for
income from long-term savings, and a system of national savings
associations designed to encourage small savings. Financial
Associations and other agricultural credit institutions were also
compelled to purchase national bonds to help raise funds for the

TABLE 82 Loans by Industry Provided by All Banking Institutions, 1937–1942
(million yen)

Year	Agr., Forestry & Fisheries		Mining & Manu-facturing		Commerce		Other		Total	
	Amount	Ratio	Amount	Ratio	Amount	Ratio	Amount	Ratio	Amount	Ratio
1937	447	32.5	237	17.2	315	22.9	377	27.4	1,377	100.0
1938	453	30.9	325	22.2	386	26.3	303	20.6	1,446	100.0
1939	546	30.0	317	17.4	593	32.6	364	20.0	1,820	100.0
1940	644	25.9	692	27.8	666	26.8	485	19.5	2,487	100.0
1941	718	25.7	847	30.3	768	27.6	458	16.4	2,792	100.0
1942	792	24.0	977	29.5	943	28.6	525	15.9	3,297	100.0

Source: NACF, 20-year History of Agriculture (1965).

war effort (see Table 83), and enforced such a new plan as the "deduction savings system." Provision of agricultural credit was thus made a secondary function of the agricultural financing institutions.

TABLE 83 Funds Operation Status of the Federation of
Financial Associations, 1940–1945
(million yen)

Year	Draft	Loan	Bond Purchase	Borrowing	Deposit
1940	210	120	49	31	101
1941	307	113	95	40	167
1942	367	109	278	24	89
1943	751	118	556	28	165
1944	1,392	119	1,235	22	118
1945.7	1,610	120	1,488	22	80

Source: NACF, *Agricultural Yearbook.*

FROM LIBERATION TO THE EARLY 1950s

Up until the time of Korean Liberation in 1945, modern banking systems were introduced and developed in Korea primarily by external elements in order to help carry out Japanese colonial policies. Only a very few banks were founded and operated with native capital. Consequently, Korea's banking system almost ceased to function for a time after the Japanese had withdrawn from the country. Accordingly, even specialized banks came to concentrate on commercial banking, and they converted themselves de facto into ordinary banks. This led to confusion involving irregular bank lending practices, stoppage of industrial financing, and other ill effects. Moreover, because of the absence of currency management, there was no effective way to prevent galloping inflation. It became imperative to reorganize the banking system in order to enforce strict currency management and to normalize the financial situation.

The Financial Associations, which had played the principal role in providing agricultural credit, degenerated into savings institutions toward the end of Japanese rule in Korea. In the wake of Liberation, even the Financial Associations' savings function was greatly weakened, not to speak of their agricultural credit function. In 1946, the Financial Associations, which had specialized exclusively in rural banking, were charged with the additional task of serving as government agencies in a number of fields. First, they were given the task of procuring, storing, and distributing under a rationing system seven types of goods of daily necessity, including cotton cloth, and in 1949 the associations took over the task of handling fertilizer, which had previously been carried on by the Korean Farming Society.

The handling of government grain, which had been carried on by the Korea Food Corporation for the account of the government, was also transferred to the Financial Associations. The task involved the purchase of food-grains, their storage, acceptance, transportation, processing and distribution. In December 1949, even the handling of straw products was transferred from the Korean Farming Society to the Financial Associations.

The Financial Associations thus came to carry out, in addition to credit services, various economic services for the account of the government. This resulted in improvement of the financial situation of the associations on account of expansion of their business activities, with their previous financial strain significantly eased. On the other hand, the relative share of credit services in the overall activities of the associations declined.

Before long, the associations' government-assigned economic services were again taken away and transferred to other agencies, causing the business activities of the associations to shrink drastically. Meanwhile, the functions of the Korean Industrial Bank, whose speciality had been the provision of long- and medium-term loans to the agricultural sector, also were greatly reduced, with the result that the bank turned itself into a de facto commercial bank. As Korea entered an era of economic rehabilitation following the signing of a truce for the Korean War in 1953, there arose an immense demand for industrial capital. To help meet the huge financial requirements of rehabilitation and reconstruction, it was felt necessary to establish a special bank by reorganizing the Korean Industrial Bank. This led to the founding of the Korea Reconstruction Bank (later renamed the Korea Development Bank) in January 1954.

Following Liberation, it also became necessary to reorganize the Oriental Development Company. The company was renamed the New Korea Company, Ltd., in November 1945, and its legal status was established by Decree No. 52 issued by the U.S. Military Government. The New Korea Company, Ltd., absorbed the Forestry Development Corporation in March 1946, thereby enlarging its scope of operations. However, the New Korea Company, Ltd., became a target of popular censure because it continued to charge high rents to tenant farmers, just as the Oriental Development Company had done. As a result, it was dissolved in March 1948, and its land improvement programs were transferred to the Korean Federation of Irrigation Associations, and its farmland management functions to a newly established Central Land Administration Office.

The conditions of agricultural credit deteriorated severely following Liberation, and no government support at all was provided. Agricultural credit availability, instead, depended solely on whatever loans the weakened Financial Associations could make. It was not until 1949 that the government finally took measures to boost agricultural credit in the form of a line of rediscount amounting to 25 million wŏn (25,000 million wŏn in terms of the unit of currency in use at that time) established at the Bank of Chosun (the then central bank, later the Bank of Korea). Thus the Financial Associations were able to borrow money under government guarantee in order to finance agricultural production credits. After 1950, the Financial Associations continued until 1955 to borrow money from the Bank of Korea (founded in December 1950) and also raised some money on their own, and these monies were loaned out to farmers (Table 84).

TABLE 84 Agricultural Loans by Financial Associations, 1949–1955
(million wŏn)

Year	Total Loan	Financing Sources	
		Borrowing	Own Funds
1949	25	25	–
1950	25	25	–
1951	100	80	20
1952	358	358	–
1953	10	6	4
1954	30	19	11
1955	60	27	33

Source: NACF, 20-year History of Agriculture.

THE LATE 1950s

Following Liberation, agricultural credit was provided mainly through the transitional or stop-gap operations of the Financial Associations, and yet the associations were unable to provide farmers with adequate levels of credit because their lendable financial resources were quite limited. Nor were the commercial banks or the Korea Reconstruction Bank able to make any significant amount of agricultural credit available. All they could do was lend short-term marketing funds to agricultural organizations for the purpose of financing collection of agricultural products. Provision of credits to farmers through normal banking channels was so inadequate that most farmers were compelled to rely on usurers to meet their financial needs, and the usury indebtedness of farmers kept mounting at an ever-accelerating pace.

In view of such pressing needs, a series of legislative steps began to be taken in order to institutionalize improved agricultural credit facilities, to expand financial sources of agricultural credit, to unify channels for fund supply, to reinforce supervised credit, and to make available long-term, low-interest loans for agricultural financing. In 1955 the government invited a team óf American experts in agricultural credit and farm cooperative systems led by Edwin C. Johnson, Deputy Governor of the U.S. Federal Land Mortgage Bank, to come to Korea in order to advise the government on how to reform the agricultural credit system. The contents of the report prepared by the team, however, were found not to be suitable for the rural economy in Korea, and the recommendations were not accepted. Later, J. R. Cooper, an official of the U.S. International Cooperation Administration who participated in the drafting of the Japanese law establishing agricultural cooperatives, was invited to Seoul to study the matter again. On the basis of Cooper's report, the National Assembly, the Ministry of Finance, and the Ministry of Agriculture and Forestry drafted for discussion an agriculture bank law, a credit union law, an agricultural cooperative law, and other related measures. Through the resulting discussions, only the idea of establishing an Agriculture Bank evolved in a concrete form. In the meantime, however, the spring season of peak agricultural activities approached and the government adopted a new policy of selling fertilizer to farmers on a cash instead of a credit basis. Because of this urgent state of affairs, the government made a basic policy decision to disband the Financial Associations and their federation, and at the same time, to establish an Agriculture Bank (in May 1956) under the Commercial Banking Law as an interim measure so as to have in operation a financial institution specializing in agricultural financing.

The principal objectives of the bank were: 1) providing loans maturing in not more than five years for the production and processing of agricultural products; 2) making available loans maturing in not more than one year for the purchase and processing of articles of necessity for farm households; 3) extending loans maturing in not more than fifteen years for the reclamation, purchase, and improvement of agricultural land; 4) making available loans maturing in not more than five years for the establishment of joint-use facilities for farmers; and 5) handling domestic exchange, borrowing monies from the government and the Bank of Korea, and issuing debentures.

The Agriculture Bank, Ltd., was intended to serve as a transitory organization pending the establishment of a government-corporation-type Agriculture Bank under special legislation. The Agriculture Bank, Ltd., had insufficient equity capital; its ability to borrow long- and medium-term funds was limited; and it was not easy for the bank to obtain government funds for its lending operations. Thus, the bank encountered great difficulty in raising lendable funds and engaged mostly in providing very small loans to as many borrowers as possible. Furthermore, it was unable to get irrigation project loans transferred to it from the Korea Reconstruction Bank; nor was it able fully to take over agricultural loans handled by commercial banks. Thus the Agriculture Bank, Ltd., failed to unify lending channels for agricultural credit. It was also unable to divest itself of the characteristics of a commercial bank resulting from the fact that it was a joint-stock company under obligation to seek profits and avoid undue risks. Therefore, it could not boldly lower interest rates on its loans; nor could it make unsecured loans or otherwise ease collateral requirements for its clients. In view of such problems facing the bank, the government hastened to establish a special Agriculture Bank with the aim of overcoming the shortcomings in the stop-gap arrangement and thus reinforcing the foundation for agricultural credit activity. A new Agriculture Bank was therefore founded in April 1958, under special legislation, and this founding led to a great expansion in the availability of agricultural credit, because the special bank was able to borrow money from the Bank of Korea much more easily and also because the government actively supplied the bank with loans out of its own fiscal resources.

With the founding of the special rural bank under the Law Establishing the Agriculture Bank, a solid institutional basis was laid for agricultural financing in Korea. During the life of the new Agriculture Bank, agricultural credit facilities were measurably improved and expanded under a series of positive agricultural policies initiated by the government—including the unification of channels for provision of agricultural loans and the extension of new government loans to the bank. More specifically, the developments featured the following (supporting data are in Table 85 to 89):

1) The administration of various types of agricultural credit, which had been handled by a number of separate banking institutions, was brought under the sole jurisdiction of the special Agriculture Bank. Previously, long-term irrigation loans had been handled by the Korea Development Bank; farm product collection loans by commercial banks; and agricultural production credits by the Financial Associations. But, following the promulgation of the Law Establishing the Agriculture Bank, irrigation loans were transferred from the Korea Development Bank to the Agriculture Bank in August 1958, thereby completing the process of unifying lending channels for agricultural credit.

2) The government began in earnest to increase the availability of its own financial resources for agricultural financing purposes. Up until 1956 only long-term irrigation loans were financed with government money, but, beginning in 1957, the government also utilized proceeds from the Vested Property Disposal Special Account, the Counterpart Fund, the Grain Management Special Account, and the Industrial Reconstruction Bonds in helping to build up the lendable resources of the Agriculture Bank.

The amount of government funds loaned to the Agriculture Bank rose from 547.3 million wŏn (5,473 million hwan in the monetary unit at that time) in 1957 to 7,007.7 million wŏn

TABLE 85 Agricultural Loans by Financial Institution, 1952–1960
(million wŏn)

Year	Agriculture Bank		Reconstruction Bank		Other Banks		Total		
	Total Loan	Agricultural Loan	Total Loan	Agricultural Loan	Total Loan	Agricultural Loan	Total Loan A	Agricultural Loan B	B/A
1952	79	57	–	–	719	131	798	188	23.6
1953	151	117	–	–	1,649	316	1,800	433	24.1
1954	476	403	736	296	1,789	142	2,999	841	28.0
1955	696	544	3,119	727	3,566	368	7,381	1,639	22.2
1956	1,811	1,514	3,604	1,873	5,252	190	10,667	3,577	33.5
1957	4,982	4,671	9,822	2,586	6,170	29	20,974	7,286	34.7
1958	8,256	7,830	10,543	–	7,629	41	26,428	7,871	29.8
1959	8,962	8,170	14,132	–	9,297	46	32,391	8,216	25.3
1960	12,842	11,484	14,852	–	11,473	52	39,167	11,536	29.5

TABLE 86 Agriculture Bank Loans Outstanding at the End of Each Year, 1956–1960
(million wŏn)

	1956	1957	1958	1959	1960
Credit Sector					
Agricultural Production Loan	9,600	10,508	10,364	13,036	13,734
Agriculture and Forestry Loan	2,115	5,614	5,701	8,109	8,580
Rice Lien Loan	–	15,366	10,385	288	19,758
Fertilizer Loan	3,662	10,725	7,002	–	–
Usury Debt Settlement Loan	–	–	–	–	2,463
General Fund Loan	2,576	3,279	3,728	7,601	9,809
Small and Medium Industry Loan	–	–	–	–	1,419
Sub-total	17,953	45,492	37,180	29,034	55,763
Government Sector					
Counterpart Fund	–	3,044	2,884	3,702	11,894
Agricultural Production Loan	–	1,055	998	973	1,342
Grain Management Special Account	–	–	42	1,447	2,077
Rice Lien Loan	–	266	8,265	9,853	658
Irrigation Loan	–	–	33,195	44,607	54,573
Small and Medium Industry Loan	–	–	–	–	2,112
Sub-total	–	4,365	45,384	60,582	72,656
Total	17,953	49,851	82,564	89,616	128,419

Source: Natural Agricultural Cooperative Federation.

TABLE 87   Agricultural Loan Made by Agriculture Bank
by Financing Source, 1957–1960
(million wŏn)

Classification	1957	1958	1959	1960
Agricultural Credit Fund	45,492	37,180	29,034	55,763
Government Fund (A)	4,359	45,384	60,582	72,656
Total (B)	49,851	82,564	89,616	128,420
A/B (%)	8.7	55.0	67.6	56.6

Source: National Agricultural Cooperative Federation.

TABLE 88   Government Funds Supplied to Agriculture Bank,
1957–1960
(million wŏn)

Classification	1957	1958	1959	1960
Counterpart Fund	3,973	6,321	19,371	21,677
Agricultural Fund	(3,973)	(4,321)	(17,355)	(18,120)
Irrigation Fund	–	(2,000)	(2,016)	(2,012)
Small and Medium Industry Fund	–	–	–	(1,545)
Vested Property Fund	1,500	5,400	8,370	11,470
Agricultural Fund	(1,500)	(1,500)	(1,500)	(1,500)
Irrigation Fund	–	(3,900)	(6,870)	(9,970)
Small and Medium Industry Fund	–	–	–	–
Agricultural Special Fund (Irrigation Fund)	–	1,800	2,462	2,462
Grain Management Special Fund	–	6,000	11,908	17,108
Agricultural Fund	–	(2,000)	(2,300)	(2,300)
Irrigation Fund		(4,000)	(9,608)	(14,808)
Industrial Reconstruction Bond (Irrigation Fund)		28,014	27,966	27,913
UNKRA Small and Medium Industry Fund	–	–	–	60
Total	5,473	47,535	70,077	80,690

Source: National Agricultural Cooperative Federation.

TABLE 89 Agricultural Loan Made by Agriculture Bank, 1957–1960
(million wŏn)

Year	Credit Fund		Government Fund		Total	
	Annual Loan	Loan Outstanding at the End of Year	Annual Loan	Loan Outstanding at the End of Year	Annual Loan	Loan Outstanding at the End of Year
1957	4,576	10,508	4,554	4,099	9,130	14,607
1958	3,843	10,368	2,201	3,295	6,043	13,663
1959	9,286	13,036	6,449	6,121	15,735	19,157
1960	10,178	13,734	16,890	15,312	27,068	29,046

(70,077 million hwan) in 1959, and further to 8,069.1 million wŏn (80,691 million hwan). As a result, agricultural loans to farmers financed with government funds increased from 435.9 million wŏn (4,359 million hwan) in 1957 to 7,265.6 million wŏn (72,656 million hwan). The proportion of government-funded agricultural loans in total agricultural credit jumped from 8.7 percent in 1957 to 56.6 percent in 1960.

3) With the sources of money for agricultural production credit diversified to include bank loans, counterpart funds, and deposits in the Vested Property Disposal Special Account, the volume of agricultural production credit made available to farmers expanded sharply.

4) In 1957, a rice liens program was instituted to provide short-term loans to cash-short farmers in order to enable them to hold their rice crop off the saturated grain market during the harvest season. This type of loan helped reduce seasonal fluctuations in food-grain prices while at the same time increasing farmers' incomes. In 1959, Agricultural Finance Bonds were issued for the first time to raise money needed to replace the usurious loans being taken by farmers with low-cost credits, and the system of supplying fertilizer to farmers on credit was improved. Thus, the rural credit system in Korea made great advances following the founding of the Agriculture Bank under special legislation, principally with the help of increased access to government funds.

THE 1960s

The agricultural credit system underwent a major overhaul again in the course of the 1960s. In the latter half of the 1950s, a basic policy was adopted providing for the separation of credit services from economic services to the agricultural sector and the placing of the two types of service under the jurisdiction of different organizations. The special Agriculture Bank was created to specialize in credit services, while economic services were assigned to newly-established agricultural cooperatives; the two organizations operated independently of each other. This separation tended to defeat their common objectives of increasing agricultural production and improving the economic and social status of farmers.

The agricultural cooperatives at that time were established under the Agricultural Cooperatives Law. Village-level cooperatives were grouped under kun (county) associations, and even a National Agricultural Cooperative Federation was organized. Outwardly, the agricultural cooperatives, therefore, had a fully developed organizational structure, but the cooperatives did not have credit-providing functions and, moreover, received rather meager government support. Thus, the financial resources for carrying out their economic services were far from adequate. Nor was the Agriculture Bank able to forge close ties with the cooperatives as a means of filling this financial gap.

The Park Military Government that came into being through the May 16, 1961 Military Revolution decided to correct the defects in the agricultural credit system as quickly as possible through a merger of the agricultural cooperatives with the Agriculture Bank. On July 29, 1961, the government repealed the old Agricultural Cooperatives Law and the Law Establishing the Agriculture Bank, and enacted in their place a new Agricultural Cooperatives Law (Law No. 670), which marked a major turning point in the history of agricultural credit systems in Korea. Under the new law, reorganized agricultural cooperatives were launched on August 15 of the same year.

The new cooperatives, which had been merged with the Agriculture Bank, first streamlined their regional and national organizational structure. The new structure was founded on ri (village) and tong (group of villages) cooperatives designed to serve farmers' needs more effectively. The village-level cooperatives had a three-tiered upper structure, namely, kun (county) cooperative associations, provincial branches of the National Agricultural Cooperative Federation (NACF), and the NACF headquarters. To ensure effective cooperative activity, a groundwork was laid to make possible close coordination of the cooperatives' credit service with their procurement and marketing service, mutual financial assistance service, and agricultural extension service. In lending operations, the cooperatives shifted from the previous methods that emphasized loan security by means of collateral on commercial banking principles to a cooperative system of mutual financing in which trust in the borrowers and their guarantors was taken into account.

As for the source of funds for the newly organized cooperatives, government lending to them increased from 8,915 million wŏn as of August 15, 1961, when they were founded, to 48,541 million wŏn by the end of 1970. Loans made by the cooperatives to farmers increased from 16,688 million wŏn in 1961, the year of their founding, to 105,360 million wŏn in 1970. The major portion of the increase in loans made by the cooperatives was financed from their own increasing deposits. During the first half of the 1960s, the ratio of private to public sector loans channeled through the agricultural cooperatives remained at around 40 to 60, but the ratio was reversed to 67 to 33 by 1970, indicating that the cooperatives' own financial resources had expanded greatly. The changing composition of fund sources for loans made by the cooperatives was well reflected in their borrowing activities. During the first half of the 1960s, external resources, such as loans from the government and the Bank of Korea, represented approximately 70 percent of the financial resources in the hands of the cooperatives. The corresponding percentage declined to 53 in 1970, as the result of increases in their own financial resources, due chiefly to expansion of deposits with them. Agricultural cooperative credit extended through the private banking sector increased from 834 million wŏn in 1961 to 70,988 million wŏn in 1970, rising faster than the public-sector credit.

Related to this expansion of rural credit was the counter-usury program aimed at ending the longstanding and widespread practice of exploitative moneylending in rural communities. Highlights of the counter-usury program were as follows:

1) "Usurious loans" were defined as debts incurred not later than May 25, 1961, which bore interest rates of 20 percent or more per annum.

2) The creditor and the debtor were required to file, separately or jointly, reports of the debts involved with the usurious debt liquidation committees in the areas of residence of the debtors.

3) If and when a reported debt was ruled to be usurious, the NACF issued to the creditor Agricultural Finance Bonds in an amount corresponding to the sum of the debt to effect debt settlement in subrogation while the debtor was required to recompense the NACF for the debt payment made on his behalf.

4) Agricultural Finance Bonds issued to a creditor were retired over a period of four years after a grace period of one year, with an annual interest of 20 percent paid on them, except that bonds with a face value of not more than 1,000 wŏn reached maturity in not more than one year.

5) The debtor repaid the NACF in equal installments over a period of not more than five years with an annual interest of 12 percent paid on the debt taken over by the NACF.

The filing of indebtedness reports under the law for liquidating usurious loans was almost fully completed by the end of December 1961. The number of usurious loans reported by farmers stood at 1,145,120, while those declared by fishermen numbered 25,521. In money terms, the farmers' indebtedness amounted to 4,599 million wŏn; the fishermen's, 207 million. Of the reported total, loans ruled to be usurious amounted to 2,927 million wŏn and the total sum of verified debts covered by Agricultural Finance Bonds issued by the NACF came to 2,663 million wŏn, encompassing 781,766 usurious loans.

Although the counter-usury program was launched with high hopes, it failed to live up to expectations for the following reasons. First, although it was necessary to increase drastically the availability of low-interest agricultural credit following the liquidation of usurious farm debts in order to prevent the continued practice of private moneylending in rural communities, no satisfactory steps were taken to attain such a goal. Second, the NACF failed to retire on schedule the Agricultural Finance Bonds that it had issued to the creditors.

This situation developed primarily because the government was unable to mobilize adequate financial resources to back up the counter-usury program, even though it did increase the volume of public funds loaned to the NACF in order to ease the financial stringency that came about in farm communities owing to a halt in private moneylending the counter-usury program caused. Loans made by the agricultural cooperatives did increase, and government funds loaned to the cooperatives rose from 8,000 million wŏn at the end of 1960, to 14,000 million wŏn in 1962, to 15,100 million wŏn in 1963, and further to 15,880 million wŏn by the end of 1964. Further, during the last half of the 1960s, Korea's agricultural credit system was operated in a more effective manner, with loans made through the system linked to such specific projects as the development of viable farms, the development of specialized agricultural production districts, and the Special Projects for Increasing Rural Income. However, the chronic problem of the short supply of lendable funds relative to the financial requirements of agricultural development remained largely unresolved.

Although the financial support of the government for agricultural development was expanded in absolute terms, it was still insufficient to meet the rapidly rising demands for medium- and long-term farm loans brought about by the acceleration of agricultural development. In other words, the necessity of expanded and reinforced development financing became even more acute than formerly, and in 1968 the government began to grant interest subsidies to the NACF in order to enable it to make low-cost, medium-term loans out of the deposits it received. Also, foreign loans began to be introduced to finance agricultural projects. Until 1968, the proportion of medium- and long-term agricultural development loans in total agricultural credit made available had remained very low. The percentage of development loans exclusive of irrigation project loans stood at a mere 4.1 percent in 1961. But the figure rose to 11.3 percent in 1964 and further to 37.8 percent in 1968. Nevertheless, it can be said that Korea's agricultural credit system was still oriented toward short-term financing for farming activities. Supporting data for the discussion in this section are presented in Tables 90 to 95.

AGRICULTURAL CREDIT IN THE 1970s

An institutional basis for agricultural credit in Korea was consolidated over the period from the latter half of the 1950s to the early 1960s. The strengthened institutional framework of rural credit has continued to be maintained through the 1970s. Major developments in the agricultural credit field since the turn of the 1970s include growth of mutual financing through basic agricultural cooperatives; the strengthening of cooperative lending functions based on the cooperatives' enhanced ability to raise funds on their own; and increased provision of medium- and long-term development loans.

Mutual financing through basic cooperatives refers to savings and loan arrangements made by economically weak farmers to promote mutual financial help for cooperative members trying to overcome their fund shortages on their own. This mutual financing system was put into practice as part of a program to strengthen the ability of basic cooperatives to stand on their own financial feet in conjunction with the enactment of the Credit Union Law in August 1970. The mutual financing system is an extension of traditional cooperative financing practices. It is aimed at pooling small sums of idle funds in rural communities and private capital hitherto used for usurious lending operations; the pool is then utilized to increase the provision of farming loans, thereby facilitating the efforts of basic cooperatives to become financially self-supporting by building up their own financial resources, instead of depending on government help.

Deposits are accepted into the mutual financing systems, and

TABLE 90 Supply of Government Funds by Use, 1961–1970
(million wŏn)

Classification	1961	1962	1963	1964	1965	1966	1967	1958	1969	1970
Irrigation Fund	6,198	6,672	7,168	7,714	6,436	6,803	7,906	9,836	11,927	14,216
Agriculture Fund	3,235	7,010	7,590	7,621	7,833	9,156	9,733	12,168	20,103	25,672
Warehouse Fund	228	220	220	220	220	220	211	202	198	184
Fisheries Fund	–	226	147	147	–	–	–	–	–	–
Usury Debt Settle-ment Fund	–	–	–	178	178	1,378	1,378	1,378	1,378	1,378
AID Loan Fund	–	–	–	–	–	–	–	–	–	–
Farm Products Collec-tion Fund	–	–	–	–	–	–	–	–	3,228	3,482
Livestock Fund	–	–	–	–	–	–	–	–	–	3,609
Forestry Fund	–	–	–	–	–	–	–	–	–	–
Total	9,661	14,128	15,125	15,880	14,666	17,557	19,228	23,584	36,834	48,541

Source: National Agricultural Cooperative Federation.

TABLE 91 Funding Sources of NACF Credit Funds, 1961–1970
(million wŏn)

Classification	1961	1962	1963	1964	1965	1966	1967	1968	1969	1970
Government Fund	9,661	14,128	15,125	15,757	14,666	17,557	19,228	23,584	40,655	48,541
Borrowing from Bank of Korea	3,054	600	3,442	2,256	16,311	20,626	24,202	30,347	33,377	32,672
Issue of Agricultural Credit Bonds	250	2,686	2,136	1,970	898	465	223	166	121	83
Deposit Received	3,471	4,238	5,673	6,450	10,641	20,948	27,774	46,783	75,909	95,416
Loan Funds	–	–	–	–	–	–	–	–	5,752	8,117
Own Funds	475	847	1,351	1,535	1,684	1,831	1,964	2,207	2,407	7,354
Total	16,911	22,499	27,727	27,968	44,200	61,427	73,391	103,087	158,221	192,183

Source: National Agricultural Cooperative Federation.

TABLE 92 Foreign Loan Funds

Classification	Total Amount	Use	Loan Terms	
			Interest Per Annum	Repayment Period
			%	
USA	US$19,870,000	Irrigation	2.0	10-year grace
			2.5	30-year repayment
Canada	CN$999,000	Milk Cows Pasture Development	3.0	7-year grace
				23-year repayment

Source: National Agricultural Cooperative Federation.

TABLE 93 Deposit Received by NACF, 1961–1965

(million wŏn)

Classification	1961	1962	1963	1964	1965	1966	1967	1968	1969	1970
(Type of Deposit)										
Demand Deposit	1,831	2,946	4,323	4,702	6,567	7,586	12,673	20,144	31,976	39,209
Saving Deposit	1,640	1,304	1,350	1,748	4,074	13,362	15,101	26,639	43,925	56,207
Total	3,471	4,250	5,673	6,450	10,641	20,948	27,774	46,783	75,901	95,416
(Depositor)										
Farmers	985	644	1,198	816	1,233	1,895	3,461	11,159	15,330	20,913
Non-farmers	2,486	3,606	4,475	5,634	9,408	19,053	24,313	35,624	60,571	74,503

Source: National Agricultural Cooperative Federation.

TABLE 94 Trend of Lending, 1961–1970

(million wŏn)

Year	(A) Total Lendings	Credit Sector (B) Amount	B/A	Government Sector (C) Amount	C/A
1961	16,688	8,134	48.7	8,554	51.3
1962	18,611	7,418	39.9	11,193	60.1
1963	17,704	7,509	42.4	12,195	68.9
1964	23,139	10,882	47.0	12,257	53.0
1965	23,259	11,360	48.8	11,899	51.2
1966	27,108	14,195	52.4	12,913	47.6
1967	34,377	18,265	53.1	16,112	46.9
1968	52,814	33,974	64.3	18,840	35.7
1969	84,413	53,360	63.2	31,053	36.8
1970	105,360	70,988	67.4	34,372	32.6

Source: National Agricultural Cooperative Federation.

TABLE 95 Borrowings From the Bank of Korea, 1961–1974
(million wŏn)

Year	Agricultural Credit Program	Marketing Program	Total
1961	2,854	200	3,054
1962	100	500	600
1963	800	2,642	3,442
1964	3,156	2,100	5,256
1965	4,011	12,300	16,311
1966	2,126	18,500	20,626
1967	4,102	20,100	24,202
1968	2,447	27,900	30,347
1969	1,377	32,000	33,377
1970	2,673	30,000	32,673
1971	8,765	32,000	40,765
1972	8,500	34,000	42,500
1973	15,770	43,000	58,770
1974	78,565	20,000	98,565

Source: National Agricultural Cooperative Federation.

loans are made out of them. The types of deposit accounts include temporary deposits, time deposits, mutual installment deposits, and free installment deposits. Two kinds of loans are made by the system—ordinary loans and mutual loans. A basic agricultural cooperative can borrow money from the NACF as well as from the kun (county) cooperative association to which it belongs. Borrowings by a basic cooperative are classifiable into temporary borrowings to meet temporary cash shortages, and mutual financing promotion loans that the NACF provides to it at low interest for use as working capital in the operation of its mutual financing system. Whereas the credit service of the NACF and kun associations, which are classed as banking institutions, is governed by the General Banking Law, mutual financing operations of basic cooperatives are governed by the Credit Union Law. Being a form of cooperative financing, the mutual financing system, by nature, is engaged, in principle, in making short-term loans alone, although the credit service of the NACF and kun associations makes medium- and long-term loans financed with government funds. Only members of cooperatives are eligible for loans from the mutual financing system; no such restrictions are imposed on eligibility for loans by the credit service of the NACF and kun associations.

In mobilizing financial resources for agricultural credit, Korea's agricultural cooperatives have been increasing their capabilities to raise funds on their own since the turn of the 1970s. Deposits continue to grow; a campaign to attract a trillion wŏn in rural savings has been conducted; the share capital of the cooperatives has been increased; and larger foreign loans have been introduced. As of the end of 1974, deposits in the cooperatives totaled 215,852 million wŏn; their share capital stood at 13,862 million wŏn; and their foreign loans received came to 9,454 million wŏn. With the increasing capacity of the cooperatives to raise money on their own, loans made by them out of their own resources reached a total of 105,200 million wŏn, accounting for 64 percent of their total loans, a large improvement in their rate of financial self-support.

The scale of medium- and long-term loans made by the cooperatives has also been expanding in keeping with their increasing ability to supply rural credit. Such loans have gone principally to the growth sectors of agriculture, such as livestock raising, sericulture, and special cash crops, which are given high priority under the Special Projects for Increasing Rural Income and the Saemaul (New Community) Projects.

Previously, the cooperatives relied chiefly on government funds in financing medium- and long-term loans for capital investment purposes. Government funds alone have become insufficient, however, to respond to the increasingly strong emphasis placed on agricultural development under the economic policy of the government. Thus, the cooperatives have begun to supplement government funding with their own financial resources in providing medium- and long-term development loans. To facilitate such long-term lending, the government has been granting subsidies to the cooperatives to offset losses that they take on low-interest medium- and long-term loans made out of their deposit accounts.

Such loans financed with the cooperatives' own resources, which are eligible for government subsidy for interest difference compensation, included the following: loans for sericulture, for Special Projects for Increasing Rural Income, for rehabilitation from natural disasters, for livestock development, and for the mushroom-processing industry. In 1974, the scope of eligibility for interest subsidy was expanded to cover loans for Saemaul Income Increasing Projects, for rehabilitation from natural disasters, for increased agricultural production, for livestock raising, for sericulture, for special cash crops, for the forest industry, for cottage industry, for non-agricultural enterprises of farmers for additional incomes, for agricultural warehouse construction, and for farm mechanization. Because private banking sector medium- and long-term loans are thus provided for the financing of such government-programmed agricultural development projects as Saemaul Income Increasing Projects, the development of sericulture and livestock and agricultural mechanization, as well as such public relief projects as rehabilitation from natural disasters, the government makes the decisions as to the amounts of loans, lending procedures, and the eligibility criteria for borrowers with respect to this particular category of lending operations. Government controls in this area are the same as those applicable to government funds made available for rural credit. At the end of 1974, medium- and long-term agricultural development credits provided amounted to 79,800 million wŏn, of which 35,500 million wŏn, or 44.4 percent of the total, was funded with deposits with the agricultural cooperatives.

In keeping pace with the expansion of the physical volume of agricultural credit, institutional improvements have continued in the rural credit system. Such institutional changes include the implementation of a credit-worthiness guarantee system for farmers and fishermen designed to supplement the inadequate ability of economically weak farmers and fishermen to provide collateral for their loans. This system was put into effect in 1972 under the Law on Guaranteeing the Credit-worthiness of Farmers and Fishermen enacted in 1971. The NACF has been designated as the agency for administering the credit-worthiness guarantee fund. Decisions on important matters relating to the fund operation are made by the Credit-worthiness Review Committee, composed of the Ministers of Finance, and Agriculture and Forestry, the Governor of the Bank of Korea, and the Presidents of the NACF and the Central Federation of Fisheries Cooperatives (CFFC). Guarantees are extended by the NACF and the CFFC alone. Loans guaranteeable are farm production loans and development loans for livestock, orchards, afforestation, and so on, which are provided to farmers and agricultural organizations, plus fishery loans provided to fishermen and fisheries organizations. The fund may guarantee a total amount of loans not more than 10 times the sum of money paid into the fund.

The sources of money for the fund are contributions by the government, the NACF, and the CFFC, and earnings through operation of the fund. At the end of 1974, the fund's financial resources totaled 1,846 million wŏn, of which 59.6 percent had been contributed by the government. Contributions by the NACF accounted for 33.9 percent. The ceilings on the amounts of loans guaranteeable per borrower were increased in 1975, in step with raised ceilings on individual unsecured loans and increases in the total availability of rural credit. The ceiling on the value of credit-worthiness guarantee per individual was increased from three million wŏn to five million wŏn; that for a forest or fishery development club was upped from 5 million wŏn to 10 million; and that for an agricultural or fisheries organization was hiked from 10 million wŏn to 15 million wŏn. Loans actually guaranteed by the fund increased from 122 million wŏn at the end of 1972 to 1,634 million wŏn by the end of 1973, and further to 8,927 million at the end of 1974.

Another major change in the agricultural credit system of the country has been the introduction of a farmland mortgage system. The agrarian reform of 1950 had made it de facto impossible to mortgage agricultural land, causing confusion and difficulty in the rural credit system for some time. In preparation for such problems, the government had instituted a system of making unsecured loans to farmers. The unsecured lending system, however, soon reached its inherent limits, and the need to break through the impasse became increasingly acute during the latter half of the 1960s. So in 1966 an Agricultural Land Mortgage Law was enacted with the aim of allowing farmers to obtain more adequate agricultural loans by offering their land as security. The farmland mortgage system went into effect in 1970 when the Executive Ordinance implementing the law was promulgated.

Farmland may be mortgaged only when the farmer who owns it wants to obtain an agricultural loan. No one other than an agricultural or fisheries cooperative is authorized to be a mortgagee, and charging advance interest payments or compound interest on mortgage loans is prohibited. As of the end of 1974, the agricultural cooperatives had extended 35,746 million wŏn in mortgage loans.

In the early 1970s, a group lending system was instituted with the aim of improving the efficacy of loans. Instead of providing loans to individual farmers, as had been the case previously, the new system was designed to make loans to productive organizations, such as crop cultivation teams and farm complexes, with the aim of linking the production and marketing of farm products to agricultural credit. Also, a system of advance announcement of credit availability has been put into force, and lending procedures have been greatly simplified to facilitate the utilization of agricultural credit.

Supporting data for this discussion of the 1970s are presented in Tables 96–104.

TABLE 96 Growth of Mutual Credit, 1969–1974

	Deposit				Lending			
			Average Deposit				Average Deposit	
Year	Balance	Rate of Increase	Per Co-op	Per Member	Balance	Rate of Increase	Per Co-op	Per Member
	(Billion wŏn)	%	(Million wŏn)	(1,000 wŏn)	(Billion wŏn)	%	(Million wŏn)	(1,000 wŏn)
1969	0.3	–	2	0.13	0.3	–	2	0.14
1970	2.4	735	7	1	2.8	812	9	1
1971	6.6	279	10	3	7.3	259	12	3
1972	13.0	197	13	5	12.9	176	13	5
1973	27.7	213	21	11	19.2	149	14	8
1974	57.9	209	38	29	34.1	178	23	14

Source: National Agricultural Cooperative Federation.

TABLE 97   Compensation of Differences in Interest,
1968–1974
(million wŏn)

Year	Loan Outstanding	Interest Difference Compensation
1968	1,427	49
1969	7,507	742
1970	18,149	1,358
1971	25,267	2,140
1972	25,268	2,121
1973	26,843	1,626
1974	32,493	1,802

Source: National Agricultural Cooperative Federation

TABLE 98   Establishment of Credit Guarantee and
Actual Amount Guaranteed, 1972–1974
(million wŏn)

Classification	1972	1973	1974
Fund			
Government Contribution	100	1,100	1,100
NACF Contribution	155	351	625
CFFC	–	3	9
Others	–	15	112
Total	255	1,469	1,846
Total Guarantee			
Agriculture and Forestry Fund	80	984	7,500
Fisheries Fund	42	650	1,427
Total	122	1,634	8,927

Source: National Agricultural Cooperative Federation.

TABLE 99 Funding Sources of NACF Credit Funds, 1971–1974
(million wŏn)

Classification	1971	1972	1973	1974
Government Fund	50,143	59,717	64,540	78,691
Borrowing from Bank of Korea	40,765	42,500	58,770	98,565
Issue of Agricultural Credit Bonds	81	4,404	3,952	3,385
Deposit Received	108,926	132,741	169,371	215,852
Loan Funds	4,956	4,808	4,724	9,454
Own Funds	7,765	8,529	11,532	13,862
Total	212,636	252,699	312,889	419,809

Source: National Agricultural Cooperative Federation.

TABLE 100 Supply of Government Funds by Use, 1971–1974
(million wŏn)

Classification	1971	1972	1973	1974
Irrigation Fund	12,746	14,748	16,899	19,308
Agriculture Fund	29,720	36,968	36,474	36,216
Warehouse Fund	176	2,397	158	149
Fisheries Fund	–	–	–	–
Usury Debt Settlement Fund	1,378	900	450	–
AID Loan Fund	–	–	4,538	4,538
Farm Products Collection Fund	1,452	407	1,645	5,459
Livestock Fund	4,671	4,297	4,376	4,236
Forestry Fund	–	–	–	500
Total	50,143	59,717	64,540	70,406

Source: National Agricultural Cooperative Federation.

TABLE 101 Borrowings From the Bank of Korea, 1970–1974
(million wŏn)

Year	Agricultural Credit Program	Marketing Program	Total
1970	2,673	30,000	32,673
1971	8,765	32,000	40,765
1972	8,500	34,000	42,500
1973	15,770	43,000	58,770
1974	78,565	20,000	98,565

Source: National Agricultural Cooperative Federation.

TABLE 102 Deposit Received by NACF, 1971–1974
(million wŏn)

Classification	1971	1972	1973	1974
(Type of Deposit)				
Demand Deposit	42,804	56,516	70,998	98,529
Saving Deposit	66,122	76,225	98,373	117,323
Total	108,926	132,741	169,371	215,852
(Depositer)				
Farmers	26,882	26,611	34,295	38,408
Non-farmers	82,044	106,130	135,076	177,444

Source: National Agricultural Cooperative Federation.

TABLE 103 Trend of Own Funds, Accumulation by NACF and
Kun Co-ops, 1961–1974
(million wŏn)

Classification	1961	1965	1968	1971	1974
NACF	385	957	1,114	2,819	5,229
Kun Co-ops	64	1,072	1,462	5,546	15,564

Source: National Agricultural Cooperative Federation.

TABLE 104 Trend of Lending, 1970–1974

(million wŏn)

Year	(A) Total Lendings	Credit Sector (B)		Government Sector (C)	
		Amount	B/A	Amount	C/A
1970	105,360	70,988	67.4	34,372	32.6
1971	123,068	87,648	71.2	35,420	28.8
1972	148,245	102,984	69.5	45,262	30.5
1973	164,400	116,894	71.1	47,506	28.9
1974	252,149	191,282	75.9	60,867	24.1

Source: National Agricultural Cooperative Federation.

EIGHT

Farm Price Policy

Due to the dominant role of food-grains in the Korean economy, the policy decisions pertaining to food-grain prices, particularly rice prices, have been the central issue in the farm price policy and the term "food-grain price policy" has been virtually synonymous with "farm price policy" in Korea. It was only recently that a few selected items of non-cereal products have been, in a limited way, included in the government support programs.

The objectives of farm price policy have been stated in many different ways with shifting emphasis depending on the general economic situation. The market system has also gone through many changes over time, such as from free-market transactions to complete control, from complete control to partial control, back to free market, and then partial control again.

THE EVOLUTION OF
GRAIN MARKET OPERATIONS

Prior to 1939, there were no government controls on the grain market. Essentially the prices of all agricultural products including staple food-grains were determined by market forces. Because of a serious crop failure due to drought in both Korea and Japan in 1939 and increased demand created by the outbreak of the Sino-Japanese War in the same year, the Japanese colonial authorities initiated various measures of control over the grain market. Japan's subsequent involvement in World War II in 1941 led to a further increase in the military demand for grain and a drastic deterioration in the overall food situation. A series of regulatory measures was designed to secure adequate food supplies and to continue exportation of rice to Japan. In 1943, the Food Control Law was promulgated to close down completely the free market for grain. Compulsory grain delivery quotas were assigned to all farmers, and supplies to all consumers were rationed. Low purchase prices paid to Korean rice producers under this program had a negative effect upon production, resulting in a further shortage in food-grain supply.

After Liberation in August 1945, the U.S. Military Government, in keeping with its liberal ideas, discontinued the food-rationing system and restored free-market transactions for all grain. The U.S. Military Government authorities apparently believed that Korea had a substantial surplus of food-grain, an opinion based no doubt on the past record of large rice exports. The result of liberalization, however, was an aggravation of the imbalance between supply and demand. Repatriation of approximately 1.2 million Koreans from abroad, the influx of refugees from North Korea, and the sudden freedom from tight food controls caused a sharp rise in rice consumption and prices. At the time the food decontrol ordinance was issued, transportation, communication, banking, and market facilities were totally inadequate to meet the needs of a free economy.

Fearing the political, economic, and social confusion subsequent to a sudden switch from a wartime system to a peacetime system, the U.S. Military Government was obliged to reconsider the free-market system. With a view to stabilizing overall prices and protecting the urban consumers' standard of living, the U.S. Military Government adopted the ceiling-price system at the retail level for eleven major consumer items, including rice and cotton cloth. But the government-set ceiling prices were never honored in the market. Hoarding and black-marketing prevailed, and consumers who had been completely dependent upon government rations for their staple food supplies suddenly found the rationing eliminated with no adequate alternative source of supply available.

Having recognized the failure of the ceiling-price system, the U.S. Military Government revised its grain policy. The U.S. Military Government believed at the time that direct government control of a sufficient grain stock would be more effective in alleviating inflationary pressures than a price control system. The free grain market was once again closed, and compulsory rice collections from farmers and a complete ration system for urban consumers were enforced.

After the U.S. Military Government operations were terminated and the Government of the Republic of Korea was established in August 1948, the same policy was pursued. Grain producers and landowners were required to sell to the government all their grain other than that required for home consumption and seed. Free-market transactions and transportation of grain except for home use were prohibited. However, the government could not procure sufficient grain to implement overall rationing. The main reason was an unreasonably low purchase price, much lower than the market price, which in turn was ascribable to an underestimation of farm production costs and limited government funds available for grain purchase. Moreover, due to administrative inability and lack of adequate statistics on rice production, it was difficult to allocate fairly the government purchase quotas among different localities and farmers.

In this situation, a fundamental change in policy was inevitable. Overall control was lifted, free-market transactions were allowed, and the overall rationing system was changed into a priority ration system. First claim to the limited government-controlled grain supply was given to the military, police, government employees, and workers in critical industries such as coal mining. Urban consumers not receiving rations were able to procure their requirements from the open market.

While continuously enforcing the partial rationing system, the government enacted the Grain Management Law in February 1950, which remains today the basic legal authority for the government food-grain policy. The primary intent of this law was to enable the government to secure sufficient grain from farmers so as to stabilize the national economy by exercising adequate control over grain distribution and consumption through manipulation of government stocks. Later in 1963 and 1967, the main provisions of the law were reaffirmed, and additional authority was given to the government, but the basic direction remained the same.

Since the enactment of the Grain Management Law, the grain market in Korea has been characterized until today by a dualistic system, combining free-market transactions and government control, though the degree of market control by the government has varied from year to year. Free-market transactions apparently exist today only by sufferance of the government and not from lack of legal authority to control the grain market. The government has full legal authority for virtually complete regulation. It has authority to import or export grain and can give orders to hotels, restaurants, grain dealers, transporters of grain, and grain processors when it considers it necessary. The government directly conducts the procurement, transport, storage, milling, and sales for the government-controlled grain.

The government acquires grain from farmers through various programs at government-set prices during or after the crop harvest season. The acquisition programs include: 1) direct purchase; 2) rice-fertilizer barter; and 3) collection of farmland tax in kind. In the early years, the government acquisition program was mainly centered on rice, but recently the share of barley has been substantially expanded. During the 1950s, the market share of government rice in total marketings was less than 10 percent but had expanded to over 50 percent by 1975. In the case of barley, the government handled more than 90 percent of the total quantity marketed in 1975.

During the 1950s, the primary function of the government grain program was to secure an adequate supply of grain for the armed forces, for government institutions such as police forces and public hospitals, and for other ministries for relief and work programs. Whether intentionally or unintentionally, this procurement action undoubtedly had the effect of providing support for prices at harvest time when grain prices normally fell.

Entering the 1960s, the government expanded the scale of grain operations through increased purchases from farmers and imports. In addition to the original function of supplying grain for institutional uses, the government began to put emphasis on seasonal price stabilization through direct sales in the market during the non-harvest season when prices normally would fall. The amount of these direct sales accounted for about three-quarters of the total government grain supply in 1975.

LEVELS OF PURCHASE PRICES

The problem of determining government purchase prices for major grains, rice and barley, was closely related to such aspects of the economy as farm income, consumer welfare, the general price level, and rural incentives for increased production. During the 1950s, the government's primary efforts were placed upon the rehabilitation of the war-wrecked economy and the alleviation of the inflationary spiral. Policy-makers were particularly sensitive to the effects of grain prices on urban consumers' living costs and the general price level. Therefore, the government's major emphasis was directed toward maintaining low prices for urban consumers rather than toward farm income support. In particular, the government desired to provide grain at less than open-market prices to wounded veterans and their families, workers in critical industries, and to other special categories assumed to be less able to buy food at open-market prices than normal consumers. Another motive the government had for setting official grain prices at relatively low levels was the desire to minimize annual payments to landlords forced to sell land under the land reform program instituted in 1949.

Above all, the availability of United States grain under the U.S. Public Law 480 program after 1955 helped make it possible for the government to pursue a low-price policy for staple food-grain. The acquisition from aid sources of grain other than rice for government use became relatively easy with the quantity of grain imported under the PL 480 program on the average equaling approximately 9 percent of annual domestic grain production during the 1956–1960 period. Apart from the significant contribution of PL 480 grain to general economic stability, the easy availability of aid grain gratis undoubtedly provided a disincentive to policy-makers to increase domestic production by means of a high-price policy.

As shown in Table 105, government purchase prices for rice were lower than the estimated costs of production in almost every year until 1960. Government grain was, in effect, requisitioned from farmers through local administrative channels. Table 106 gives the estimated loss on the part of farmers due to low pricing of rice during the years 1950–1960, and indicates how low grain prices contributed in a major way to rural poverty.

The Farm Products Prices Maintenance Law was promulgated in June 1961. The purpose of the law was "to maintain proper

TABLE 105 Government Purchase Prices and Market Prices
versus Cost of Production for Rice, 1948–1975
(wŏn per 80 kg)

Year	Purchase Price (A)	Cost of Production (B)	Market[a] Price (C)	A/B (%)	A/C (%)
1948	2.47	3.72	7.10	66.3	34.8
1949	2.67	6.71	13.21	39.9	20.2
1950	16.40	15.88	52.30	103.6	31.4
1951	65.37	n.a.	157.50	–	41.5
1952	200.62	329.09	447.50	61.0	44.8
1953	200.62	330.94	350.00	60.6	57.3
1954	308.33	330.94	581.00	93.2	53.1
1955	390.56	838.44	962.00	46.6	40.1
1956	1,059.00	1,134.00	1,591.00	93.4	66.6
1957	1,059.00	1,384.00	1,311.00	76.5	80.8
1958	1,059.00	1,297.00	1,157.00	81.6	91.5
1959	1,059.00	1,300.00	1,368.00	81.4	77.4
1960	1,059.00	1,313.00	1,687.00	80.7	62.8
1961	1,550.00	1,377.00	1,768.00	112.6	87.7
1962	1,650.00	1,422.00	2,801.00	116.3	58.9
1963	2,060.00	1,373.00	3,470.00	149.7	59.4
1964	2,967.00	1,936.00	3,324.00	153.3	89.3
1965	3,150.00	2,672.00	3,419.00	117.9	92.1
1966	3,306.00	2,495.00	3,750.00	132.5	88.2
1967	3,590.00	2,735.00	4,289.00	131.2	83.7
1968	4,200.00	3,403.00	5,140.00	123.4	81.7
1969	5,150.00	3,565.00	5,784.00	144.5	89.0
1970	7,000.00	4,642.00	7,153.00	150.8	97.9
1971	8,750.00	4,682.00	9,844.00	186.9	88.9
1972	9,888.00	6,115.00	9,728.00	161.7	101.6
1973	11,377.00	6,578.00	12,175.00	173.0	93.4
1974	15,760.00	7,959.00	17,821.00	198.0	88.4
1975	19,500.00				

Sources: MAF, *Grain Statistics Yearbook 1967–1975.*
MAF, *Cost of Production Survey, 1967–1975.*

Note: [a]November-January average prices.

TABLE 106 Estimated Farmers' Losses[a] Due to Sales to Government, 1950–1955

Year	Quantity of Sales (1,000M/T)	Market[b] Price (wŏn/80kg)	Government Purchase Price (wŏn/80kg)	Loss[a] Per Bag (wŏn/80kg)	Total Loss (million wŏn)
1950	538	52.30	16.40	35.90	241.4
1951	266	157.50	65.37	92.13	306.3
1952	268	447.50	200.62	246.88	827.0
1953	400	350.00	200.62	149.38	746.9
1954	333	581.00	308.33	272.67	1,135.0
1955	246	962.00	390.56	571.44	1,757.2

Source: Computed from MAF, Grain Statistics Yearbook 1964.

Notes: [a]The concept of loss used in making this calculation is a rough approximation at best. If the data were available, it would be better to subtract the government price from what the market price would have been in the absence of government intervention in the market.

[b]November–January average price.

prices of agricultural products to insure the stability of agricultural production and the rural economy." The expressions "equitable income distribution between rural and urban families" and "increased food production" have been frequently used in official documents as the objectives of farm price policy. As seen in Table 105, the government purchase price for rice has exceeded the estimated cost of production since 1961, but these same prices were in most years lower than the market prices, and the old practice of grain delivery quotas continued to be implemented.

Because development strategy was centered on rapid industrial growth in the formulation of the First Five-Year Plan (1962–1966), the industrial sector not only received high priority in sharing investment resources, but general economic stability, particularly general price stabilization, was also listed as a top priority. Farm price support continued to receive less attention, as higher prices for major food-grains led by higher grain support prices were believed to cause a rise in the general price level. The high weight of grain in both the wholesale and consumer price index (Table 107) made the maintenance of low grain prices a key anti-inflationary instrument.

The low-price policy for food-grain, however, also hindered efforts to increase food production and, at the same time, stimulated rice consumption, resulting in a widening food gap. When a large portion of the food-grain shortage was met by local currency purchases under the U.S. PL 480 program, the food gap itself did not impose a serious burden on the country's foreign exchange position. But in the face of changes in United States policy to cash or credit sales in U.S. dollars in the late 1960s, the food-grain situation became directly related to the country's balance of payments position. The total value of imported food-grain amounted to $70 million (U.S.) in 1962, of which $34 million worth, or approximately 50 percent, was secured under the U.S. PL 480 program. But since 1970, grain imports have been dependent upon the country's own foreign

TABLE 107 Weight of Grain and Rice in Price Index
(%)

	Wholesale Price		Consumer Price	
	All Grains	Rice	All Grains	Rice
1947–1954	39.00	35.29	35.70	30.50
1955–1959	24.94	20.37	22.60	21.20
1960–1964	14.71	12.19	18.62	17.16
1965–1969	13.06	10.50	23.15	13.86
1970–	11.17	8.83	14.04	13.20

Sources: BOK, *Price Survey 1947–1970*; EPB, *Annual Report on the Price Survey 1970–1973.*

exchange reserves, and in 1975 almost $700 million worth of grain was imported under cash or credit terms.

With the increasing food shortage and, at the same time, the growing income disparity between urban and rural households, policy-makers were obliged to give serious consideration to expanding food-grain production and to a more equitable income distribution. Starting from the 1968 crop, the government took the initiative to improve the terms of trade in favor of farm producers by raising the purchase prices for rice and barley. Table 108 gives the movement of the terms of trade between agricultural products and non-farm products for the period 1959–1975 and shows that, since 1968–1969, the prices of agricultural products and rice received by farmers rose more rapidly than the prices of non-farm products paid by farmers. This initiation of high purchase prices for major grains clearly reflects a major change in farm price policy and is part of an effort on the part of government to stimulate domestic production of food-grain and to upgrade farm income.

TABLE 108 Terms of Trade for Agricultural Products, 1959–1975

Year	Index of Prices Received by Farmers		Index of Prices Paid by Farmers (C)	A/C	B/C
	All products (A)	Rice (B)			
1959	17.4	18.0	24.8	70.2	72.6
1960	20.9	21.8	26.6	78.6	82.0
1961	24.6	27.1	28.7	85.7	94.4
1962	27.1	28.7	31.8	85.2	90.3
1963	40.1	45.8	35.3	113.6	129.8
1964	50.2	57.0	44.8	112.1	127.2
1965	52.2	55.3	51.8	100.8	106.8
1966	55.4	56.5	58.1	95.4	97.3
1967	63.5	62.2	65.8	96.5	94.5
1968	74.3	73.2	78.8	94.3	92.9
1969	84.8	90.8	86.8	97.7	104.6
1970	100.0	100.0	100.0	100.0	100.0
1971	121.4	125.6	114.4	106.1	109.8
1972	147.9	159.5	130.5	113.3	122.2
1973	164.2	167.7	143.1	114.7	117.2
1974	215.6	242.8	192.5	112.0	126.1
1975	267.6	305.7	237.9	112.5	128.5

Source: Computed from an NACF 1975 Rural Price Survey.

FIGURE 20 Movements of Terms of Trade for Agricultural
Products, 1959–1975

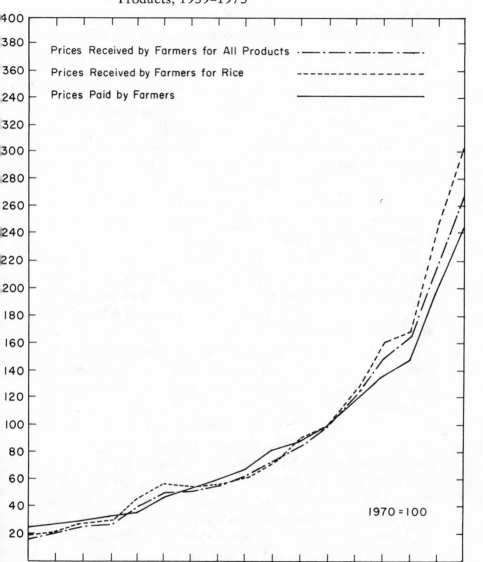

THE TWO-PRICE SYSTEM

Under the single-price system, if the government attempts to increase production and reduce consumption of rice in order to keep the aggregate demand and supply of rice in balance while importing less foreign rice, both the purchase and selling prices must be maintained at a relatively high level. The resulting high price of rice certainly contributes to increasing farm revenue as well as to saving foreign exchange through reduction in rice consumption, but high rice prices also will cause an upward pressure on the general price level and an adverse effect upon urban consumer welfare. A two-price system for rice and barley, a higher price for farmers and a lower price for urban consumers, was one means of resolving this dilemma and attaining simultaneously these apparently conflicting objectives.

The two-price system for barley was put into effect beginning from the 1969 summer crop, and for rice beginning from the fall of 1969. The comparisons between government purchase prices and selling prices for rice and barley are presented in Table 109 and 110 respectively.

Until 1969, the selling prices for rice were determined by adding intermediate handling costs to the original purchase prices, resulting in no financial loss due to price differentials. Beginning from 1970 (the 1969 crop), the selling prices have been, with the exception of 1972, below the costs of acquisition and intermediate handling. The government purchase price for the 1969 rice crop was 5,150 wŏn per 80-kilogram bag (polished) but the selling price for the 1970 marketing period was 5,470 wŏn, a differential of 320 wŏn per bag. This differential was not sufficient to cover the actual handling costs of 572 wŏn. For 1971, the selling price was lower than the acquisition price, causing the government to bear not only the full handling costs but also the price difference. For 1972, the selling price was high enough to cover the acquisition and handling costs. Since 1973, the difference between the purchase prices and selling

TABLE 109 Government Purchase versus Selling Prices for Rice, 1956–1975

(wŏn per 80 kg)

	Purchase Price (A)	Selling Price (B)	Price Difference (B) − (A)	Handling Costs	B/A (%)
1956	1,059	1,216	157	157	114.8
1957	1,059	1,216	157	157	114.8
1958	1,059	1,216	157	157	114.8
1959	1,059	1,216	157	157	114.8
1960	1,059	1,216	157	157	114.8
1961	1,550	1,792	242	242	115.6
1962	1,650	1,888	238	233	114.4
1963	2,060	2,312	252	251	112.2
1964	2,967	3,450	483	346	116.3
1965	3,150	3,350	200	394	106.3
1966	3,306	3,900	594	446	118.0
1967	3,590	4,100	510	507	114.2
1968	4,200	5,200	1,000	496	123.8
1969	5,150	5,470	320	572	106.2
1970	7,000	6,500	−500	662	92.9
1971	8,750	9,500	750	738	108.6
1972	9,888	9,500	−388	785	96.1
1973	11,372	11,264	−108	905	99.1
1974	15,760	15,850	90	1,265	100.6
1975	19,500	16,800	−2,700	1,560	86.2

Source: Food Bureau, Ministry of Agriculture and Forestry (Fisheries).

prices continued to widen and the loss incurred by the government amounted to more than 1,000 wŏn per bag.

In the case of barley, the price differences were even wider, reflecting a heavier price subsidy than for rice. The government efforts to keep barley prices at a low level for consumers were motivated by a desire to encourage consumers to substitute barley for rice in their diet. Prior to 1968, the market price of

TABLE 110 Government Purchase versus Selling Prices for
Barley, 1964–1975
(wŏn per 80 kg.)

Year	Purchase Price (A)	Selling Price (B)	Price Difference (B) – (A)	B/A (%)
1964	1,147	1,377	230	120.1
1965	2,295	2,463	168	107.3
1966	2,295	2,463	168	107.3
1967	2,490	2,632	142	105.7
1968	2,640	2,750	110	104.2
1969	3,348	2,750	–598	82.1
1970	3,850	3,100	–750	80.5
1971	4,890	4,300	–590	87.9
1972	6,357	4,800	–1,557	75.5
1973	6,993	6,000	–933	85.8
1974	9,091	8,320	–771	91.5
1975	11,100	8,320	–2,780	75.0

Source: Food Bureau, Ministry of Agriculture and Forestry (Fisheries).

barley had been maintained at the level of over 65 percent of the rice price (see Table 111) but, with the increasing subsidy for barley prices after 1969, the price of barley for urban consumers was lowered to almost 40 percent of the rice price.

Although an accurate estimate of the net effect of the barley price subsidy is lacking, comparisons of the quantity of government-controlled grains released to the market indicate that the low price ratio of barley relative to rice was somewhat conducive to diversion of consumption away from rice (Table 112). As per capita income grows, however, the average consumer will respond less sensitively to changes in the relative price of barley due to a higher income elasticity of demand for rice and other higher-quality foodstuffs. It is doubtful, therefore, that the two-price policy for barley will have a continuous effect on the marginal rate of substitution between rice and

TABLE 111 Rice versus Barley Consumer Prices, 1960–1975

Year	Rice Price (A) (wŏn/80 kg bag)	Barley Price (B) (wŏn/76.5 kg bag)	B/A (%)
1960	1,520	1,005	66.1
1961	1,835	1,360	74.1
1962	1,885	1,450	76.9
1963	3,010	2,310	76.7
1964	3,680	3,130	85.1
1965	3,515	2,535	72.1
1966	3,650	2,525	69.2
1967	3,955	2,890	73.1
1968	4,600	2,980	64.8
1969	5,390	2,980	55.3
1970	6,320	3,710	58.7
1971	7,760	4,790	61.7
1972	10,790	5,740	53.2
1973	10,833	6,000	55.4
1974	14,970	6,350	42.4
1975	19,903	7,965	40.0

Source: BOK, *Price Survey, 1960–1975.*

TABLE 112 Government Rice and Barley Sold to the Market, 1971–1975
(1,000 M/T, polished)

	Rice	Barley
1971	1,181	137
1972	589	419
1973	606	420
1974	972	676
1975	553	461

Source: Food Bureau, Ministry of Agriculture and Forestry (Fisheries).

barley unless the price difference between the two grains is widened as income grows.

The two-price policy through price subsidy for rice and barley might have contributed to increasing barley consumption and, at the same time, to alleviating an upward pressure on consumer prices, but the implementation of the two-price policy caused government costs for grain operation to increase at an accelerating pace. Table 113 gives the actual financial loss the government incurred since 1972 due to the price subsidy.

The effect of the government financial deficit from grain operations depends on how the deficit is financed, that is, the source of financing. If the deficit were compensated for from the general budget account, the effect would be simply a reduction in budget expenses for other sectors. So far as the grain operation was concerned, this was not the case. A large portion of the deficit has been financed through a long-term overdraft from the central bank, causing an increase in the money supply.

Table 114 shows how the net increase in the money supply due to the grain deficit contributed to monetary expansion. It accounted for 13.3 percent of the total increase in money supply in 1972, 20.6 percent in 1973, increased to 68.5 percent in 1974, and rose again to 98.2 percent in 1975, implying that nearly all the increase in money supply that occurred resulted from money creation in the grain sector.

In view of the importance of enhancing the economic position of farmers, policy decisions regarding grain prices must be looked at from a perspective different from that of monetary policy alone. However, it is undeniable that the increase in money supply caused an upward pressure on the general price level. In effect, the expanding scale of the government deficit due to the two-price system has emerged as one of the serious constraints on farm price policy.

TABLE 113 Financial Loss in the Grain-Management Funds, 1972–1975

Year	Item	Purchase Price (wŏn)	Handling Costs (wŏn)	Breakeven Price (wŏn)	Actual Selling Price (wŏn)	Profit or loss per Bag (wŏn)	Total Loss (billion wŏn)
1972	Domestic rice	8,750	738	9,488	9,500	12	1.0
	Imported rice	$310		8,325	9,500	1,175	3.9
	Barley	4,890	672	5,562	4,300	-1,262	-5.1
	Other cereals						-2.0
	net loss						-2.2
1973	Domestic rice	9,888	792	10,680	9,500	-1,180	-1.9
	Imported rice	$331		9,190	9,500	310	2.8
	Barley	6,357	796	7,153	4,800	-2,353	-8.8
	Imported wheat	$116		1,485	1,130	-355	-16.5
	Other cereals						-1.0
	net loss						-25.4
1974	Domestic rice	11,377	1,039	12,416	11,264	-1,152	-12.2
	Imported rice	$550		18,728	11,264	-7,464	-14.1
	Barley	6,993	1,040	8,033	6,000	-2,033	-45.1
	Imported wheat	($223)		2,797	1,805	-932	-54.8
	Other cereals						-0.7
	net loss						-126.9
1975	Domestic rice			18,280	16,065	-2,215	-13.2
	Imported rice			16,685	14,865	-1,820	-3.1
	Barley			10,821	8,029	-2,792	-22.0
	Imported wheat						-54.6
	Other cereals						-0.7
	net loss						-93.6

Source: Food Bureau, Ministry of Agriculture and Forestry (Fisheries).

TABLE 114 Effect on Money Supply of Financial Deficit Due to Grain Operation, 1970–1975
(billion wŏn)

Year	Long-Term Overdraft	Repayment	Balance (A)	Outstanding Balance	Total Money Supply	Increase in Total Money Supply (B)	A/B (%)
1970	2.0	2.0	0	0	307.6	55.6	–
1971	0	0	0	0	358.0	50.4	–
1972	50.0	14.0	36.0	36.0	519.4	161.4	22.3
1973	50.0	0	50.0	86.0	730.3	210.9	23.7
1974	160.0	0	160.0	246.0	945.7	215.4	74.3
1975	231.7	0	231.7	477.7	1,181.7	236.0	98.2

Sources: Food Bureau, Ministry of Agriculture and Forestry (Fisheries); BOK, *Economic Statistics Yearbooks 1970–1975.*

SEASONAL PRICE STABILIZATION

As already stated, the government began to place much emphasis on reducing the seasonal variation in grain prices with a view to protecting urban consumers during the non-harvest season. Whether or not the objective of seasonal price stabilization could be achieved has been largely dependent upon the availability of grain in the government stock.

As shown in Table 115, when the market share of the government-controlled grain was relatively small before 1968, the rate of seasonal variations in rice prices was relatively large, ranging from 15 to 30 percent. With the increased market share of government-controlled rice due to increased acquisition and imports, the seasonal fluctuation in rice prices was reduced to 5–8 percent, but rose again in 1974 and 1975.

From a pricing point of view, the government was in a position to achieve market prices of a desired level when it held at its disposal roughly from 1 to 1.2 million metric tons. But whether or not the flattening of the seasonal prices of rice has been an effective policy has long been a subject of public debate. It has been persistently argued that price policy which aimed at flattening seasonal prices provided farmers with little or no incentive to retain their rice for later sale because it did not allow farmers and wholesalers to cover their storage costs. The consequences may have been not only a discouragement of investment in storage facilities but a stimulation of rice consumption in rural areas, since the relative price of rice would otherwise have been higher during the summer when new barley supplies were becoming available.

BUFFER STOCK OPERATION FOR
SELECTED AGRICULTURAL PRODUCTS

Because agricultural products are produced only during limited periods of the year and are perishable, their prices fluctuate

TABLE 115 Scale of Government Operation in the Rice Market
and Seasonal Price Variation, 1960–1975

Year	Production[a] (1,000M/T)	Government[a] Acquisition (1,000M/T)	Government Sales (1,000M/T)	Market Share (%)	Seasonal Price Variation (%)
1960	3,149	198	76	5.4	25.2
1961	3,046	141	20	–	25.0
1962	3,463	309	94	–	24.0
1963	3,015	278	264	22.2	23.2
1964	3,758	224	74	5.4	21.4
1965	3,954	240	94	5.0	18.3
1966	3,501	302	217	12.2	15.1
1967	3,919	355	285	13.5	11.7
1968	3,603	286	442	21.7	9.0
1969	3,195	156	681	90.4	7.6
1970	4,090	326	749	29.6	6.7
1971	3,939	365	1,181	44.3	6.7
1972	3,998	517	589	29.5	6.7
1973	3,957	507	606	30.3	2.3
1974	4,212	480	972	46.3	41.2
1975	4,524	735	553	24.6	22.0

Source: Computed from MAF Grain Statistics Yearbooks 1964–1975.

Note: [a]Production and acquisition of the previous year

widely from season to season. Prices decline during harvest season owing to market glut. But demand is fairly constant throughout the year in most cases, with the result that farm product prices rise during pre-harvest seasons. Such a pattern of price fluctuations is repeated year after year.

In order to stabilize the prices of farm products, the government set in motion in 1970 a program to purchase proper amounts of selected agricultural commodities at government-set prices during their harvest seasons and to sell during the off seasons when their prices begin to rise. The selling prices are determined on the basis of the purchase prices plus handling and storing costs, with the prevailing market prices also taken into consideration.

The list of products stockpiled under the program encompassed red pepper, garlic, and sesame in 1970 and 1971. In 1972, apples and eggs were added to the list. Red pepper and garlic alone were stockpiled in 1974. Since 1975, garlic has been excluded from the stockpile, because it has been placed under the marketing regulation program (see Table 116).

TABLE 116 Scale of Government Purchase, 1970–1975
(M/T)

Year	Red Pepper	Garlic	Sesame	Apples	Eggs
1970	1,160	117	700	–	–
1971	1,917	151	595	–	–
1972	1,532	293	325	797	79
1973	648	133	787	900	145
1974	300	70	–	–	–
1975	1,144	–	537	–	–

Source: MAF, "Agricultural Products Purchase Program," 1975.

In light of the fact that the stockpiling program requires a lump-sum release of government funds to purchase crops in short periods of time, the government instituted a marketing-regulation program in 1972 to complement the stockpiling program. Under the new program, the government makes advance payments to producers—instead of buying crops directly from them—to enable them to hold their products off the market during the harvest seasons, when prices tend to be depressed. The marketing-regulation program was put into effect in the areas producing Chinese cabbages and radishes for preparing *kimjang*, the ingredients used in making *kimch'i* (pickled vegetables) to last the winter. In 1976, garlic, apples, and eggs were additionally placed under the program.

A study indicates that the implementation of this program

made a substantial contribution to alleviating seasonal price fluctuations for selected items, resulting in benefits for both producers and consumers. According to this study, the average rates of seasonal price changes of red pepper and sesame were as high as 88.5 percent and 55.1 percent respectively during the period 1965–1970 when there was no such program but, after the program was put in effect, the corresponding rates of fluctuation were reduced to 29.1 percent for red pepper and 38.7 percent for sesame during the period 1971–1973.[1]

The benefits accruing to producers due to price-raising effects of the government purchase activities during the harvest seasons were estimated as approximately 3.7 billion wŏn in the case of red pepper, 500 million wŏn in the case of garlic, and almost 650 million wŏn in the case of sesame (Table 117).

FORWARD PRICING SYSTEM FOR
SELECTED AGRICULTURAL PRODUCTS

A cash-crop price-support policy was launched in 1962, when the government purchase of five commodities (flax, sweet potatoes, castor beans, rapeseed, and cotton) was put into practice.

Although there was a need to encourage the production of the above five agricultural products, the instability of their prices had been discouraging to farming. Thus, the government instituted a policy of announcing their prices before harvest time so that farmers would be assured of minimum prices for their products.

During its initial stage, the government purchase program did not adequately attain its intended two-pronged goal of price stabilization and price support, because prices paid to the farmers under the program were either unduly higher or unduly lower than actual market prices. But the price-support policy soon became much more effective as a result of the creation of the Agricultural Product Price-Stabilization Fund.[2]

TABLE 117 Estimated Price-Raising Effects and Increase in Cash Income
Due to Marketing-Regulation Program

	Red Pepper		Garlic		Sesame[a]	
	Price[b] Raising %	Net Increase In Cash Income (million wŏn)	Price[b] Raising %	Net Increase In Cash Income (million wŏn)	Price[b] Raising %	Net Increase In Cash Income (million wŏn)
1971	5.4	624	11.9	327	–	–
1972	37.6	3,789	9.8	436	32.6	563
1973	34.9	3,726	11.9	534	34.5	650

Source: MAF, "Agricultural Products Purchase Program," 1975.

Notes: [a]No government purchase in 1971

[b]Rate of price raise $= \dfrac{P_R - P_E}{P_E} \times 100$

where P_R = Actual prices received by producers during harvest season

P_E = Estimated prices during harvest season without program

To elaborate, with the fund providing money to cover part of the costs of the purchase program, the list of agricultural commodities eligible for price support was greatly expanded. It became possible for the government even to post its purchase prices for certain crops before their planting time.

In the mid-1970s, farm commodities covered by the forward pricing system included: 1) industrial raw material crops, such as sesame, rapeseed, and castor beans; 2) import-substitute crops, such as barley for brewing beer; 3) export crops, such as silk cocoons, mushrooms, rushes, and ramie. The list of farm products supported by the system has been revised from time to time. For example, prices of soybeans and sesame were announced in advance of harvest in 1972, but these two were removed from the list thereafter. As for barley, its government purchase price was announced before planting time only in 1973. Rapeseed was dropped from the list in 1976.

A system of contract farming has been introduced to encourage farmers to grow certain crops even when the markets are volatile. The contract farming system has been applied to sweet potatoes, ramie, corn (maize), and beer barley since 1964. Currently, the system covers sesame, rape, mushrooms, rushes, loofah, ramie, castor beans, sweet potatoes, and so on.

Broadly speaking, three types of farming contracts are currently in practice. The first are arrangements between agricultural cooperatives and individual farmers. The second are cultivation contracts concluded between farmers and private firms engaged in export business. The third are agreements whereby farmers sell their crops directly to the end-users. The kinds of crop covered by cultivation contracts vary from year to year, and the contents of such contracts vary from crop to crop.

LIVESTOCK PRODUCT PRICING POLICY

The government policy for pricing livestock products has featured various administrative actions intended to hold down the prices, rather than being centered on measures to support prices in order to protect the incomes of livestock farmers. And yet the policy has served to lower livestock production costs by concentrating on distribution of animal feeds and stabilization of their prices, because of the importance of feed to the livestock industry.

It can thus be said that the principal component of the livestock product pricing policy of the government has been a feed policy. This tendency is clearly reflected in the enactment of the Feed Management Law in 1963 and the annual overall feed demand-supply program of the government. Under the program, the government makes projections of feed demand and supply situations each year. It then purchases and stocks feeds—including those produced domestically—in consideration of both the seasonal feed requirements of the livestock industry and seasonal availability of feeds. The stock is released when shortages of commercial feed supply are anticipated. The aim is to ensure timely feed supply at stable prices.

In more recent years, Korea's livestock industry has become increasingly dependent on imports to meet its feed requirements. This necessitates an effective program to achieve long-term stability in the prices of compound feeds in spite of fluctuations in world market prices of feed-grains. In view of such a need, the Feed Price Stabilization Fund was established in 1975.[3]

Currently, consumer prices of meats (beef and pork) are determined by what are supposed to be agreements among members of the livestock industry association. But the consumer prices are not closely tied to production costs. In other words, consumer prices remain rigid, not much influenced by changes in prices received by producers. This poses a problem in the marketing of livestock products.

NINE

Local Government and Rural Development[1]

Local governmental agencies in South Korea are tightly integrated into a highly centralized, bureaucratic administrative system directed from Seoul. Very little authority or initiative is delegated to provincial and county levels, where self-administration and the carrying out of directives from above are the principal functions. The ideology and attitudes of officials towards their work and their "constituencies" are largely consistent with this organizational structure; by and large they do not see themselves as representing or reflecting the opinions, desires, and needs of villagers in their districts. Rather they tend to be overwhelmingly concerned with finding ways of handling pressures from higher echelons for the fulfillment of predetermined plans and quotas.

Historically in Korea, there has been a very high degree of concentration of prestige and power in the hands of a central bureaucratic elite that administered local areas through rigidly hierarchical organizations. Such traditions were reinforced and rationalized under colonial rule, and a good deal of the organizational framework and administrative procedure existing today has been derived from models imposed on Korea by the Japanese. Since 1945, efforts have been made to promote varying degrees of local autonomy, but at present there is virtually no popular participation either in the choice of local officials or in the formulation of the policies they implement.

In the late 1950s, towards the end of the Syngman Rhee regime, a significant change took place in the upper echelons of the Korean national bureaucracy. Until then, most high officials had carried out their duties in a formalistic, almost ritualistic manner, their primary concerns usually the cultivation of connections necessary to stay in power and the promotion of their own personal status and wealth. After about 1956, such men were gradually replaced by more highly trained professional "technocrats," who brought an entirely different perspective to public administration. Subsequently, under the military and then civilian regimes of Park Chung Hee, there has been a reinforcement of that trend, as younger men (both military officers and civilians) with a relatively pragmatic, problem-solving, achievement orientation took over key positions in the government.

While lip service had been paid to developmental goals for rural areas since the Republic was founded in 1948, it is only since 1962 that the Korean government has made any determined sustained effort to promote increases in agricultural productivity and farm incomes. And throughout most of the 1960s, as indicated in previous chapters, such efforts were relatively limited, with a large part of the developmental momentum and resources concentrated in industrial projects.

During this period, traditional patterns, which emphasized formalistic attention to detailed bureaucratic procedures, maintenance of the status quo, and control functions, continued to prevail in local government agencies and, as a result, standards of administrative effectiveness lagged far behind those of the central ministries in Seoul. But as the aspirations and energies of the nation—particularly the governing elite—became more and more closely bound up during the 1960s with the formulation and implementation of five-year economic plans, there were increasingly insistent demands for improved performance at local levels. The centralized authoritarian nature of the government, in particular the fact that it has virtually complete control of local finance, enables it to exert great pressures on local administrators to produce immediate dramatic and concrete results. Such pressures were sharply stepped up after the inauguration of the New Community Movement (Saemaul Undong) in 1971, which reflected President Park's greatly increased concern with regional development and the improvement of rural living standards.

Inevitably a certain amount of spurious bureaucratic sleight of hand and wheel-spinning has resulted, as county governments struggled to meet heavy burdens of quotas and deadlines on meager budgets. Nevertheless, a transformation in attitudes, morale, and standards of performance has been taking place in recent years that is perhaps as far-reaching and momentous in its implications for rural development as the earlier changes in the national bureaucracy have been for the development of the national economy.

STRUCTURE OF LOCAL ADMINISTRATION

Although a number of ministries such as Agriculture, Education, Commerce and Industry, Health and Social Affairs, Construction, and Transportation are deeply involved in local affairs, the Ministry of Home Affairs has primary jurisdiction. Provincial governors, county heads, and township (sub-county) chiefs all operate under the direction of the Bureau of Local Administration of the Ministry of Home Affairs, and as a rule they have pre-eminent rank, prestige, and power within their own territorial jurisdictions. The other bureau of the Ministry of

Home Affairs is the Police Bureau, which also has an extensive organizational network reaching down to the sub-county level. Both branches of the Ministry, the police and the regular regional and local bureaucracy, are more tightly organized, have better communications, and report to more influential and more highly placed officials in Seoul than any other local government agencies.

The county office or "local autonomous government" is the focal point of local administration and directly oversees the work of the sub-county offices and through them the village chiefs. The county executive also has some supervisory functions over the other government agencies in his district and is responsible for coordinating their activities. Within each county office there is an Agriculture and Forestry Section. The other agencies that are primarily concerned with agriculture at the county level are the Rural Guidance Offices (RGO agricultural extension service) of the Office of Rural Development, and the branch offices of the National Agricultural Cooperative Federation (NACF).

The Office of Rural Development (ORD) is attached organizationally to the Ministry of Agriculture and Fisheries, but it consists of a separate centralized bureaucracy with its own headquarters in Suwŏn and more than 10,000 extension workers stationed at the county and sub-county level all over rural Korea. The ORD also operates several agricultural experiment stations in various regions of the country and some well-equipped research laboratories in Suwŏn, where the extensive work in developing new varieties of seeds, plants, and other agricultural products mentioned in previous chapters has been done.

The National Agricultural Cooperative Federation is also a centrally directed bureaucracy, and local members have no voice in determining policies or the manner of operation. The Central Federation directs and supervises all activities, which are conducted mainly through branch offices at the county level and 1,545 "primary" cooperatives with 2,300,000 farmer members. The NACF has a monopoly on fertilizer distribution to farmers;

it also sells them insecticides, tools, and other agricultural supplies, as well as consumer goods on a non-profit basis. In addition to providing credit and banking services for their members, the cooperatives purchase surplus rice and other products, handling storage, transportation, and marketing where necessary.

Figures 21 through 23 illustrate these institutional relationships.

THE COUNTY (KUN)

The assigned administrative functions of the county government that have a direct bearing on agricultural development are economic planning, farm development, highway maintenance, flood control, and reforestation. But, since the county head now has primary responsibility for promoting the New Community Movement (NCM), his office is necessarily involved in every aspect of community development including the following: irrigation projects, the replacement of thatched roofs with metal tile or composition; improvements to village roads, bridges, wells and sanitation facilities; the construction of village meeting halls; the establishment of mutual credit associations; various technical projects to introduce commercial crops or off-season activities that will raise farm income; and the promotion of local industry.

Since the programs of other local agencies are either supervised, coordinated, financed, or evaluated and reported on by officials of the county government, it has become the administrative focus for nearly all developmental activities. While there is some direct involvement in community projects by county personnel, particularly as a result of the intensification of activity that has accompanied the New Community Movement, in general county officials are too exalted for working-level contacts with ordinary farmers. The more influential and better-educated village leaders sometimes visit friends or relatives in the county bureaucracy, however, in order to seek advice or favors.

FIGURE 21 Ministry of Home Affairs and Local Administration

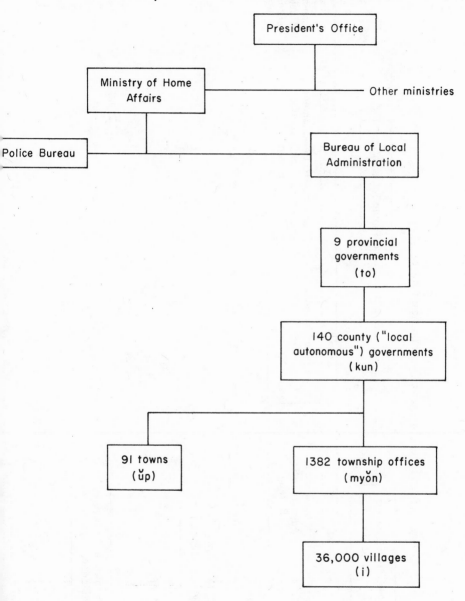

FIGURE 22 Office of Rural Development Organization Chart

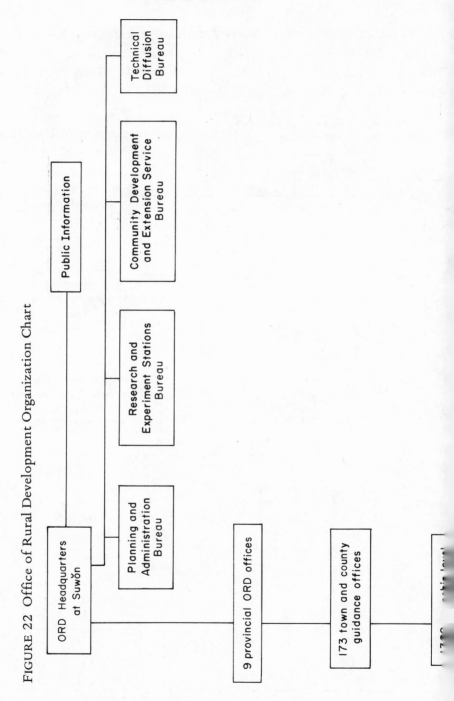

FIGURE 23 County (Kun) Administration Organization Chart

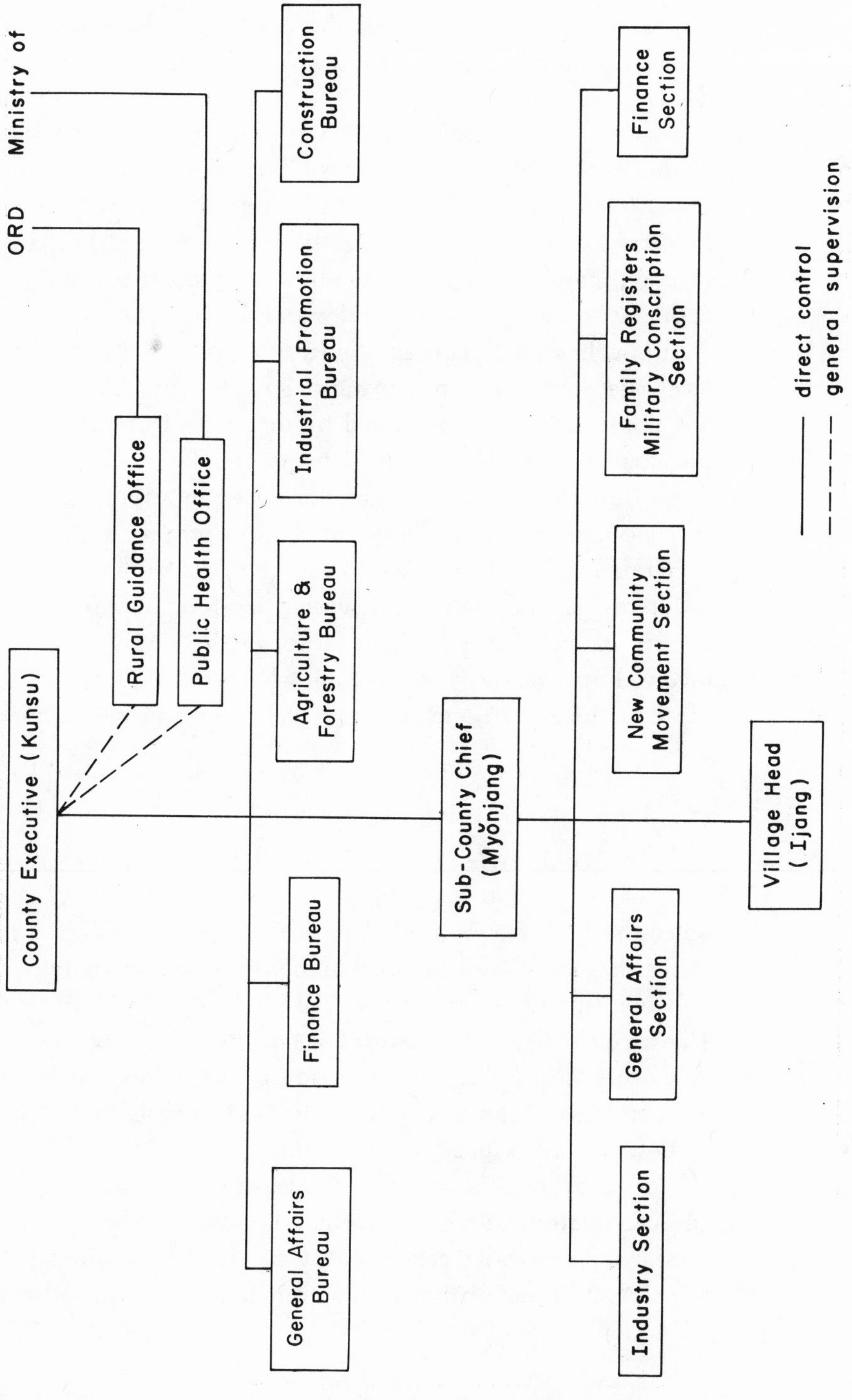

Most of the work done at the county level consists of the transmission of instructions from provincial or higher levels down to the townships and the evaluation and reporting of results of development activities up through bureaucratic channels. Possibly the most crucial role played by the county is that of institutional linkage, both vertically up and down the Ministry of Home Affairs chain of command, and horizontally to coordinate the activities of the township offices, the branch cooperatives, and the extension workers.

Although annual plans and budgets are prepared at the county level and submitted to provincial governments, their contents are almost entirely determined by directives from above. Since the county office is heavily dependent (about 80 percent) on funds from the central government for its operations, there is not much opportunity for local initiative in proposing agricultural development projects. Nevertheless, a considerable number of proposals generated at the township level are submitted to the county office each year and, although the amounts of money involved are relatively minor, anywhere from about 20 to 60 of these local projects may be approved and given financial assistance.

THE TOWNSHIP (SUB-COUNTY, MYŎN)

Since the Park regime took over in 1961, the township or myŏn offices have been downgraded in importance, and many of their previous functions are now vested in the county governments. The myŏn office is the lowest level of the bureaucracy, and it is here that the village leaders must come for instructions, indoctrination, and the presentation of complaints or special requests for assistance.

Like other local government institutions, the township office has been jolted out of its former somnolence by the intense pressures from above generated by the New Community Movement. As the organization in most direct contact with farmers, it is heavily involved in promoting, supervising, and reporting on local cooperative self-help projects. The myŏn budget contains only small amounts for development activities, however, so that, in addition to passing on instructions from above, officials at this level are mainly engaged in exhorting villagers to make greater contributions of their own towards the fulfillment of NCM goals. Village leaders must channel their requests for financial support of pet schemes for community development through the township office, and a few projects requiring relatively small amounts of money (under about $1,200) are funded each year. Well-to-do farmers occasionally seek special favors through their personal influence among myŏn officials. In cooperation with the Rural Guidance Office the township agency is also engaged in agricultural extension work. Village leaders are summoned to the myŏn office where they listen to lectures and watch demonstrations of new agricultural methods. Then they return home and attempt to persuade the local residents, more or less forcefully, depending on their individual styles of leadership and the degree of conviction they hold concerning the practices in question.

OFFICE OF RURAL DEVELOPMENT (AGRICULTURAL RESEARCH AND EXTENSION)

The elaborate agricultural research and extension service operated by the Japanese Colonial Government was largely dismantled after Liberation in 1945. The Rhee regime did almost nothing to replace it, and Korean farmers generally expressed no regrets because of their dislike of the previously mentioned coercive measures used by the Japanese to enforce their programs. In the late 1950s, several efforts were made to establish agricultural guidance activities in connection with local development programs. These were based on the voluntary, community self-help doctrines that were currently fashionable. By the early

1960s, AID agricultural officials were convinced on the basis of previous performance that the Ministry of Agriculture and Forestry was not capable of running an aggressive extension service effectively, partly because of its Seoul-centered, elitist bureaucratic character, and also because a substantial amount of United States aid intended for rural development had been consistently diverted to other, often "informal," purposes. In 1962 under the new Park Military Government, the Office of Rural Development was established with the energetic backing and financial assistance of AID as a separate, centralized, and largely independent agency attached to the Ministry of Agriculture and Forestry (renamed in 1973 the Ministry of Agriculture and Fisheries).

Since its establishment, the scope of operations of the ORD has been gradually expanded, and it is now trying to run an extremely ambitious program with severely limited resources. The Rural Guidance Offices are currently involved in implementing the following "policy directions": 1) development of food resources; 2) utilization of idle land resources; 3) technical dissemination to increase farm income; 4) innovation of farming techniques; 5) improvement of rural life; 6) training of rural leaders; 7) strengthening of cooperation among institutions (collaboration among ORD, agricultural colleges, and agricultural high schools).

There are Rural Guidance Offices (RGO) in every county, and additional branches are being opened at the sub-county level as well. Rural Guidance workers are in frequent and direct contact with farmers, and as a result they are usually more aware of local problems and attitudes than most other officials. They are also somewhat more likely to be trusted and listened to. On the other hand, ORD policy is determined at its central headquarters in accordance with national agricultural objectives, so that the advice of the field worker, who is closely supervised and controlled, is not always well adapted to local conditions or to the best interests of the farmers themselves.

Although it was launched with great energy and dedication, the rural guidance program has had some trouble holding on to well-trained, effective workers. The low pay, menial nature of the work (from a conventional bureaucratic perspective), and lack of opportunity for promotion have resulted in a high rate of turnover. Large numbers of graduates of agricultural high schools go into the service, but their youth and lack of experience tend to undermine their influence with farmers.

Two other problems have arisen that adversely affect the operation of the extension program. The administrative weakness of the Rural Guidance Office at the county level, where most of the authority and responsibility for implementing development policies is in the hands of the regular county government (Ministry of Home Affairs), leads to a situation in which the extension workers often have nothing to offer but advice, although various kinds of other resources may be needed if the advice is to be acted on. Another difficulty is the marked variation in the amount of attention given to different villages. Rural Guidance workers generally travel by bicycle to their work sites, and the more remote communities inevitably tend to be neglected. Also, since the Guidance Office is interested in demonstrating to superior authorities the success of its efforts, there is often a tendency to concentrate on showcase projects in villages particularly favored by good soils, progressive attitudes, and a record of cooperative accomplishment. As a result, existing differences in the relative prosperity of different communities may be reinforced, particularly where they are related to geographic accessibility. An effort has been made to meet this problem by stationing extension workers for long periods in certain villages as Residential Guidance workers. Unfortunately there has been a particularly high turnover rate within this group, since the living conditions are poor, and no provision has been made for an extra "hardship" allowance.

Since 1971, the RGO has focused most of its efforts on developing and promoting the cultivation of new varieties of "miracle" rice seeds, with dramatically successful results. In the process, however, many of its other research and extension programs have had to be cut back or eliminated.[2]

AGRICULTURAL COOPERATIVES

The National Agricultural Cooperative Federation (NACF) and its local branches constitute a large, growing, and crucially important institution for Korean agricultural development. Over 90 percent of Korean farmers belong to the local or "primary" cooperatives. In addition, there are 140 county-level cooperative branch offices and 145 special cooperatives (for horticulture, livestock breeding, or other cash crops), organized as shown in Figure 24.

FIGURE 24   National Agricultural Cooperative Federation Organization Chart

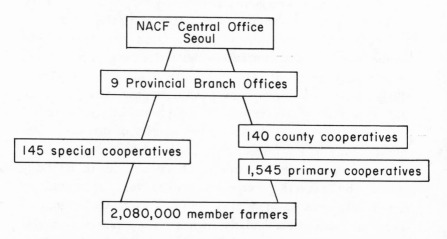

The only thing cooperative about the organization, however, (in addition to the name) is the fact that farmers must buy one or more shares in their primary cooperatives in order to belong. The NACF is a centrally controlled and directed bureaucracy engaged in planning and implementing policies, either on behalf of the government or to achieve its own institutional goals. Partly because the federation has had a long record of corruption and abuses in the past, supervision of local operations and finances is strict. In addition to the NACF's own hierarchical controls, representatives of the Ministries of Finance and Agriculture and Fisheries ride herd on the cooperatives at all levels.

The stated objectives of the NACF to increase agricultural productivity, to improve the social and economic status of farmers, and to enhance their cultural betterment, are quite similar to those of the New Community Movement. Since the NACF is an independent organization, there is a certain element of competition with the regular bureaucracy in its efforts to supply services to farmers that are also the responsibility of various government agencies. For example, the NACF has constructed 85 township community halls that are used for weddings, nursery schools, farmer education, and general meetings. Another 1,500 halls are to be built by 1980. Some local cooperatives operate health clinics, barber shops, and public baths. Primary cooperatives are also engaged in organizing groups of farmers who specialize in a particular crop in order to improve the planning and management of production.

While the central federation is engaged in a wide variety of activities, some of which, particularly banking and insurance, have little to do with agriculture, the primary cooperatives are of vital importance to the farmer, since they are his only source of fertilizer and of credit at low interest rates. By the 1970s, slightly over half the NACF funds available for loans were being generated locally from farmers' savings deposited with the local cooperative banking facilities (see Table 99). Large infusions of government money have substantially increased the amounts available for rural credit, and consequently borrowing at very high interest rates from such traditional sources as relatives, friends, and usurers has declined (see discussion in Chapter 7).

The rural credit program is confronted with a choice between economic and humanitarian goals. There is great competition among farmers for the limited supply of credit available through the cooperatives at interest rates of from 8 to 15 percent, which compares to 50 or 60 percent and even higher for loans from private sources. While the cooperative administration prefers to make loans to the wealthier farmers who are better risks, and who can be expected to use the money for productive investments in agriculture, there is great pressure from village leaders for a more equitable distribution of loans to poor farmers, many of whom need the money to feed their families or to overcome some other crisis. As farm income has increased in recent years, the amounts loaned for consumption purposes have declined somewhat, but such small loans are still made, and in most local cooperatives a certain number are each year quietly forgotten. In the past, farmers regularly had to give kickbacks to cooperative officials of 5–10 percent of the loan principle, but the rigorous anti-corruption campaigns of recent years have nearly eliminated such practices.

Local cooperatives also sell other agricultural chemicals (particularly insecticides) as well as tools and machinery, and they purchase grain at prices determined by the government. In addition, they operate non-profit retail stores selling consumer goods. The transportation and storage of agricultural products beyond the township level is handled by the county cooperatives, which also conduct farm guidance services and marketing research, as well as dealing with such matters as crop insurance, feed mixing and distribution, and artificial insemination. County cooperatives are also increasingly engaged in promoting a variety of specialized commercial agricultural projects. In addition to providing technical advice and loans, they are developing storage, processing, and marketing facilities for seri-culture, fruit cultivation, and the raising of pigs and poultry.

There is widespread recognition within the cooperative organization that close personal contacts with the farmer and the adjustment of policies to suit his needs and desires have been sacrificed to large-scale organization and central direction. But it is argued that this was necessary in order to achieve maximum efficiency, particularly with regard to implementing the government's insistence on rapid increases in food production.

SAEMAUL UNDONG
(THE NEW COMMUNITY MOVEMENT)

Economic planners of the Park Chung Hee regime concentrated during the 1960s on investment in industry and social overhead, while agriculture and rural development were consistently neglected. The increasing rural-urban imbalance had by the end of the decade confronted the government with some major challenges: 1) A low rate of increase in agricultural productivity and the stagnant rural economy could not cope with higher food consumption standards and the growing population, so that large amounts of foreign exchange were required for grain imports. 2) Extremely large and chaotic rural-urban migration had added greatly to administrative burdens in the cities (particularly Seoul) and to the potential for social and political unrest. 3) The presidential election of 1971 demonstrated that a substantial erosion of support for the regime had taken place in the predominantly rural areas of the southwest.

President Park responded in the winter of 1971–1972 by launching the New Community Movement (NCM), a highly organized, centrally directed effort to mobilize farmers on a national scale. From the start, the movement has had political overtones. Its stated purpose was to upgrade the quality of village life by promoting cooperation, self-help, and the transformation of conservative rural attitudes. Although actual government investments in NCM projects have been quite small as a proportion of total budget expenditures, the administrative and propaganda effort at both local and national levels has been intense. President Park has backed the NCM with the full weight and prestige of his office and, during the period 1972–1975, its momentum was not only sustained but even intensified. Every village in the country has, to a greater or lesser extent, felt the economic, political, and cultural pressures.

The intensity of purpose with which the New Community Movement was launched and the sustained pressure of the bureaucratic follow-through have resulted in much greater developmental activity by local agencies than in the past. Determined efforts have been made to mobilize all government personnel, including even the police and locally based army units in support of rural development efforts. In addition to their regular duties, local officials, teachers, extension agents, village and hamlet leaders, co-op administrators and clerks, and others are putting in long extra hours working on self-help projects or in "raising village consciousness."

Everyone is expected to engage in the furtherance of NCM goals, regardless of his organizational niche, so that there has often been a blurring of functional roles and considerable overlap among the different agencies. As a result, the coordinating functions of the county head have become even more crucial, and in fact an enormous amount of time seems to be spent in meetings by everyone from the village leaders on up through township, county, cooperative, and ORD personnel. The county head is constantly bustling back and forth to the provincial capital, just as the village head must frequently visit the sub-county office, and the township chief must go to the county. Since individual reputations and promotions depend to a considerable extent on performance in achieving NCM objectives, there is a strong tendency for lower echelons to emphasize mainly their successful accomplishments in reporting to superiors. Ceremonial visits of encouragement and inspection by higher officials to villages where key projects are underway take place frequently. But, along with the propagandistic rhetoric, the endlessly time-consuming transmission of detailed bureaucratic instructions and reports, and the often phony overfulfillment of plans, there are signs of improved morale, a greater sense of purpose, and a new confidence in the administration's capacity to deal with practical problems. Also, standards of personal performance have improved, both in terms of upgrading administrative and technical skills, and through shutting off many of the opportunities for corruption.

Most farmers are gratified by the attention they are getting, but there has also been some puzzlement and cynicism over the frequent visits and other demonstrations of concern by officials for their welfare.

Except for a limited number of model villages where extensive investment and great efforts by the residents have been matched by government assistance, there is a gap between the glowing official descriptions of NCM accomplishments and the actual situation. Local self-help projects require considerable unrecompensed labor and often the contribution of other scarce resources by villagers, while in many cases benefits are intangible or deferred for many months or even years. Where projects have been undertaken that were not appropriate to village conditions or needs, or where lack of cooperation or inadequate leadership has resulted in the dissipation of community enthusiasm and hard work, the initial reaction of farmers has been strongly critical. Also, there have been many cases where an excess of bureaucratic zeal in carrying out the assigned mission on the part of local officials resulted in coercive pressures being applied to farmers, with widespread resentment the result. Memories of the methods used during the Japanese colonial period are still vivid for many villagers, and they are quick to make comparisons when mobilization for development becomes excessively authoritarian.

On the other hand, the New Community Movement has had far-reaching positive effects throughout the Korean countryside. Whether willingly and harmoniously or not, farmers have worked together to improve local roads, bridges, wells, washing facilities, and toilets. They have also undertaken extensive irrigation and flood-control projects, many of which have been pushed to successful completion. Large numbers of thatched roofs have been replaced with more permanent materials. Some villages have done much more than others, but no place has been untouched by the general fervor. The movement has reinforced rising expectations among rural Koreans, while their receptivity to change and technological innovation has been enhanced.

Since 1973, there has been a gradual shift away from the movement's initial emphasis on improving village physical environments and towards the promotion of greater agricultural productivity and increased rural incomes. The government's insistence that rapid agricultural development could only be achieved through the psychological transformation and mobilization of the rural population is open to question, but there is no doubt that the NCM has speeded up the pace of social and economic change in rural areas.

If the movement has sometimes been administered with excessive bureaucratic zeal, there has also been a considerable degree of flexibility and a willingness to learn through mistakes and to adjust policies accordingly. Detailed information about the reasons for success or failure in the case of specific projects does in fact flow back up through bureaucratic channels, and problems are discussed and solutions proposed at every level of the hierarchy.

In the past, while local administration in Korea had been effective in accomplishing its traditional mission—that of maintaining order and mobilizing the population for labor, military service, voting (appropriately) in elections, and the payment of taxes—it was not an effective development action instrument. Neither the organizational structure nor the performance standards of individual bureaucrats have been adequate for this role until relatively recently.

The strong vertical and hierarchical component in Korean institutional structures created a familiar administrative problem: the difficulty of coordinating efforts horizontally, both among different bureaus of the same ministry and among different ministries. Officials tend to be narrowly preoccupied with their own immediate assignments and to cater exclusively to the preoccupations of their superiors, without much concern for the end product of their efforts in terms of services to farmers. There is fierce competition and rivalry at all levels for resources, prestige, and credit for accomplishments. One result is that, at the local level where several agencies are engaged in rural development activities, and where great pressure is felt to show immediate results, bureaucratic concerns predominate, and administrative clout is more important in getting policies implemented than worthiness of objective in terms of rational economic priorities or the degree to which a program fits in with the aspirations and needs of the rural population.

It is within this special institutional context that one must interpret the dominant role of the Ministry of Home Affairs in current rural development efforts. Responsibility for rural credit, agricultural extension, farm mechanization, irrigation, technological innovation, the collection, storage, and distribution of products, rural welfare, and so on is fragmented among different agencies and branches of agencies. Extensive, if sporadic, efforts in the past to promote development in one or more of these sectors have usually had disappointing results. Many foreign observers, as well as Koreans engaged in promoting agricultural development, have generally assumed that the principal reason for inadequate performance and the slow growth of agricultural productivity has been the lack of adequate government investment through the Ministry of Agriculture and Fisheries, the Office of Rural Development, or the National Agricultural Cooperative Federation. Some planners and government leaders blamed Korean farmers for their presumed resistance to change. The point to be made here, however, is that another and possibly more critical obstacle to successful rural development has been a lack of institutional mechanisms to provide the organizational linkages necessary to bring about an effective coordination of efforts.

To put it more bluntly, neither the Ministry of Agriculture and Fisheries (MAF) nor the ORD has had enough control of the local administrative machinery to achieve its objectives. Whatever degree of priority any particular program of rural development might have had in national planning, there has not been a group of local officials with the authority, skills, prestige, degree of commitment, and organizational backing to carry it through to successful completion. For example, rural guidance workers are low-paid, low-ranking government employees whose effectiveness in promoting agricultural change is impaired by their administrative weakness; they are unable to mobilize such resources as credit, or funds for construction, in order to support the programs they have been advocating to farmers. And the NACF, which has been relatively isolated from the other bureaucratic organizations, has tended to follow policies dictated by its own special economic and organizational objectives without much coordination with other development programs.

Real power in rural areas continues to remain in the hands of officials of the Ministry of Home Affairs (MHA), including both the regular local administrative bureaucracy and the police. When it became apparent in 1971 that a major national effort with the personal backing of the President would be launched to promote rural development, an intense bureaucratic struggle between Agriculture and Home Affairs for control of the NCM ensued. The ultimate victory of the MHA, that is, the emergence of the county government as the pivotal agency for NCM administration, was criticized at the time as a subversion of the movement's economic and social objectives to political ends. But it may have been necessary in order to achieve coordination and focus of effort, given the Korean administrative system. While political aspects of the NCM are undoubtedly important, it now appears that the administrative involvement of the local "autonomous government" has been an essential factor in the considerable degree of developmental momentum achieved so far.[3]

PART FOUR

The Impact on Rural Welfare

TEN

Land Reform

The politics of Korea's land reform efforts in the 1945–1952 period is as complex and confusing as most of what was going on in that period. American combat officers with no prior experience in Korea understandably made less than ideal administrators during the first years of Liberation from Japanese rule. Koreans, with no recent experience in running their own government, weren't much of an improvement. Chaos characterized many government efforts, and land reform was no exception. And yet, when the process was completed, the Korean people had successfully implemented the redistribution of rented land from landlords to owner-cultivators.

By any standard, the land tenure system Korea inherited from the Japanese colonial period was an oppressive one. Tenancy had been increasing steadily under Japanese control (see Table 118) and, by the late 1930s, nearly 60 percent of all cultivated

TABLE 118 Owner-Tenant Distribution Before 1945

	1913–1917	1918–1922	1923–1927	1928–1932	1933–1937	1938
Owners	21.8	20.4	20.2	18.4	19.2	19.0
Part-Owners	38.8	39.0	35.1	31.4	25.6	25.3
Tenants	39.4	40.6	44.7	50.2	55.2	55.7
Total	100.0	100.0	100.0	100.0	100.0	100.0

Source: Takeo Suzuki, *Chōsen no nōgyō* (Tokyo, 1942), p. 246, as reported in Ki Hyuk Pak, p. 48.

land in what is now South Korea was owned by landlords rather than cultivators. This tenure system served the Japanese Empire reasonably well. While per capita grain output in Korea in the 1920s and 1930s actually fell slightly, Korean grain exports to Japan more than doubled.[1]

Part of the tenant-farmed land had been owned by Japanese landlords prior to 1945. Redistributing that land posed few political problems. Most of the rented land (over 80 percent), however, was owned by Koreans. In 1945 most Korean landlords were under a political shadow from having prospered under the hated Japanese rule, but in the next several years they were able to regain a measure of power. Equally important, the attitudes of the U.S. Military Government toward land reform were not very clear. Americans were reluctant to participate in a reform that involved a substantial element of expropriation of private property. To some Americans, land reform smacked of "communism" and was, therefore, suspect. And other Americans, ignorant of the Korean language and the nature of Korean society, were prey to manipulation by Koreans with their own private interests to pursue. On the other hand, Americans were carrying out a thorough land reform in Japan, one involving a large measure of expropriation, and many other Americans were aware that failure to carry out land reform was contributing to the collapse of the Kuomintang position on the Chinese mainland. In short, there was no clear and consistent American position on Korean land reform. On balance, the American position supported land reform in principle, but in practice American efforts to carry out an orderly reform probably slowed temporarily Korean political forces that were pushing the nation in the direction of a rapid and thorough reform.

A detailed history of the politics of land reform in this period is beyond the scope of this study and has, in any case, been discussed at length elsewhere.[2] The key formal steps were the creation of the New Korea Company, Ltd. in November 1945 by the U.S. Military Government to administer farmland formerly owned by the Japanese. That corporation was replaced by the Central Land Management Office, which succeeded in distributing over 240,000 hectares of former Japanese land, most of it going to the former tenant-cultivators of the land. The establishment of the Republic of Korea Government on August 15, 1948 was followed by the new Land Reform Act of June 21, 1949, amended into its final form on March 10, 1950. Under this legislation the government purchased and distributed nearly 330,000 hectares of farmland by 1952. As can be seen from Table 119, however, over 500,000 hectares appear to have been sold directly by landlords to their tenants. For reasons that will become apparent, the Land Reform Act provided both landlords and tenants with a powerful incentive to make their own separate deals.

Whatever the mechanism, there can be no serious doubt that land reform succeeded in turning most of Korea's cultivated acreage over to the families cultivating it. The number of families that owned all the land they farmed rose dramatically and the number of tenants fell to a miniscule 5 to 7 percent of the total (from 48.9 percent in 1945—Table 120). The amount of rented land fell simultaneously from nearly 60 percent of the total to less than 15 percent. There was some cheating, but surprisingly little, given the history of land reform in other nations. The land reform legislation itself allowed for certain exceptions to redistribution—clan land, land owned by educational institutions and other special groups, and reclaimed land

TABLE 119 Distribution of Former Rental Land
(in chŏngbo, 1 chŏngbo = 0.992 hectare)

1. "Vested farmlands" sold by the U.S. Military Government	245,000
2. Farmlands distributed in accordance with 1949 Reform Act	332,000
3. Land privately sold, 1945–1951	573,000
4. Land not yet distributed as of 1952	320,000
5. Total land to be reformed as of 1945 (= 1 + 2 + 3 + 4)	1,470,000

Source: Jae Hong Cho, "Post-1945 Land Reforms and Their Consequences in South Korea" (unpublished PhD dissertation, Indiana University, 1964), p. 92, is the source of the methodology used in this table, but the figures are somewhat different from his. No. 1 and No. 2 come originally from the BOK *1955 Economic Statistics Yearbook*. No. 3 is basically a residual whose precise size depends on one's estimate of total rental land in 1945 (No. 5) and the amount of land not yet distributed. No. 5 is from the BOK *Economic Statistics Yearbook* as reported by Ki Hyuk Pak, p. 88, and the land not yet distributed was estimated by assuming that the percentage of rented land in 1952 was the same as in the first farm household survey of 1962 (14% of 2,320,000 chŏngbo).

TABLE 120 Owner-Tenant Distribution of Farm Households, 1945–1965

	1945	1947 (end)	1964	1965
Full owner	13.8	16.5	71.6	69.5
Owner-Tenant	16.4	38.3	14.8	15.5
Tenant-Owner	18.2		8.4	8.0
Tenant	48.9	42.1	5.2	7.0
Farm laborer and burnt field farmers	2.7	3.1	–	–
Total	100.0	100.0	100.0	100.0

Source: Ki Hyuk Pak, pp. 87, 89, 131. The 1945 and 1947 figures are originally from the BOK *Economic Statistics Yearbook*. The 1964 and 1965 figures are from Farmland Surveys of those years of 1,136 farms (in 1965) and 922 households (in 1964).

were the major exceptions. According to the 1965 Farmland Survey, these legal exemptions in that year constituted half the total rented area.[3] The other half was privately owned and rented and hence illegally held under conditions of tenancy. Together, however, both types of land amounted to only 16 percent of the cultivated acreage in 1965 (or 8 percent each).

COMPENSATION VERSUS EXPROPRIATION

The success of a land reform, however, must be judged by more than just how much land was redistributed. Two key economic issues are whether the land reform succeeded in redistributing *income* in the rural areas as well as land ownership and whether the reform promoted or held back increases in productivity.

Generally speaking, land reforms that fully compensate the landlord for his land at free market prices[4] are not likely to lead to any significant income redistribution at all. One can go further: unless a land reform involves a high degree of expropriation, it is not likely to be successful. Little land and no income will be redistributed because tenant farmers won't be able to afford land at high market prices. If they could, they probably would have bought that land or some other long before.[5] It is not surprising, therefore, that the successful Korean land reform involved much more expropriation than compensation. Much of the expropriated income went to the former tenant and a part was used to finance the government war effort.

Initially, the U.S. Military Government did attempt to ensure that payment for land formerly held by the Japanese was to be in full (although the government, not the Japanese landlords, was to receive the money). Payment was to be equivalent to 3 times the average annual product of the land, which was roughly equivalent to the actual price of land in the 1930s in a fairly good year. At first the authorities intended to use 1930s production levels as the basis for this calculation, but in the end had to reduce 1930s levels by 40 percent because of the decline in land productivity that had occurred in the interim. Compensation was set at 20 percent of the annual harvest paid over a 15-year period. Most former Japanese land was in fact sold in accordance with this formula. Clearly, 20 percent of the harvest per year for 15 years was less burdensome than 50 percent forever. If one discounted payments in the second through fifteenth year at the comparatively modest interest rate (for rural Korea) of 10 percent, the actual price of the land was less than twice its average annual yield.

One problem with this 1948 formula was that the market price of land in that year was less than half the average annual output at the time (see Table 121). The price of land had fallen precipitously because many landlords were no longer able to collect rent on much of their land. Even those landlords who had managed to reacquire a measure of political power under the U.S. Military Government were blocked from receiving full payment because of the government grain collection program.[6]

The U.S. Military Government compensation formula disappeared with that government. The Korean politicians who led the first independent government had been elected on platforms that promised land to the tillers, and this principle was written into the Korean Constitution. No doubt many of the election promises made were insincere, and many nations have land reform laws on the books and little land reform in practice. But, unlike the situation in many less-developed countries where landlords dominate the rural political scene, in Korea political power was greatly weakened by the taint of the collaboration of many of them with the Japanese.

The 1949 Land Reform Act, therefore, reduced the compensation rate to 1.5 times the normal output of the main crop (instead of 3 times). Further, landlords were to be compensated with government bonds whose value was expressed in terms of rice (the purpose of the latter feature being to protect landlords from inflation). Originally farmers were only to pay 1.25 times the annual yield with the difference between 1.25 and 1.5 to be

TABLE 121 Paddy Land Prices in Korea, 1936–1949

Year	Average Market Price per Chŏngbo (in sŏk of rice)	Average Land Reform Compensation Rate (in sŏk of rice)
1936–1945 average	74.5	–
1945 (August)	7	–
1946 (June)	8	–
1947 (October)	6	–
1948	7	50–60
1949	–	25–30

Source: Jae Hong Cho, pp. 59, 76, 88. Cho only speaks of a 1948 compensation rate of 60 sŏk, but the formula was to reduce the 1936–1937 output base for the compensation calculation by 40% which would make the average compensation rate more like 50 than 60 sŏk.

made up by the government, but the 1.25 figure was amended upward in March 1950 to 1.5 times as well. Payment (and compensation) was to take place over five years or a rate of 30 percent of the annual harvest each year. Discounted at 10 percent, therefore, the total land payment was equivalent to 1.25 times the annual yield.

Because the government often did not redeem the bonds on time and most landlords were not allowed to use the bonds as collateral for business loans, the price of these bonds in the market fell precipitously. A bond with a face value of 30 sŏk of rice sold on the market for the equivalent in wŏn of only 3.5 sŏk.[7] The landlord's loss, however, was not the farmer's gain. Korea, after June 1950, was at war, and the government needed all the grain it could get hold of to feed the army. Under wartime conditions, this grain payment of 30 percent of the main crop each year proved to be a considerable burden to many farmers and, according to one estimate, over 200,000 Korean farm families were forced to sell the land they had recently acquired.[8]

As indicated above, however, many landlords and tenants avoided the government distribution procedure altogether. By selling their land directly, landlords could avoid being stuck with government bonds of dubious value. Tenants could afford a higher price than the market value of the bond because the alternative for them was to pay 30 percent of their main crop to the government for five years. Land yielding 20 sŏk of rice a year provided the landlord with a 30 sŏk bond actually worth only 3.5 sŏk. If the former tenant got his land through government procedures, he had to pay the government 6 sŏk a year for five years. Even if the second through fifth year payments are discounted at a rate of 30 percent, the present value of that 6 sŏk a year for five years is 19 sŏk. Between 3.5 and 19 sŏk there was a lot of room for landlord and tenant to gain at the expense of the government.

What then was the net effect of land redistribution on rural income inequality? Under the Japanese, about 4 percent of the rural population (the landlords) had received roughly half the main crop or about one-quarter of farm income. After land reform, Japanese landlords had lost everything, while Korean landlords had received compensation equivalent to perhaps one-sixth to one-fourth of their former land assets.[9] Landlords without sources of income other than land were, in effect, wiped out. Owner-cultivators with more than 3 chŏngbo (roughly 3 hectares) of land (the upper limit allowed) were also hurt, but the number of farmers involved was only about 4 percent of the total.

Ultimately, former tenants were the main beneficiaries of this change. During the period of payment, however, the government took a large cut on that land it had redistributed. Because the Korean War and post-war reconstruction were going on simultaneously with this repayment, it is likely that the government would have found some other way to get an equivalent amount of grain if the payment device had not been available. Once payment for the land had been completed, a tenant farmer who had previously paid rent equal to half his main crop or perhaps one-quarter of his income, now received that income himself, a per capita income increase of 33 percent. Tenants who were part owners of land already also received benefits but less than 33 percent, the precise amount depending on their past ratio of rented or owned land. Farmers who received no new land or were cheated out of land on which they had been tenants (a surprisingly infrequent event)[10] actually lost income. On the average, however, an 80 percent or so decline in income of the top 4 percent was matched by a 20 to 30 percent increase in income of the bottom 80 percent who had been tenants or owner-tenants before 1948 if total farm output before and after the reform was about the same.

LAND REFORM AND PRODUCTIVITY

If agriculture had been stagnating prior to land reform and in the Japanese period and then had spurted ahead in the years following the reform, an attempt to measure the quantitative impact of land reform on productivity might have been feasible, although by no means easy. Agricultural output in the late 1950s after the end of land reform and the Korean War did grow at 3.6 percent a year (1953 to 1961) as contrasted to 2.9 percent in the 1930s[11] —a rise, but not a very large one given that some of that 1950s' increase represented recovery from depressed agricultural conditions during the 1940s and the Korean War. Rice output, which should have benefited most from land reform,[12] actually grew at only 2.7 percent during the entire 1957–1969 period as contrasted to 3.7 percent in the 1930s.[13]

One reason land reform may not have had a more pronounced √ positive impact on productivity in the short run was the government's small effort to replace those rural functions that previously had been provided by landlords. Specifically, landlords in the 1930s were responsible for most major capital improvements on their land, that is, improvements involving cash outlay or more than 5 or 10 days of labor. Landlords were also a major source of seeds, commercial fertilizer, and farm implements. At least this was the landlord role where share tenancy (in contrast to a fixed rent) was practiced, and share tenancy was the dominant form on paddy land.

Thus, pre-1945 rental contracts on paddy land were in a form that discouraged tenant investment in land and other productivity improvements, since the landlord received half of any output gain. At the same time, however, share tenancy provided landlords with an increased incentive to invest. Land reform, therefore, improved tenant investment incentives while eliminating landlords' interest in the land. The net effect on productivity depended on whether increased investment by former tenants exceeded or fell short of that previously provided by landlords. Since, according to one estimate, expenses borne by landlords commonly reached 20 percent of average annual production, there was a considerable gap to be filled.[14] To fill a gap of this magnitude, former tenant farmers would have had to save most of their increased income, an extraordinarily high marginal savings rate for people so poor.[15] An alternative would have been for the government to make capital available to the countryside in the amount needed and at an interest rate that would make loans for productive purposes feasible.[16] Former landlords themselves, of course, could not make adequate amounts of loan money available because they had lost so much capital as a result of the land reform, and farmers were unable to use their land as collateral, making the prospect of receiving a loan from those who still had money to lend even less likely.

Since the Farm Household Surveys did not begin until 1962, it is impossible to say what, in fact, happened in precise quantitative terms. Some current investments (seeds, farm tools, crucial repairs, and so on) could not be postponed without dire consequences for current income. In the immediate post-Korean War period, however, rural poverty was such that farmers probably chose consumption over postponable investments in capital improvements, thus slowing the rate of growth of rural capital formation and output.

A major concern in some land reforms is that the reform will break up "efficient" large farms and turn them into "inefficient" smaller farms. The problem can become particularly acute when one is breaking up modern plantations and turning them into small holdings farmed with insufficient capital and knowledge. There were no plantations to speak of in Korea, however. There were, on the other hand, farms whose size was above the 3-chŏngbo limit set by the land reform. According to data in Tables 122, 123, and 124, although only 4 percent or so of farm families ran farms whose size was greater than the limit, these farms cultivated 30 percent or more of the total acreage. The key issue is whether there were pronounced economies or dis-economies of scale in Korean agriculture. Most attempts to esti-mate agricultural production functions for Korea or for countries with similar farming techniques get results that indicate modest (and statistically insignificant) economies of scale at best.[17] Unfortunately, none of these tests have been run on data where farms are classified by size. To get a sample that included farms over 3 chŏngbo in size, of course, one would have to use data from the 1930s, which would have questionable validity after 20 years and after major institutional and technological changes.

If there is little available statistical evidence with which to analyze the impact of the elimination of over-3-chŏngo farms, there is at least an intuitive case for arguing that their elimina-tion harmed farm output little, if at all. Farms of that size in the Korean context could not be cultivated by a single family. They had, instead, to rely heavily on hired labor. To be more efficient, therefore, the superior managerial capabilities and more efficient use of capital would have had to offset the lower efficiency of hired (as contrasted to family) labor. On individual farms this may have happened, but one doubts that it was the general rule.

Other evidence that the 3-chŏngbo (roughly 3-hectare) limit was not an important negative influence can be deduced from what has happened to the size of farms since land reform. Land reform does appear to have created a great many farms that, whatever their efficiency, were too small adequately to support a farm family. Farms in the 0.5 hectare and below category, as a result, have declined from 45 percent of the total in 1953 to

TABLE 122 Distribution of Farm Households by Size of Farm, 1937

Province	Under 0.5 Chŏngbo		0.5-1.0		1.0-2.0		Above 2.0	
	No.	%	No.	%	No.	%	No.	%
Kyŏnggi	76,223	31.7	77,801	32.4	62,556	26.0	23,900	9.9
North Ch'ungch'ŏng	84,423	62.6	33,175	24.6	13,652	10.1	3,512	2.6
South Ch'ungch'ŏng	84,256	39.5	71,318	33.4	46,437	21.8	11,435	5.4
North Chŏlla	114,200	53.1	54,093	25.2	29,990	14.0	16,610	7.7
South Chŏlla	209,333	56.1	94,913	25.4	46,521	12.5	22,359	6.0
North Kyŏngsang	176,987	51.9	97,879	28.7	50,600	14.8	15,573	4.6
South Kyŏngsang	167,875	58.8	73,669	25.8	33,455	11.7	10,466	3.7
Kangwŏn	72,533	33.4	55,563	25.5	51,376	23.6	37,982	17.5
TOTAL (of above provinces)	985,830	48.8	558,411	27.6	334,587	16.6	141,837	7.0

1 chŏngbo = 0.992 hectare

Source: Himeno Minoru, *Chōsen keizai zuhyō* (1940), p. 171. These figures were taken from an investigation of the Japanese Government General's Office.

TABLE 123 Distribution of Farm Households by Farmland Under Cultivation, 1945–1973
(%)

Farm Size	Year							
	1945	1947	1953	1955	1960	1965	1970	1973
Under 0.3 ha	⎫ 72.1	⎫ 41.2	44.9	19.0	19.7	17.2	15.7	15.4
0.3 – 0.5 ha	⎬	⎬		24.1	23.2	18.7	15.9	17.0
0.5 – 1.0 ha	⎭	33.3	34.2	31.1	30.1	31.7	31.7	31.5
1.0 – 2.0 ha	23.8 minus	18.8	16.5	20.1	20.7	25.6	25.8	26.3
2.0 – 3.0 ha	⎫ 4.1 plus	5.3	4.3	5.5	6.0	5.6	5.0	4.8
Over 3.0 ha	⎭	1.4	0.1	0.25	0.3	1.2	1.5	1.5
Non-crop farms	–	–	–	–	–	–	4.4	3.5
TOTAL	100.0	100.0	100.0	100.0	100.0	100.0	100.0	100.0

Sources: Ki Hyuk Pak, pp. 92, 95; MAF, *Yearbook of Agriculture and Forestry Statistics 1968*, pp. 46–47, and *1974*, pp. 28–29. The earlier data (1960 and before) are given in chŏngbo rather than hectares but, since 1 chŏngbo = 0.992 hectare, no attempt was made to convert the earlier figures to hectares. The break point in the 1945 data is 5 chŏngbo rather than 3, so the actual 1–3 chŏngbo statistic would be lower than that in the table while the over-3-chŏngbo percentage would be higher.

TABLE 124 Distribution of Cultivated Area by Size of Farm, 1945–1973
(%)

Farm Size	Year					
	1945	*1955*	*1960*	*1965*	*1970*	*1973*
Under 0.3 ha	⎫	5.8	5.3	3.8	3.4	3.4
0.3 – 0.5 ha	10.4	12.2	11.4	8.6	7.3	8.0
0.5 – 1.0 ha	⎭	29.2	27.9	26.7	27.8	26.9
1.0 – 2.0 ha	⎫ 40.0 minus	35.9	37.0	40.5	40.6	41.4
2.0 – 3.0 ha	⎭	15.9	17.3	15.3	13.6	13.3
Over 3.0 ha	26.4 plus	1.0	1.2	5.1	7.3	7.0

Sources: MAF, *Yearbook of Agriculture and Forestry Statistics 1968*, pp. 46–47 and *1974*, pp. 28–29; and Ki Hyuk Pak, p. 92. The break point in the 1945 figures is 5 chŏngbo (1 chongbo = 0.992 hectare) rather than 3, so the actual 1–3 hectare figure would be lower than that in the table while the over-3-hectare percentage would be higher.

32.4 percent in 1973 (Table 123). The main area of expansion, however, has not been in the large 2- and 3-hectare farms (whose share remained virtually constant) but in 1- to 2-hectare farms that rose from 16.5 percent to 26.3 percent.

Land reform was not the cause of the decline in small farms nor had it created them in the first place. Nearly 50 percent of all farms in Korea in the Japanese period were under 0.5 hectares in size, a figure even higher than that of 1953. The decline after 1953, therefore, was not a return to a pre-land-reform "equilibrium" but the result of marginal farmers leaving the land entirely to take up work in the cities.

Thus, the effect of land reform on farm size probably did little to hurt agricultural productivity and might have helped if 3-chŏngbo farms and up were less efficient. There is also little evidence to suggest that the 3-chŏngbo limit has become onerous with the passage of time. The number of farms above this limit has increased as has the amount of land held by such farms. But the number of farms and the amount of land in the 2- to 3-hectare class has declined. The rise in 3-chŏngbo farms may reflect little more than rich farmers expanding their holdings by opening up new land (to which the 3-chŏngbo limit does not apply) and hence probably says little or nothing about the efficiency of these farms. The 3-chŏngbo limit continues to serve as a useful way of preventing some forms of land accumulation at the expense of smaller farms. If the smaller farms are more efficient, the law actually promotes efficiency. The data available don't allow any firm conclusions on this score, however. They do, on the other hand, make one skeptical about arguments about the need to remove the limit in order to give the entrepreneurial talents of large farmers full play. The result of elimination would be more large farms but not necessarily more agricultural output.[18]

Overall, therefore, the direct impact of land reform on agricultural productivity was probably neither strongly positive nor strongly negative. The main impact of land reform lay instead in the area of income redistribution, a theme to which we shall return in the next chapter.

ELEVEN

Rural Living Standards

During the past thirty years, the standard of living of Korea's urban population, particularly its middle class, has gone through a dramatic transformation. As urban opportunities have expanded, rural people have received at least some of these benefits. It is farmers or their children, after all, who provided many of the migrants to the cities (refugees from the north being the other major source). Where farm families in the 1930s constituted 75 percent of the total in the eight southern provinces, by the mid-1970s their percentage had fallen below 40. If the rural areas had had to absorb the over two million families that made up the difference between 40 and 75 percent in 1975, average living conditions per family would almost certainly have been much lower than they were in fact.

The principal concern here, however, is not with what might have been, but with what has actually happened to the income and way of life of the Korean farm household, to those who have remained in the rural areas rather than those who have left. Average income figures per farm household tell part of the story. On the basis of agricultural value-added data, there is no question that farm product per household has risen substantially (by roughly one-third to one-half) between the early 1960s and the mid-1970s (Tables 125 and 127). There was also some growth in the latter half of the 1950s, but at a pace considerably below that of later years (see Table 126). Three percent a year may seem modest when compared with GNP growth rates of 10 percent but, as suggested in a previous chapter, by international standards it is a fairly high agricultural growth rate.

But is agricultural value added in constant prices a good guide to what has happened to farm household income, and what is the appropriate concept of income? Farm household income, to begin with, includes income from many non-agricultural pursuits, but the percentage from these side-business activities has hovered a bit above 20 percent, so that they do not much affect the trend. Farm income, as defined in the Farm Household Survey, also includes increases in the value of farmers' agricultural products inventory, animals, and trees but asset appreciation is excluded from the income concept used here.[1]

There is also a problem connected with the choice of a price series with which to deflate farm household income estimates. The index of prices paid to farmers comes closest to duplicating the agricultural value-added series in constant prices. (This is the index used in Table 127.) But in this section we are concerned with what has happened to standards of living; hence it makes more sense to use an index of prices that farmers paid for their consumption goods. Both kinds of price indexes are used in this section, but deflation by the latter index is the basis for most of the conclusions about standards of living.

The data are not very good and are not presented here, but there is no doubt that average farm incomes fell sharply in the late 1940s and early 1950s when compared with the 1930s. It is not entirely clear when farm income per household finally recovered to pre-1945 peak levels. Farm Household Survey estimates, deflated by prices paid to farmers (Table 127), indicate full recovery by the early 1970s, but there was

TABLE 125 Agricultural Value Added, 1953–1975
(constant 1970 prices)

Year	Agricultural Total (billion wŏn)	Per Farm Household (thousand wŏn)
1953	384.88	172.28
1954	415.68	186.07
1955	427.25	192.67
1956	396.36	180.08
1957	430.54	194.73
1958	459.52	207.18
1959	454.05	200.29
1960	450.36	191.64
1961	502.84	216.09
1962	471.65	191.03
1963	509.47	210.87
1964	587.62	239.85
1965	574.98	229.35
1966	637.04	250.80
1967	599.56	231.76
1968	610.38	236.67
1969	689.76	270.92
1970	678.16	272.57
1971	692.65	279.07
1972	688.02	280.60
1973	707.21	288.66
1974	732.04	307.58
1975	787.51	330.89

Sources: BOK, *Economic Statistics Yearbook 1967* and *1976*, and BOK, *National Income in Korea 1975*, pp. 146–147.

TABLE 126 Growth Rates of Agricultural Value Added,
1953–1974
(% per year)

Years	Agricultural Value Added		Value Added per Farm Household	
	beginning-end year calculation	3-year moving average	beginning-end year calculation	3-year moving average
1953–1960	2.2	–	1.6	–
1954–1960	1.2	2.2	0.3	1.6
1960–1970	4.2	3.9	3.6	3.1
1970–1975	3.0	–	4.0	–
1970–1974	1.9	2.0	3.1	3.0

Source: Derived from data in Table 125.

probably an upward bias in the 1930s surveys.[2] Output data suggest, instead, that full recovery was achieved in the late 1950s or early 1960s.[3]

Whenever pre-war averages were achieved, there is little doubt that all Korean farmers other than landlords were far better off in the late 1950s than in the 1930s. The reason, of course, is that a large part of "income," as that concept was defined in the 1930s surveys, went to pay rent whereas, after land reform, rent payments fell from one-quarter of income to around 2 percent according to the Farm Household Surveys. Thus, household income net of rent in the early 1960s was 30 percent or more above the 1930s level, if one reduces the 1930s figures in Table 128 by an appropriate amount to account for the upward bias.

How one appraises what happened to farm incomes in the 1960s and the first half of the 1970s depends very much on which price index one uses to deflate. A deflating with prices paid *to* farmers yields a growth rate of farm income between 1962 and 1974 of only 1.1 percent per year. A deflating with prices paid *by* farmers for consumption goods, however, yields a growth rate of 3.4 percent. The difference, of course, reflects the impact of government price policy on incomes. As is clear from the discussion in Chapter 8 and from the estimates in Table 129 (columns 5 and 6), the terms of trade moved against

TABLE 127 Farm Household Income from Survey Data, 1933–1975

	(1) Farm Household Income— Current Prices per Household	(2) Farm Product Price Index	(3) Farm Household Income— Constant 1934 Prices Paid To Farmers
	(yen)	(1934 = 100)	(yen)
1933[a]	304.96	82.3	370.55
1938[a]	545.92	134.4	406.19
	(wŏn)		
1962	59,286	18,081.2	327.89
1963	82,799	28,196.0	293.66
1964	107,913	36,424.8	296.26
1965	109,839	33,823.2	324.74
1966	118,349	35,165.1	336.55
1967	138,718	40,199.5	345.07
1968	161,096	44,780.0	359.75
1969	186,852	51,120.0	365.52
1970	230,170	60,185.4	382.44
1971	317,236	72,925.5	435.01
1972	359,120	89,014.2	403.44
1973	426,756	98,824.4	431.83
1974	465,794	129,759.7	358.97
1975	722,716	165,449.6	436.82

Sources: (1): 1933 and 1938, derived from data in Nōka keizai gaikyō chōsa, and BOK Farm Household Survey, 1962–1975.
(2): Sŏng-hwan Pan, Han'guk nongŏp ŭi sŏngjang, 1918–1971, pp. 243–244, except 1972–1975 which are derived from the index of prices paid to farmers in MAF Yearbook of Agriculture and Forestry Statistics.
(3): (1) ÷ (2).

Note: [a]1930s' estimates are based on surveys with an upward bias.

TABLE 128   Farm Household Income Net of Rent Payments,
1933–1975

	(1) *Farm Household Income per Household*	*(2)* *Rent per Household*	*(3)* *(1)–(2)*
	(constant 1934 yen)	*(constant 1934 yen)*	
1933	370.55	84.29	286.26
1938	406.19	117.44	288.75
1962	327.89	3.83	324.06
1965	327.74	6.00	321.74
1970	382.44	9.69	372.75
1973	431.83	8.03	423.80
1975	436.82	9.29	427.53

Sources: 1962–1965: MAF, *Report on the Results of Farm Household Economic Surveys 1933* and *1938* and Table 127; rent data for 8 southern provinces are given in terms of grain in Bureau of Agriculture and Forestry, *Nōka keizai gaikyō chōsa.* These rent data were converted into yen at 0.075 yen per shō of paddy in 1933 and 0.113 yen per shō in 1938. These current price figures were then converted into constant prices using the farm price index in Table 127. Net rent payments are total rent for farmland paid.

agriculture in the mid-1960s (the 1970 terms of trade are taken as parity) and then sharply in favor of agriculture in the first half of the 1970s. From the trough in the terms of trade in 1968 to the peak in 1973, farm household incomes rose by 19 percent over what they would have been if agricultural prices had remained at depressed 1968 levels. One would get slightly different results if one made these calculations using agricultural value added per farm household rather than survey data on income per household, but the basic message is clear. Changes in the agricultural terms of trade had a marked effect on rural Korea's standard of living.

The three-decade trend in the standard of living of Korean farmers, therefore, has been the product of two kinds of forces. On the one hand, the performance of agricultural production; depending on when one dates the recovery of per household output to pre-independence levels, the rise in agricultural

TABLE 129 Relative Price Changes and Their Impact on Farm Incomes per Household, 1962–1975

(1970 wŏn)

	(1) Agricultural Output per Household	(2) Agricultural Gross Income per Household	(3) Change in Real Output over Preceding Year	(4) Change in Real Income over Preceding Year	(5) Cumulative Impact of Relative Price Changes on Gross Income (wŏn)	(6) Cumulative Impact of Relative Price Changes on Gross Income (% of income)
1962	270,908	230,868	—	—	-40,040	-17.3
1963	251,683	285,907	-19,225	55,039	34,224	12.0
1964	255,124	285,875	3,441	-32	30,751	10.8
1965	222,205	223,921	-32,919	-61,954	1,716	0.8
1966	237,197	226,174	14,992	2,253	-11,023	-4.9
1967	237,787	229,476	590	3,302	-8,311	-3.6
1968	238,335	224,725	13,610	-4,751	-13,610	-6.1
1969	253,086	247,254	14,751	22,529	-5,832	-2.4
1970	248,064	248,064	-5,022	810	0	0
1971	293,713	311,684	45,649	63,620	17,971	5.8
1972	289,381	327,965	-4,332	16,281	38,584	11.8
1973	292,487	335,614	3,106	7,649	43,127	12.9
1974	308,168	345,149	15,681	9,539	36,981	10.7
1975	332,942	374,508	24,774	29,359	41,566	11.1

Source: Farm Household Surveys.
(1) Agricultural gross receipts in current prices divided by an index of prices paid to farmers (1970=100).
(2) Agricultural gross receipts in current prices divided by an index of prices paid by farmers (1970=100).

output accounts by 1975 for an increase of well over 50 percent in real farm income per household when compared with the 1930s. On the other hand, farm income has also been greatly affected by government policy decisions more or less deliberately designed to redistribute income toward the rural sector. Land reform in the late 1940s and early 1950s raised non-landlord income by one-third, and improvement in agriculture's terms of trade in the early 1970s added another 10 percent. Farm household income in 1975 (excluding landlords) as a result of both kinds of forces was more than double that of the 1930s. About 60 percent of this rise was due to increases in agricultural output, while the other 40 percent has come about because of the above-described efforts at redistribution.[4]

DISTRIBUTION OF RURAL INCOME

An increase in average income, of course, does not necessarily lead to a rise of a comparable amount in the income of the majority of farmers. If most of the increase in income accrues to the richer farmers, the poorer majority may not benefit much or at all from the general rise in production and prices. Increased purchase prices, in particular, often benefit richer farmers who market a higher percentage of their crop.

Fortunately for rural Korea, it is possible to do more than speculate about likely trends in income distribution at least for the 1960s and 1970s. For earlier years, there is no question that land reform caused a substantial shift in the direction of greater equality, but the magnitude of the change cannot be documented in any detail, nor can much be said about trends in income distribution in the late 1950s.

TABLE 129 (continued)

---

(3) Derived from column (1).
(4) Derived from column (2).
(5) Derived by subtracting column (1) from column (2).
(6) Derived by dividing column (5) by column (2).

Income distribution estimates for years after 1962 can be derived from data published in the Farm Household Surveys. Except for 1974, where the estimates were derived by going back to the original survey data,[5] the size distribution of income data in Table 130 had to be adjusted slightly to make each survey comparable to the others. These adjustments, together with the small size of the sample, suggest caution in interpreting small percentage changes from year to year, particularly in the case of the 1963 estimates.

The basic conclusion readily apparent from even a glance at Table 130 is that there has been little if any change in the size distribution of rural income, the 1963–1967 change probably reflecting errors introduced by assumption in the 1963 estimates.[6] Farm households, therefore, benefited from the rise in incomes in proportion to their original income position in the early 1960s.

This result is not really very surprising when one considers certain relevant supporting information. The distribution of farmers by size of farm (in terms of cultivated acreage) also does not show much change. Nor would the policy of raising farm purchase prices have a disequilibrating effect. Except for very small farms (under 0.5 hectares) all other Korean farms market the same percentage (about 45 percent) of their crop output, and even the smallest farms market 30 percent.

Household data to some extent exaggerate the degree of inequality, mainly because the richer households have more people. The data in Table 130, for example, indicate that the average per-household income of the top 10 percent of farm households was 7 to 8 times that of the bottom 10 percent. But the households in the upper group also had nearly 80 percent more people per household (7.10 versus 3.95) than did the lower group, and 50 percent more workers. Thus the ratio from top to bottom of per capita income was more like 4 or 5 to 1.

Equality and inequality, of course, are not solely questions of income. In some societies, equality of opportunity is seen as

TABLE 130 Distribution of Farm Household Income,
1963–1974

(%)

	1963	1967	1970	1974
top 10%	31	24	–	25
top 20%	44	40	39	41
next 40%	36	40	40	40
next 30%	17	17	18	16
bottom 10%	3	3	3	3

Sources and Methodology: All data are from the Farm Household Surveys for the year concerned. The 1974 figures are from an unpublished survey of 2,416 households as calculated by W. I. Abraham. The 1963, 1967, and 1970 surveys of 1,200 households were all published but in a form that required some adjustments to make them comparable to the 1974 estimates. Specifically, the figures are given as numbers of households per income level so one can, for example, get precise figures for the bottom 8.6% of households in one case, the bottom 11.1% in another, and so on. To convert these figures to 10% intervals, the incomes of 1.4% of the households in the next interval had to be added to the 8.5%, etc. These incomes were derived by assuming that households were distributed evenly over each interval. For the 1963 survey there is the added problem that we had to assume that average income in each interval was equal to the midpoint of each interval. For later years when data are available this assumption appears in general to be valid. We also do not know the average income of the top 1.8% in the 1963 survey and the entire difference between total income (652 households X average income) minus the incomes of households in all other income intervals was attributed to this sector. This assumption may lend an upward bias to the estimate of income of the top 10% in 1963 although downward bias is also conceivable but less likely.

The 1967 distribution is only broken down into 5 income groups but data on pp. 71 and 77 allow one to break this down further into 13 groups, greatly reducing the potential error introduced by the assumptions used in constructing the table.

being of greater importance than equality of result. In traditional Korea before the Japanese occupation, opportunity to rise up in the income ladder was restricted to some degree by the existence of a rigid class structure. Although this class system was beginning to break down at the end of the nineteenth century, the upper class, or *yangban*, remained basically a hereditary elite that constituted perhaps 10 percent of the total population.[7] Individuals not of the *yangban* class could become wealthy as landlords or traders, but they could not achieve equality of power with the *yangban*, and hence their position near the top tended to be very insecure.

When the Japanese took over after 1910, they contributed to the rapid decline of the *yangban* class but replaced them with a Japanese elite that was even more closed to new entrants from the point of view of the Korean farmer. The end of the Japanese occupation in 1945, together with the land reform which eliminated what was left of a rural Korean "ruling group," therefore, markedly widened the opportunities for advancement within the rural sector for Korean farmers. Industrialization in the 1960s further increased those opportunities, particularly for the young and the educated, by creating hundreds of thousands of new jobs in the modern sector.

If one takes a long-term perspective, therefore, there have been dramatic improvements in the distribution of income and opportunity within rural Korea. Within the past fifteen years, there has been no further improvment in the distribution of income, and most of the increase in opportunity has taken the form of a greater availability of jobs in the modern sector. But one would not expect much further progress toward greater rural income equality. By international standards, rural Korean income distribution was already among the most equal rural distributions (see Table 131). Even in the People's Republic of China, the top 10 percent of farm families probably earned around 5 times the income of the bottom 10 percent.[8]

Up to this point, consideration has been given only to the distribution of income within the rural areas. But of comparable significance to the farm household is the income differential between the urban and rural households. As is apparent from the data in Table 132, this differential widened substantially in the late 1960s and has been gradually closing ever since. In fact, data that received considerable publicity in Korea suggested that, by the mid-1970s, farm households actually enjoyed higher incomes than their urban counterparts. This conclusion, however, was based on rural income data that included a form of asset appreciation which, for reasons explained in Appendix C, has been removed from the income concept used in this study.

The lack of a rural-urban income differential in 1963 reflects

TABLE 131 International Comparisons of Rural Income
Distribution

	Ratio of Average Income of Top 10% to that of Bottom 40%
Rural Korea (1974)—families	5.26
Rural India (1950)—families	8.08
Rural Ceylon (1952-1953)—recipients	10.28
Rural Puerto Rico (1953)—families	5.32
Rural Sweden (1959)—recipients	6.04
Rural U.S.A. (1950-1953)—families	8.56

Sources: Table 130 and Simon Kuznets, "Quantitative Aspects of the Economic Growth of Nations: VIII Distribution of Income by Size," *Economic Development and Cultural Change* (January 1963), pp. 50, 51, 56.

two conditions. On the one hand, poor harvests in the early 1960s led to a substantial improvement in the agricultural terms of trade in 1963 (see Chapters 5 and 8). On the other hand, modern economic growth in the urban areas was just barely under way. The widening of the gap in the late 1960s resulted from both a sharp rise in urban incomes and a substantial worsening in the rural-urban terms of trade. In the 1970s, as already discussed, the government instituted policies that led to a marked change for the better in the rural-urban terms of trade, a change that did not simply reflect a poor harvest in preceding years.

The rural-urban terms of trade, therefore, played a central role in the distribution of income in the nation as a whole. In fact, as a recently published study makes clear, increases in agricultural output in isolation may actually lead to a decline in the income of the agricultural sector because inelastic demand for food will cause prices to fall sharply.[9] Only if the urban sector and hence urban demand for food is growing rapidly (or if food imports are being cut back) is a rise in farm output likely to be combined with improved terms of trade. In Korea, these basic market forces were also supplemented by subsidies to grain producers.

TABLE 132 Rural and Urban Household Incomes Compared,
1963–1975
(wŏn in current prices)

Year	Urban Income per Household	Average People per Household	Rural Income per Household	Average People per Household	Ratio (1)/(3)
	(1)	*(2)*	*(3)*	*(4)*	*(5)*
1963	80,160	5.56	82,799	6.39	0.97
1967	248,640	5.46	138,718	6.12	1.79
1970	381,240	5.34	230,170	5.92	1.66
1974	644,520	5.21	465,794	5.66	1.38
1975	859,320	5.15	722,716	5.83	1.19

Sources: MAF, *Report on the Results of Farm Household Economic Survey* various issues, and Economic Planning Board, *Annual Report on the Family Income and Expenditure Survey, 1975.*

THE PROVISION OF EDUCATION
AND HEALTH SERVICES

An economic study of Korean agriculture is not the place to attempt a full description of the changing lifestyle of the "typical" Korean farmer. But, even in an economic study, income and its distribution are an inadequate guide to what has happened to the standard of living. Certain basic services, health care and education, for example, are provided in part by the government, and that part supplied by the government is not included in household income. More to the point, improvements in health (and other services) can often make a difference in the quality of life out of all proportion to the expenditures on those services. In cost-benefit analysis, one handles this problem by measuring the "consumer surplus" provided by these services, but such an approach is neither feasible nor desirable here. A brief description of what has happened to certain key services over time, however, is in order.

Education has always been valued by the Korean villager. Even in the late nineteenth century, large numbers of individuals in the countryside could read materials written with the basic Korean alphabet (Han'gŭl)[10] although only the elite were fluent in the written language of the court, classical Chinese. During the Japanese colonial period, there was some progress in modernizing schools and spreading literacy and, by the late 1930s, it was a rare farm family (only 9 percent of the total—Table 133) that possessed no member who could read at a level that included more than just writings in Han'gŭl. Still, in the 1930s it would appear that less than half the rural population was literate in anything except the Korean alphabet,[11] and relatively little was written exclusively in that alphabet.

Since independence, and particularly after the Korean War, there has been a dramatic expansion of formal education in rural Korea. By the early 1960s, virtually all children in the appropriate age groups were in primary schools, and increasing numbers were going on to middle school (see Table 134). The big spurt in middle school education, however, appears to have coincided with the rise in incomes in the 1970s. Middle school education (9 years of formal education) is not yet universal in rural Korea, but it is rapidly approaching that point.

Unfortunately, there are not many good statistics on rural health conditions. The most virulent killer communicable diseases have been largely eliminated throughout the country, although there were still 600 to 800 cases of typhoid fever in 1973 and 1974 (down from several thousand in the 1960s). But the really important diseases in rural Korea today, important, that is, in that they have a major impact on the quality of rural Korean life, are tuberculosis and parasite infections, and for these diseases the data are poor. There is no question that both kinds of diseases were widespread in rural areas in the past and continued to be prevalent in 1975. Although rates per 1,000 of these diseases were undoubtedly much higher in the Korean War years and the late 1940s than in the 1960s, there is little or no solid evidence to support a further decline in these rates during the 1960s and early 1970s.

TABLE 133 Literacy in Rural Korea, 1938
(in 8 provinces of present-day South Korea)

| | | Literate Families (above Han'gŭl) | | |
	Illiterate Families	under 20% of family literate	20–50%	above 50%
Number of families	115	148	637	375
Percent of total	9.0	11.6	50.0	29.4

Sources: Based on Farm Household Survey of 1,859 families (1,225 in the 8 southern and central provinces). Chōsen Sōtokufu, Nōrin-kyoku, Nōka keizai gaikyo chōsa (1940).

TABLE 134 Rural Literacy and Education, 1963–1974

| | | | Percent of Farm Population (excluding preschool children) Who Are | | |
Year	Illiterate	Literate	Primary Education or More	Middle School Education or More	High School Education or More
1963	16.6	83.4	61.7	12.0	3.7
1967	11.7	88.3	67.5	13.2	3.9
1970	12.3	87.7	65.1	14.2	3.9
1973	10.0	90.0	73.1	20.6	5.9
1974	11.8	88.4	75.8	21.8	6.5

Source: MAF, Report on the Results of the Farm Household Economic Surveys.

There is evidence, however, that rural health conditions were improving in other areas. Infant mortality rates are often considered a useful indicator of the level of health for more than just infants, and sample estimates of the change in these rates are presented in Table 135. The rate by 1970 had fallen to less than half that of the late 1940s, but the rate was still high compared with the most advanced countries. In Japan, for example, the rate in 1970 was 13.1 per 1,000 whereas in the United States it was 19.8 per 1,000. There is also evidence that rural people are making greater use of health facilities. The number of

TABLE 135 Rural Infant Mortality, 1944–1970

Year	Rural Infant Mortality (per 1,000 live births)	Size of Sample
1944	160.9	230
1945	258.7	201
1946	143.7	327
1947	125.0	280
1948	168.9	302
1949	85.8	338
1950	87.6	331
1951	119.3	285
1952	92.5	346
1953	105.5	256
1954	–	–
1955	–	–
1956	59.2	359
1957	63.8	470
1958	59.3	337
1959	67.1	432
1960	68.5	394
1961	66.7	1,289
1962	61.2	1,372
1963	55.7, 67.6	1,508, 1,375
1964	60.7	1,317
1965	59.6	6,645
1966	26.5	226
1967	63.3	5,952
1968	49.3	223
1969	–	–
1970	42.2	2,083

Source: Various sample survey results by D. J. Yun, S. J. Lee, and others were compiled by Ok Ryun Moon and Jae Woong Hong, "Health Services Outcome Data: A Survey of Data and Research Findings of the Provision of Health Services in Korea" (unpublished paper, December 1975), pp. 18–20.

hospitalizations per 1,000 population in the rural areas, for example, appears to have doubled between 1963 and 1974, although the rate is still well below that in urban areas.[12] Health centers are a more common source of rural health care, and herbal doctors are also widely used. In general, although there have been considerable improvements in Korean rural health care, the overall performance in this area has not matched the more dramatic improvements in education. New Community Movement activities in the early 1970s designed to improve public health through better protection of well water from contamination, for example, may indicate a changing attitude on the part of the government as may the emphasis on health in the Fourth Five-Year Plan; but in 1975 there was still a great deal to be done.

OTHER INDICATORS
OF A CHANGING RURAL LIFE

There have been many other changes in rural life over the thirty years since 1945 that also are not captured adequately by household income statistics. The size of the farm family itself has declined significantly. In the 1930s, there were about 6.4 people per rural family, and in 1963–1964 there were still 6.4 people. But by 1972–1974, there were only 5.7 people, the difference being almost entirely due to the declining number of children per family. The rural population was also older, not solely because of fewer children, but also because of the disproportionate migration, both temporary and permanent, of men and women in their twenties.

Contact with the world outside the village had also increased in many different ways. By 1974, Korean farmers were buying half their family living needs (food, clothing, and so on) from the market rather than with their own production as contrasted to one-third only ten years earlier. The great expansion in the number and quality of roads, together with rural bus services, has brought a high percentage of farmers within a few hours' traveling time from Seoul. As recently as 1965 or even 1970, Korean rural roads were narrow, strewn with large rocks, and often impassable in a rain storm. Even in the early 1960s, there were only 4,000 buses in the whole country, most in Seoul and a few other large cities. By 1975, there were over 20,000 buses, many of them connecting regularly with all but the most remote parts of the country.

Electricity as late as 1966 reached only 15.9 percent of Korea's villages, and the big push in this area did not begin until the early 1970s; but, by 1975, 64.9 percent of rural villages had electricity, most for home as well as productive purposes.[13] One could go on in this vein, but the basic point is clear. Korean rural households in 1975 were still poor by the standards of Japan or North America. An annual family income of U.S. $1,000 is not large.[14] But not only was this income much higher than it was one or two decades earlier, but rural life was in the process of a transformation of a fundamental kind. In speaking of the rural standard of living, it was still possible in 1975 to speak of the way of life of the farm household as traditional, but the distinction between traditional and modern was becoming blurred.

TWELVE

Off-Farm Migration

In 1961 Simon Kuznets noted that

> the magnitude of the migration from the agricultural sector to the nonagricultural sector and of the factor contribution involved may not have been given the attention it deserves . . . this transfer of workers from the A to the non-A sectors in the process of modern economic growth means a sizable capital contribution because each migrant is of working age and represents some investment in past rearing and training to maturity . . . And granting that the "contribution" in question depends upon the employment capacity of the non-A sector we could still argue that the internal migration of labor from agriculture represents a large transfer of valuable resources to the non-A sectors and a large contribution to the country's economic growth.[1]

Since this was written, the literature on the patterns, determinants, and consequences of inter-regional migration and

316

inter-sectoral labor mobility during the early stages of economic development has expanded enormously,[2] and the rather mechanical treatment of off-farm migration in the earlier models of dualistic development has been fleshed out in more recent efforts to simulate the development process. Although the author's perspective on the relationship between off-farm migration and economic growth has been deeply influenced by the literature on dualistic development, no attempt is made here to "measure" the contribution or costs of off-farm migration during Korea's recent economic history from one or another of the perspectives subsumed within this very diverse body of theory. Instead, answers are sought to more basic and prosaic questions. What has been the trend of farm household and farm population growth? What has been the rate and volume of off-farm migration? What have been the demographic characteristics of this movement? How have the growth and structure of the agricultural and non-agricultural labor forces been affected? How has off-farm migration and farm household formation varied among regions? From what categories of farm households have migrants been drawn, and into which non-agricultural activities have they typically moved?

Studies of inter-regional labor mobility in Korea have confirmed the expectation that movement has been from lower-income regions to higher-income regions[3] but, since these do not focus specifically on off-farm migration, the elasticity of off-farm migration to inter-regional income differences, levels of unemployment, and other variables cannot be properly adduced from the findings. However, these studies and the results of several sociological studies that have included questions on the motivations of migrants[4] suggest that economic considerations, broadly defined, provide the principal reason for voluntary migration from farm villages, or indeed for inter-regional migration in general. Expressed motivations for past actions may have only a tenuous relationship to the actual decision-making process, however, and "autonomous" decision-makers make up only a part of any migration stream, a large fraction of which consists of dependents with a limited role in the decision to migrate. Thus, while the author shares the belief that economic motivations have been primary in the migration decision, it is felt that, for the moment, an examination of the structural aspects of the process and consequences of off-farm migration will yield greater rewards.

Off-farm migration is to be distinguished from rural-urban migration, and is defined here as movement out of the farm household population and into the non-farm household population. Such farm–non-farm mobility does not necessarily entail physical migration from the rural sector. The general tendency to equate farm–non-farm with rural-urban movement results not only in mismeasurement of the former, but also obscures fundamental social and demographic processes which play a central role in the long-term shift of labor from the agricultural sector to the non-agricultural sectors. It also obscures the fact that movement from agriculture into the non-agricultural sectors is generally more responsive to a positive gap between anticipated earnings in the non-agricultural sectors and agriculture than is movement in the reverse direction responsive to higher earnings in agriculture.

In addition, in both developing and developed countries, labor appears to leave the agricultural sector too slowly to bring about a significant reduction of the inter-sectoral gap in per capita productivity and income. Recent studies tend to see this sluggishness of response to inter-sectoral wage and income differentials as more than a simple resistance to change or a preference for the agrarian life. Emphasis is now placed on the limited short-run transferability of human capital between very diverse economic activities. Thus, while for persons in the rural non-agricultural sectors there are corresponding urban occupations to which their past training and experience can be applied, peasant farming entails skills and knowledge—embodied human capital—which are of little relevance to non-agricultural activities. Even when farm incomes are very low, the salvage value of a farmer's skill in the urban sector may be less than its value in the agricultural sector.[5] In peasant agriculture, farm skills may also be specific to the soil, micro-climate, and marketing structure of a closely circumscribed locale. Insofar as an individual farmer is dependent on labor-sharing arrangements and communal irrigation facilities, a large component of his stock of human capital is vested in his place in the community where production decisions take place. For these reasons, and the difficulty of securing a new homestead,[6] peasant owner-farmers as an occupational class are typically much less mobile inter-locally than other workers.

By the same token, the salvage value of many urban occupational skills in the agricultural sector is also limited, and the constraints on movement into agriculture by those born and raised in the non-agricultural sector are, if anything, even more severe than the constraints on off-farm migration. As a result, the farmers of the future are drawn with few exceptions from among the farm boys of the present, or from among former off-farm migrants who have a measure of farm skills and experience and access to agricultural property through inheritance, marriage, or customary rights.

POST-KOREAN WAR TRENDS
IN THE FARM HOUSEHOLD POPULATION

Rural-urban and off-farm migration in Korea became numerically and socially important during the 1920s with a substantial outflow of population from the rural areas of the south to the industrial centers of Japan. The pace of movement out of the southern provinces quickened during the 1930s in the wake of Japanese development programs in Manchuria and northern Korea.[7] At the root of this movement lay grinding rural poverty, due in part to a rapidly deteriorating man-land ratio, a legacy of technical backwardness, and an oppressive and unstable tenancy system, but also greatly aggravated by a Japanese colonial agricultural policy narrowly attuned to maintaining stable prices for agricultural wage goods in urban Japan.

Real wages in the industrial sector rose very little for native Korean workers during the colonial period and were well below those received by Japanese workers in Japan and Manchuria; the fact that the colonial authorities found it necessary to resort to forced recruitment of industrial workers toward the end of the period suggests that industrial employment offered few positive attractions even in the face of desperate rural conditions. Because the demand for industrial labor in Korea was increasingly concentrated in the northern half of the peninsula during the late colonial period, the initial non-agricultural work experience of many southern Korean men was received not in the regional cities of the south but in northern Korea or the cities and mines of Japan and Manchuria.

At the time of the Liberation in 1945, approximately 3.5 million Koreans were residing outside the peninsula, over 75 percent of whom are estimated to have emigrated from the southern provinces. In the immediate post-Liberation period, 1.38 million overseas Koreans were repatriated to the American Zone, to which were added 740,000 repatriates and refugees from the Soviet Zone. The Korean War years brought a new influx of refugees from the north, estimated by Tai Hwan Kwon[8] to have totaled 646,000 persons, and resulted in the temporary displacement of several million persons in the south, and the death or disappearance of approximately 1.9 million South Koreans as a direct or indirect result of the war. Because of these population movements during the second half of the colonial period, the post-Liberation years, and the Korean War, the South Korean farm population had been subjected to substantial dislocation which had served to loosen the hold of the traditional agrarian village on its members even prior to the rapid industrialization of Korea during the 1960s.

An absence of appropriate data regrettably precludes detailed consideration of patterns and trends in off-farm migration prior to 1960, although the situation during the preceding five years can be deduced from an examination of trends in the growth of farm households and the farm population. In any case, the turmoil of the first post-Liberation decade and the problematic reliability of population and other statistics for this period make it preferable to limit the examination of recent farm population trends to the period after 1955.

Even for the period after 1955, analysis of the annual data series is complicated by changes in the definition of *farm household,* changes in the method and date of data collection or estimation, changes in the reported age categories, and an uneven degree of accuracy and completeness. Important discontinuities are noted in Figure 25 by vertical division lines. The potential difficulties arising from changes in the official definition of *farm household* appear to have been ameliorated by a lesser degree of precision in the field, with the result that marginal categories nominally excluded under the earlier definitions may have been largely included in the figures reported by local village heads. Under the more precise definitions employed in the agricultural and population census enumerations of 1960, 1966, and 1970, the total farm household population shortfalls the MAF administrative figure for the same year by 2.5–5.0 percent.

Figure 25 indicates that the total reported farm population grew from 13.3 million in 1955 to a maximum of 16.1 million in 1967, an average annual rate of increase of 1.6 percent for the period, at a time when the total population was growing at the rate of 2.6 percent per year. From 1967, the farm population declined steadily to 13.2 million in 1975, an annual rate of decrease of 2.4 percent, while the national population grew at 2.0 percent per year. Examination of inter-censal growth rates for the total and farm populations clearly indicates that the growth of the farm population began to lag behind that of the nation as a whole after 1955. This gap broadened rapidly during the 1960s and has apparently changed less rapidly during the 1970s.

FIGURE 25 Growth of Farm Households and Farm Population, 1955–1976

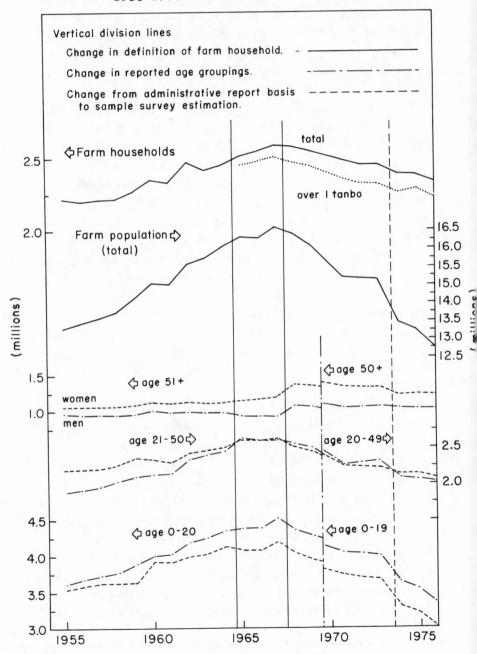

After declining in the immediate post-Armistice period with the return of war refugees to the cities, the number of farm households grew from 2.22 million in 1955 to 2.59 million in 1967 (an annual rate of increase of 1.3 percent) and declined from this maximum to 2.38 million in 1975 (an annual rate of decline of 1.1 percent). Figure 25 indicates that an upturn in the number of farm households began after 1957–1958, in part a reflection of the reunification of families and the creation of new households following the general post-war demobilization of the armed forces in 1957.

The trends in farm household growth at the provincial level (see Figure 26) generally parallel the aggregate national pattern insofar as an upward trend is evident in every province through the late 1960s, followed by a decline thereafter. In every province but South Ch'ungch'ŏng and South Chŏlla, the maximum number of farm households was recorded in 1967 and, in these two provinces, the peak was reached a year later. For the provincial-level farm populations (not shown) the spread of peak years was somewhat broader. Seoul and Pusan and the surrounding provinces of Kyŏnggi and South Kyŏngsang showed an absolute decline in the farm population after 1965, while in most other provinces the decline began after 1967. The farm population continued to increase through 1968 in South Chŏlla and through 1969 on Cheju Island.

From the broad trends in the decline of the farm population after the mid-1960s it may be noted that between 1965–1966 and 1975–1976 the decline in the farm population age 0–19 accounted for approximately two-thirds of the total decline of 2.8 million persons. Although the decline in the farm population age 20–49 accounted for little over one-third of the total decrement between 1965–1966 and 1975–1976, the actual population in this age group declined by one-fifth, the same proportionate decline as occurred in the age group 0–19. In general, the decline in the farm population was probably smoother than the annual population and household figures would suggest.

FIGURE 26   Provincial Trends in Farm Household Growth and
Decline, 1956–1976

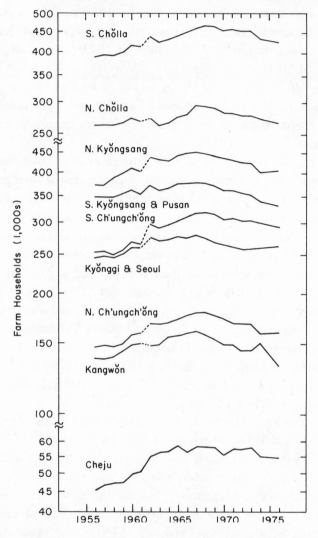

Notes: 1. Based on annual administrative enumeration of farm households, except
that 1970 figures based on MAF, *Agricultural Census* 1970. No adjustment has been
made for changes in the definition of *farm household* or the date of enumeration.
Nominal enumeration date is December 31 for 1956–1969, October 1 for 1971–
1973, and December 1 for 1970 and 1974–1976.
2. Dashed line between 1961 and 1962 data points indicates that the provincial
boundaries were altered as of January 1, 1963. These changes were apparently
reflected in the year-end figures for 1962.

THE VOLUME AND STRUCTURE OF
OFF-FARM MIGRATION IN THE 1960s

Direct data on the volume of off-farm migration are unavailable, and the estimates of the volume and annual rate of net off-farm migration during 1960–1970 have been calculated by the forward census survival ratio method. Table 136 summarizes these estimates, which exclude net migration among infants born during the period and absences directly associated with military conscription. The restriction of the analysis to the period after 1960 is dictated by the availability of appropriate data. The estimates are for the two inter-censal periods, 1960–1966 and 1966–1970. Fortunately, this periodization corresponds reasonably well to the major phases of aggregate farm population growth and decline already described, although the fit is obviously better for some provinces than others (being least good for South Chŏlla and Cheju Island).

Annual net out-migration from farm households rose from an average of 243,000 persons per year during 1960–1966 to 568,600 persons per year during 1966–1970 under the 1960 definition of farm households (with infants born to migrant mothers during the period excluded from the estimate). Under the broader 1966 definition of farm households, estimated average annual net off-farm migration during the late 1960s rose to 623,000 persons (see Table 137). Net off-farm migration among the working-age population between 15 and 60 years old averaged 169,700 persons per year during the first half of the decade and 372,000 or 402,700 persons per year during the second half, based on the 1960 and 1966 farm household definitions, respectively.

Both male and female off-farm migration rose sharply in every province but Kangwŏn between the two periods, and nationwide the rate of net migration from farm households rose from 1.8 percent per year during 1960–1966 to 4.2 percent per year during 1966–1970,[9] an increase reflected in the absolute decline of the farm household population after 1967. During the early

TABLE 136 Net Off-Farm Migration and Average Annual Net Off-Farm Migration Rates, 1960–1966 and 1966–1970

Province	1960–1966 Net Off-Farm Migration[a] (1,000s)	1966–1970 (I) Net Off-Farm Migration (1,000s)	1966–1970 (II) Net Off-Farm Migration (1,000s)
Kyŏnggi & Seoul	158.3	315.3	319.8
Kangwŏn	103.0	69.7	79.6
North Ch'ungch'ŏng	38.0	214.4	225.3
South Ch'ungch'ŏng	205.3	276.0	290.6
North Chŏlla	151.2	244.6	279.7
South Chŏlla	243.3	370.2	437.0
North Kyŏngsang	248.8	421.1	442.9
South Kyŏngsang & Pusan	265.6	321.4	368.8
Cheju Island	-4.1	38.1	43.9
Whole Country	1,409.3	2,270.8	2,487.5

TABLE 136 (continued)

Average Annual Off-Farm Migration Rates by Sex[a]

Province	1960–1966 (%)		1966–1970 (I) (%)		1966–1970 (II) (%)	
	Men	*Women*	*Men*	*Women*	*Men*	*Women*
Kyŏnggi & Seoul	1.67	2.02	5.00	5.57	4.95	5.56
Kangwŏn	2.32	2.49	1.90	2.59	2.16	2.86
North Ch'ungch'ŏng	0.56	0.83	5.18	5.51	5.38	5.70
South Ch'ungch'ŏng	1.90	2.29	3.74	4.23	3.88	4.38
North Chŏlla	1.97	1.65	4.04	4.06	4.51	4.54
South Chŏlla	1.87	1.71	3.71	3.78	4.29	4.35
North Kyŏngsang	1.87	1.76	4.09	4.54	4.24	4.70
South Kyŏngsang & Pusan	2.40	2.26	3.95	4.10	4.42	4.58
Cheju Island	-0.36	-0.34	4.61	3.77	5.13	4.31
Whole Country	1.84	1.85	4.01	4.30	4.31	4.62

TABLE 136 (continued)

Notes: [a]Negative values indicate net in-migration to farm households.

[b]Estimation and Definition of Farm Households:
1) Net off-farm migration is estimated by the forward census survival ratio (CSR) method. The estimates refer to net migration among the population surviving to the end of the period in question, and the annual net migration rate corresponds to an exponential rate of net decrement (or increment) in which only those members of the initial farm household population surviving to the end of the period are included in the population at risk. The estimates provided here exclude migration among children born to migrant mothers, since estimates for this group are almost solely dependent on the fertility rate assumed for migrant women.

Formally, net off-farm migration (i.e., with a positive sign on net out-migration) between the dates T_0 and T_1 is calculated as:

$$NM_{i+1}^{T_0, T_1} = F_{i,j}^{T_0} \cdot \frac{P_{i+1}^{T_1}}{P_i^{T_0}} - F_{i+1,j}^{T_1} \quad ,$$

where

$NM_{i+1}^{T_0, T_1}$ = net off-farm migration during $T_0 - T_1$ for the birth cohorts of the $(i+1)$th age group at time T_1.

$F_{i,j}^{T_0}$ = the farm population in the ith age group in region j at T_0.

$F_{i+1,j}^{T_1}$ = the farm population in the $(i+1)$th age group in region j at time T_1, where the age groupings in each period are arranged in such a manner that the $(i+1)$th age group at T_1 includes the same group of birth cohorts as in the ith age group at T_0.

$P_i^{T_0}, P_{i+1}^{T_1}$ = the total national census populations in the ith age group at T_0, and in the $(i+1)$th age group at T_1, respectively; dividing the second by the first gives the intercensal "census" survival ratio for the group of birth cohorts age i at T_0 and age $i+1$ at T_1.

The annual net off-farm migration rate between T_0 and T_1 for the birth cohorts of the $(i+1)$th age group at T_1 is calculated from the algorithm:

$$F_{i+1,j}^{T_1} = (1 - nm_{i+1}^{T_0, T_1})^t \cdot F_{i,j}^{T_0} \cdot \frac{P_i^{T_1}}{P_{i+1}^{T_0}} \quad ,$$

where

$nm_{i+1}^{T_0, T_1}$ = net average annual migration rate.

t = length of the intercensal period in years.

The accuracy of CSR estimates of net migration depends on the viability of the following implicit assumptions of the method: a) the national population is closed to net immigration or emigration; b) age-sex specific mortality does not differ between regions; c) in each region the ratio of enumeration coverage to the national coverage rate remains unchanged between censuses for each cohort. (See K.C. Zachariah, 1962). During the period in question these assumptions were not violated to a

TABLE 136 (continued)

degree sufficient to affect our analysis (see Tai Hwan Kwon, *Demography of Korea: Population Change and its Components,* Seoul, 1977, pp. 220–223).

2) The figures indicated are summarized from estimates made for male and female quinquennial age groups, and are based on revised estimates of the farm population in 1960, 1966, and 1970 living within 1970 provincial boundaries. The total population and farm population enumerated in the agricultural censuses of 1960 and 1970 and the population censuses of 1960, 1966, and 1970 were adjusted for documented under-enumeration, exclusion of the population currently in military service, and changes in provincial boundaries. The provincial farm populations were adjusted to be consistent with the KASS-POP adjusted total farm population developed for use with a farm–non-farm population growth simulation model. The estimated number of regular farm household members currently away in military service was added back into the farm population in each year to avoid the distortion of young male migration rates that is caused by the inclusion of service related moves.

3) The figures for 1960–1966 and 1966–1970 (I) are based on the 1960 census definition of farm household and include only households directly working at least 300 p'yŏng of land. The 1966–1970 (II) figures reflect the slightly broader 1966 census definition of farm household (see Appendix A). For the purposes of these estimates farm laborer households are considered part of the non-farm population.

1960s the highest rates of off-farm migration occurred in Kangwŏn and South Kyŏngsang provinces, while at the other extreme farm households on Cheju Island experienced net in-migration. In the second period, North Ch'ungch'ŏng and Kyŏnggi provinces showed the highest rates of off-farm migration, while Kangwŏn had the lowest.

Overall, the rapid increase in off-farm migration in the late 1960s was associated with a surprising shift in the inter-provincial pattern of these rates. Although all provinces except Kangwŏn showed a substantial increase, this was particularly marked in the provinces that had experienced below-average rates in the preceding period. The shift was sufficiently dramatic in the case of male migration to produce a large negative rank-order correlation among provincial rates between the two periods (Spearman's r_s = −0.71). The four provinces away from the metropolitan zones in which paddy field agriculture prevails (that is, South Ch'ungch'ŏng, North and South Chŏlla, and North Kyŏngsang) remained clustered at the middle of the range of migration rates, with the most dramatic shifts evidenced in the upland provinces of Kangwŏn and North Ch'ungch'ŏng, the island province of Cheju, and the metropolitan provinces of Kyŏnggi and South Kyŏngsang. While the rate of female off-farm migration generally increased more than among males of the same province, for women there was no statistically significant change in the inter-period rank order of these rates (Spearman's r_s = -0.03). Whereas the rate of male off-farm migration exceeded the rate of female off-farm migration in five of the nine provinces during 1960–1966, in the second period female migration rates were higher everywhere but on Cheju Island.

Before any consideration of the demographic structure of the net migrant population and cohort-specific net migration rates, it must be clearly understood that estimates of net migration afforded by the census survival ratio method pertain to specific birth cohorts, which are of changing and not constant age during the period. That is, migration during 1966–1970 among the cohort age 20–24 in 1970 involved individuals who were anywhere between 16 and 24 years of age at the time of migration. Moreover, migration at a given age can include incidents of migration attributable to either of the two five-year birth cohorts that have individuals passing through this age during the period. For example, migration at age 22 can involve members of either the cohort age 20–24 at the end of the period or the cohort age 25–29.

In the discussion that follows, all cohorts are referred to by the age of the cohort at the end of the inter-censal period in question. So that the migration behavior of specific birth cohorts can be compared over two successive intervals, the age groupings employed for 1966 differ from the age groupings used for 1960 and 1970. This is made necessary by the unequal length of the inter-censal periods.[10] Both sets of age groups as well as the (unvarying) mid-period age of the cohort are indicated in Table 137 and Figure 27.

In examining patterns of cohort-specific net migration, it is imperative that both the cohort-specific volume of migration and the cohort-specific migration rate be considered. Taken together, migration volume and migration rate respectively represent the stock and flow dimensions of what is essentially an *ex post* accommodation to *ex ante* supply and demand conditions.

Table 137 indicates that, during both inter-censal periods, the net inter-sectoral transfer of population was dominated by persons under 35, a pattern entirely consistent with most studies of migration. In both periods, just over one-half of net inter-sectoral migration was contributed by the cohorts between age 15 and 30.[11] The cohorts age 30–40 contributed 15.5 percent of the total net movement in the first period and 12.7 percent during the second period. In contrast, the share contributed by cohorts under age 15 (excluding infants) increased from 27.3 percent to 35.1 percent, while the annual number of such migrants tripled between the two periods. If infants born to migrant mothers are included, it seems likely that migration before age 15 accounted for approximately one-third of total net off-farm migration during 1960–1966 and a somewhat larger proportion during 1966–1970. Although it is necessary to keep clear the distinction between net migration and individual movement, an implication of these figures is that a large fraction of the individuals leaving agriculture were too young to have received any meaningful exposure to farm work prior to their departure and are hence nearly as unlikely as non-farm-born youths to enter farming at a later age.

Among working-age cohorts, females age 15–19 and 20–24 show the greatest increase in the volume of net out-migration between the two periods. Net off-farm migration in the former cohort increased from 20,700 per year during 1960–1966 to 56,300 per year during 1966–1970, while in the latter cohort the increase was from 17,400 per year to 58,100 per year, a 234 percent increase. The increase in net female off-farm migration in these cohorts far exceeded the increase among males of the same age, leading to a sharp reversal of the slight male predominance in net migration among young adults observed in the first period. For working-age cohorts (age 16–60/15–59) as a whole, the sex ratio shifted from 107 male net migrants per 100 female net migrants to 92 males per 100 females in the second period. Among male working-age cohorts, the greatest increase in annual net off-farm migration was

TABLE 137 Cohort-Specific Average Annual Net Off-Farm Migration, 1960–1966 and 1966–1970

Age of Cohort 1966/1970	Mid-Period Age	1960–1966 Period					1966–1970 Period (I)				
		Men		Women		Sex Ratio	Men		Women		Sex Ratio
		1,000s	%	1,000s	%		1,000s	%	1,000s	%	
6-10/ 5- 9	3- 7	16.5	6.8	18.5	7.6	89	61.6	10.8	50.5	8.9	122
11-15/10-14	8-12	15.4	6.3	15.7	6.4	98	42.6	7.5	45.0	7.9	95
16-20/15-19	13-17	22.5	9.3	20.7	8.5	109	44.1	7.8	56.3	9.9	78
21-25/20-24	18-22	17.9	7.4	17.4	7.2	103	37.0	6.5	58.1	10.2	64
26-30/25-29	23-27	20.7	8.5	24.7	10.2	84	50.8	8.9	53.6	9.4	95
31-35/30-34	28-32	19.9	8.2	6.5	2.7	306	40.0	7.1	16.4	2.9	245
36-40/35-39	33-37	6.4	2.6	4.9	2.0	131	9.8	1.7	6.2	1.1	158
41-45/40-44	38-42	0.6	0.3	0.8	0.3	80	-2.9	-0.5	-3.8	-0.7	78
46-50/45-49	43-47	0.3	0.1	1.6	0.6	17	2.9	0.5	2.8	0.5	107
51-55/50-54	48-52	-0.4	-0.2	1.8	0.7	-25	-2.7	-0.5	1.2	0.2	-219
56-60/55-59	53-57	-0.2	-0.1	3.6	1.5	-5	-1.2	-0.2	3.2	0.6	-37
61-65/60-64	58-62	0.6	0.3	3.2	1.3	20	-2.2	-0.4	0.5	0.1	-455
66-70/65-69	63-67	-0.2	-0.1	0.9	0.4	-21	-2.4	-0.4	1.0	0.2	-233
71+/70+	68-70+	0.9	0.4	1.9	0.8	48	-1.1	-0.2	1.1	0.2	-97
Total		121.1	49.8	122.0	50.2	99	276.4	48.6	292.2	51.4	95
Working Ages: 16-60/15-59		87.8	36.1	81.9	33.7	107	177.9	31.3	194.1	34.1	92

TABLE 137 (continued)

Age of Cohort 1966–1970	Mid-Period Age	1966–1970 Period (II)					Percentage Increase in Annual Net Migration 1960–1966 : 1966–1970 (I)		
		Men		Women		Sex Ratio	Men %	Women %	Both Sexes %
		1,000s	%	1,000s	%				
6-10/ 5- 9	3- 7	66.8	10.7	55.3	8.9	121	272.4	172.6	219.7
11-15/10-14	8-12	47.2	7.6	49.4	7.9	96	176.4	187.0	181.7
16-20/15-19	13-17	47.6	7.6	59.9	9.6	79	95.9	171.7	132.3
21-25/20-24	18-22	39.7	6.4	60.9	9.8	65	107.0	234.1	169.7
26-30/25-29	23-27	53.2	8.5	55.8	8.9	95	144.8	117.3	129.8
31-35/30-34	28-32	42.1	6.7	17.8	2.9	236	100.8	151.6	113.3
36-40/35-39	33-37	10.9	1.8	7.5	1.2	147	52.5	27.1	41.5
41-45/40-44	38-42	-2.2	-0.4	-2.7	-0.4	82	-557.3	-569.5	-564.1
46-50/45-49	43-47	3.8	0.6	3.9	0.6	97	1011.3	76.2	211.7
51-55/50-54	48-52	-1.9	-0.3	2.4	0.4	-79	511.1	-29.7	-211.6
56-60/55-59	53-57	-0.3	-0.0	4.3	0.7	-7	521.2	-10.4	-39.9
61-65/60-64	58-62	-1.5	-0.2	1.4	0.2	-105	-450.6	-84.6	-145.6
66-70/65-69	63-67	-1.8	-0.3	1.8	0.3	-105	1215.3	19.7	-302.5
71+/70+	68-70+	-0.3	-0.0	2.4	0.4	-11	-218.6	-41.1	-98.8
Total		303.3	48.7	320.1	51.3	95	128.3	139.5	133.9
Working Ages: 16-60/15-59		193.0	31.0	209.7	33.6	92	102.7	137.0	119.2

Notes: See notes to Table 136 for method of estimation, data sources, and definition of *farm household* employed in estimates I and II for 1966–1970. The estimates in this table have been adjusted for mortality among migrants during the inter-censal period.

experienced by the cohort age 25–29, and in the second period this cohort provided the largest number of net male migrants of working age, so that in the second period, unlike the first, net off-farm migration among males in their late twenties exceeded that among males in their late teens. In both periods, the male cohort age 21–25/20–24 accounted for less net migration than any other male cohort between the ages of 15 and 35. In one sense, this is a consequence of the adjustments made to eliminate the effects of moves in conjunction with military service from the estimates of cohort-specific off-farm migration. Beyond this, however, the nearly universal conscription of men between the ages of 20 and 25 for a period of two to three years has a dampening effect on the mobility of men of this age. Not only is the effective period of risk of migration shortened, but the vulnerability to immediate conscription makes it difficult to find employment or take other actions that require a commitment of several years. Much of the intended migration among this cohort is apparently postponed until military obligations are completed, while skills and experiences gained in the course of military service may themselves stimulate migration among men recently discharged from the armed forces.

Low levels of off-farm migration or net in-migration among older males are observed in both periods. In both periods, the greater amount of net off-farm migration among older women than older men is most likely a reflection of the higher mobility rates of widows, who both outnumber widowers and remarry much less frequently. Widows, and also aging widowers, may leave the farm for reasons other than the absence of a male heir—every son may have decided in favor of a non-agricultural career, or one outside of agriculture may be better able to support an aging parent and so take on this responsibility even if not the principal heir. Widows in particular may find it difficult to secure a livelihood from agriculture unless they have sons old enough to carry out the heavier agricultural tasks and maintain the prerogatives of the household within the community. It is noteworthy, however, that net in-migration among male cohorts age 50–54 and older became the dominant pattern in the late 1960s, while at the same time net out-migration among women in these ages declined both relatively and absolutely. This change in the pattern of migration between 1960–1966 and 1966–1970 may seem puzzling in light of the very great general increase in off-farm migration. Equally unexpected is the observation of net in-migration to farm households during the second period among cohorts in their late thirties and early forties. Several reasons can be suggested for these patterns but are best considered together with changes in cohort-specific net migration rates.

From estimates of net off-farm migration and the KASS-POP adjusted farm and non-farm household populations the contribution of migration to the growth of the farm and non-farm civilian working-age population can be assessed. Although the correspondence is inexact, the farm and non-farm civilian population age 15–59 may be roughly identified with the potentially economically active population in agriculture and non-agricultural sectors. Between 1960 and 1966, the farm civilian working-age population increased by 1.3 percent (2.9 percent for men), while in the absence of migration it would have grown by 15.7 percent (18.6 percent among men). Hence, net off-farm migration resulted in a 92 percent reduction in the expected growth of the farm working-age population, and this inter-sectoral transfer accounted for 59 percent of the 30.8 percent increase in the non-farm working-age population during the period. The late 1960s saw off-farm migration exceed the expected growth of the farm working-age population by 48 percent (27 percent for men), resulting in a 6.7 percent decline in the farm working-age population compared to a 14.0 percent increase that would have occurred in the absence of any migration. However, net farm–non-farm migration still accounted for approximately the same proportion, 60 percent, of the increase in the non-farm working-age population, which grew by 33.4 percent between 1966 and 1970.

Cohort-specific average annual net off-farm migration rates

for 1960–1966 and 1966–1970 are presented in Figure 27. A comparison of these rates with the age-sex composition of net off-farm migration presented in Table 137 reveals, first, that the large volume of net migration among cohorts less than 15 years of age is a function of the relative size of these cohorts rather than of particularly high migration rates. Second, although in both periods the rates of off-farm migration among adolescent and young adult women between 15 and 30 exceeded the rates for the corresponding cohorts of men, the volume of net male migration exceeded net female migration in the cohorts age 16–20 and age 21–25 during 1960–1966. This reflects the substantial sex imbalance in the adolescent farm population in 1960 and suggests that off-farm migration was substantially greater among young adolescent females in the late 1950s than among males. Third, it appears that, although net migration among the cohorts over 40 had a very small influence on the volume of net farm–non-farm migration, the *rates* of net migration among these cohorts are in some instances not insignificant. The inter-period change in the migration rate for a particular birth cohort is interpretable as the combined response of the birth cohort to both the effects of aging on mobility and to changes in exogenous circumstances. As seen in Figure 28, the change in annual net migration rates was generally less than one percentage point among cohorts over 41 in 1966, with the possibly spurious exception of the oldest male cohort. In spite of the growth of the non-agricultural sector, which greatly accelerated the general pace of off-farm migration in the late 1960s, the inter-period change in the migration rate for the same birth cohort is less among these older cohorts than the difference in the migration rate between adjacent groups in the second period.

Young birth cohorts moving into the ages of highest mobility show the greatest inter-period increase. This is to be expected, not solely because individuals in these ages face the fewest constraints on mobility, but also because these cohorts have experienced little previous self-motivated migration and hence

FIGURE 27 Average Annual Age-Specific Net Off-Farm
Migration Rates, 1960–1966 and 1966–1970

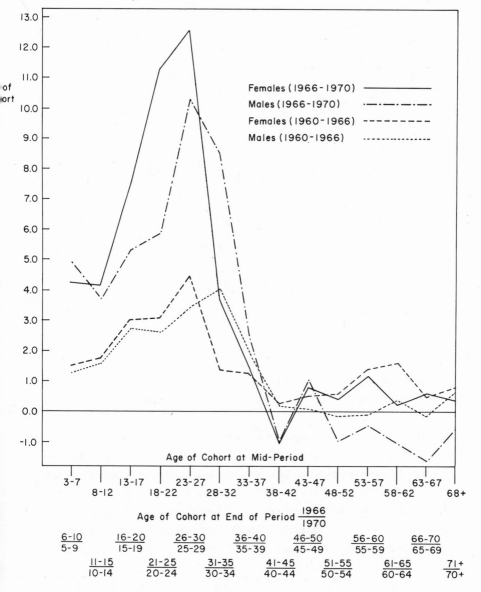

FIGURE 28 Inter-Period Percentage Point Change in Average
Annual Net Off-Farm Migration Rates by Birth
Cohort, 1960–1966 to 1966–1970

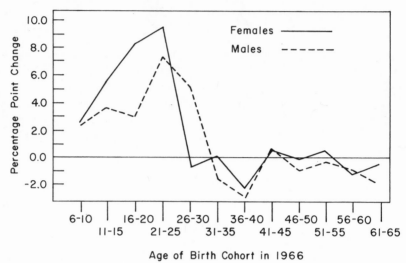

Source: Figure 27

have not yet been depleted of those individuals with the
greatest incentives to leave agriculture—that is, those facing
particularly good prospects in the non-agricultural sector or
particularly poor prospects in farming.

The highest cohort-specific migration rates in each period occur
for the female cohort age 26–30/25–29, that is, among women
who passed from their early to late twenties during the period in
question. In the second period, the annual net migration rate
for this cohort was 12.5 percent, corresponding to a 40 percent
net loss through out-migration between 1966 and 1970. In both
periods, however, the cohort age 26–30/25–29 began the period
—at age 20–24/21–25—with the majority of its members already
married. Information on marital status is not separately
available for the farm population but, for rural women as a

whole, 72 percent of those age 20–24 in 1960 were married, as were an estimated 70 percent of those age 21–25 in 1966.[12] It seems safe to assume that no more than 40 percent and perhaps as few as 30 percent of the farm women in these cohorts were single at the beginning of either migration period. Since Korean women typically give up or are excluded from wage employment after marriage, we find, paradoxically, that the highest net off-farm migration rates occur among women who are largely withdrawing from the general labor market. Moreover, the pattern of female net off-farm migration rates is very different from the pattern of net rural-urban migration rates among women, which in both periods shows maximum values for the cohort age 16–20/15–19. A great deal of off-farm migration during the 1960s, therefore, appears to have occurred in direct conjunction with marriage and the simultaneous formation of new non-farm household units. This process and its relationship to the changing circumstances facing potential off-farm migrants becomes clearer in an examination of the migration behavior of three separate female birth cohorts.

Let us first consider the behavior of the birth cohort age 30–34 in 1970. In 1966, at the beginning of the second migration interval, it is estimated that over 90 percent of this cohort were already married and, between the two inter-censal periods, the net off-farm migration rate of this birth cohort declined from 4.5 percent to 3.7 percent per year, while the annual volume of net off-farm migration declined by one-third.

In contrast, the female birth cohort age 25–29 in 1970 showed a much different migration history during the 1960s. For this cohort, the rate of net off-farm migration rose from 3.1 percent per year during 1960–1966 to 12.5 percent per year during 1966–1970, while the annual volume of off-farm migration tripled.

This increase took place in a specific historical context. During the late 1950s and the early 1960s when this birth cohort was passing through the ages between adolescence and marriage, off-farm employment opportunities for unmarried young women were limited, and largely restricted to domestic service, other menial occupations, or less respectable employments. In addition, more tractable, young teen-age girls were generally preferred as domestic workers, and the pattern of female off-farm migration typically entailed leaving for off-farm employment before age 15 and returning to the farm household a few years later. Young adult farm-born males also found it extremely difficult to move into the non-agricultural sector, although it was equally clear that agriculture could not absorb (that is, provide a viable level of livelihood for) a large proportion of the new biological families being generated by the farm population. As a consequence, it is likely that the intended formation of new non-farm households by young farm-reared couples was postponed, as perhaps was marriage itself in some instances. During the early 1960s, then, the net movement of this cohort of farm women into the non-agricultural sector either as individuals or as wives in newly formed biological families was inhibited by the limited capacity of the non-agricultural sector to absorb workers from the farm household sector. As the possibilities for long-term integration into the non-agricultural sector improved in the second half of the decade, the potential for off-farm migration that had gone unrealized in the early 1960s contributed to the surge in off-farm movement.

Finally, the birth cohort age 20–24 in 1970 began the second period as unmarried older adolescents and included a large proportion of potential off-farm migrants who had not been siphoned off in the early 1960s. As potential off-farm wage workers, farm women of this birth cohort were greeted by a non-agricultural labor market which was rapidly expanding and seeking young workers for manufacturing and modern service industries. The structure of non-agricultural labor was also changing, as the most rapidly growing sectors sought young female and male workers who were physically and mentally more mature and better educated than the workers previously sought for domestic service and similar employments. At the same time, members of the birth cohort who had previously left agriculture had less need and fewer incentives to return. The non-agricultural sector was also more capable of absorbing new biological families desiring to leave agriculture, while aspirations in this direction were undoubtedly stimulated as well. However, with women of this birth cohort marrying later than farm women of the preceding birth cohort, no doubt partly as a consequence of more off-farm employment opportunities, the importance of marriage-related off-farm migration probably declined relative to individual employment-oriented migration during the late 1960s.

To the extent that off-farm migration is immediately associated with the formation of new non-farm households by young biological families generated by the farm population, the underlying determinant of female off-farm migration is the pattern of occupational choices of farm-born men. The changing pattern of female off-farm migration rates thus helps to bring the changing pattern of off-farm migration among young adult male cohorts into sharper focus. Much that has been said about the dynamics of female migration also applies to male off-farm migration, and only a few supplemental comments are needed.

The youngest male birth cohort to show a declining rate of migration between 1960–1966 and 1966–1970 was the cohort age 35–39 in 1970. As with the female birth cohort age 30–34, this indicates the dampening influence on mobility of marriage, family formation, and—in the case of men—occupational maturation. The difference in the age at which male and female inter-sectoral mobility sharply declines reflects the difference in the age at which men and women generally marry in Korea and, of course, is related to the average difference in age between husband and wife. Among young couples marrying during the 1960s, husbands were typically two to five years older than their spouses.

The strong positive sanctions toward marriage in Korea mean that, while unfavorable economic circumstances may cause men to delay marrying for a few years, there is still considerable pressure to marry before age 35, even when prospects for an independent livelihood are poor. In 1970, 93 percent of all Korean men age 30–34 were married, including 86 percent of those age 30.

That the peak male rate of off-farm migration during the 1960–1966 period occurred in the cohort age 31–35 in 1966 (that is, the birth cohort age 35–39 in 1970) suggests not merely that non-agricultural opportunities for younger farm males were very limited in this period, but also that men in their late twenties and early thirties cannot indefinitely put off family formation and the necessity of choosing between farming and a non-farming occupation. For minor sons and principal heirs facing very marginal long-term prospects in agriculture, this "choice" may in fact have been no more than the final and long anticipated separation from family holdings at the time of marriage.

Like the cohort age 31–35 in 1966, the cohort age 26–30 had also faced limited opportunities for entering the non-agricultural sector during 1960–1966. Unlike the older cohort, however, it still retained considerable potential for off-farm migration by marriage, family formation, and five to ten years of commitment to farming. Of perhaps equal importance, the birth cohort age 26–30 in 1966 consisted of men born during 1936–1940 who had benefited from the expansion of primary and secondary education after Liberation and had generally been too young for military service during the Korean War—fortuitous advantages largely denied to men five years older.

The male birth cohort age 21–25 in 1966 was in a particularly favorable position to respond to the increased labor absorption capacity of the non-agricultural sector in the late 1960s, since this period of expansion found that cohort between the ages of military service and marriage, the age when the search for regular employment is most intense. Slightly over half this birth cohort were married by the end of the second period, including one-fourth the men age 25, and the rapid absorption of farm-born males of this cohort into the non-agricultural sectors undoubtedly was a major cause of the high off-farm migration rates during the second period that are indicated for the female cohorts age 20–24 and 25–29 in 1970.

The interpretation given to young adult migration patterns is reinforced by the pattern observed among older cohorts. The net in-migration in the cohort age 40–44 during the late 1960s, amounting to a 4 percent gain over the entire period, suggests that some young adults who were involuntarily squeezed out of the agricultural sector during the early 1960s were able to achieve reintegration into agriculture during the latter half of the decade. Net in-migration among both men and women indicates that this was not principally the result of return migration by married males who had temporarily left for wage employment outside agriculture.

Because of the particular historical and social circumstances which influenced the pattern of off-farm migration during the 1960s, it is likely that the pattern of off-farm migration during 1970–1975 will turn out to have been quite different when appropriate data for this period become available. With a smaller backlog of unrealized migration among farm-born young adults, the male cohort age 30–34 and the female cohort age 25–29 (in 1975) will probably show a somewhat lower rate of off-farm migration relative to younger cohorts. Because of the effective reduction in the period of military service and the broader exemptions which have followed upon the rapid increase in the number of service-age males as the post-Korean War "baby boom" cohorts come of age, off-farm migration among men age 20–24 is likely to have increased substantially and, in general, a greater proportion of total inter-sectoral movement probably occurred among unmarried employment seekers.

The contrasting pattern of cohort-specific net off-farm migration rates and net rural-urban migration rates has already been noted. Although these two phenomena often occur simultaneously, it is a serious mistake to assume that they are essentially identical. For instance, among migrants moving from the urban sector to the rural sector a highly disproportionate number move into the rural non-agricultural sector. Consider, for example, reported in-migration to the rural sector of South Chŏlla province during 1965–1970. Although the non-primary sector accounted for only 16 percent of total rural employment in South Chŏlla in 1970, it absorbed 41 percent of the in-migrants from other provinces (from both rural and urban areas) and 31 percent of those coming from urban areas of South Chŏlla. Even among men age 40–44, the cohort showing the highest rate of net in-migration to farm households, more than 45 percent of the in-migrants from other provinces or from cities within the province were found in non-primary occupations, compared to 22 percent for the cohort as a whole.[13] It is important to both the conceptualization and analysis of rural-urban migration that a clear distinction between these two types of mobility be maintained.

The estimated volume of net off-farm migration during 1960–1966, 1.41 million persons age 6 and over, is two-thirds higher than the estimated volume of net rural-urban migration during the same period. During 1966–1970, the estimate of 2.49 million persons for net off-farm migration in the population age 5 and over based on the 1966 definition of farm household is 13 percent above the estimate for net rural-urban migration. Age-specific rates of off-farm and rural-urban migration during the second period are juxtaposed in Figure 29, and it may be observed that rates of off-farm migration generally exceeded rates of rural-urban migration among cohorts under 35, while net rural-urban migration rates are typically higher at older ages. Perhaps the most striking contrast in Figure 29 is the large gap between off-farm and rural-urban migration rates for female cohorts age 20–29 (or 21–30) and male cohorts age 20–34 (or 21–35). This strongly suggests that a large fraction of the off-farm migration associated with marriage and the formation of new households does not entail migration out of the rural area and indeed may largely take place within the original community. It might be noted in this regard that, contrary to the common assumption that the rural non-agricultural population is largely confined to ŭp (towns), 68 percent of rural workers in secondary and tertiary industries in 1970 were in myŏn (township) areas. Moreover, fully 18 percent of employed workers in myŏn were enumerated in non-primary industries.[14]

Labor absorption constraints in the urban sector during the early 1960s may have led many off-farm migrants to remain in the rural sector, especially those who had been "pushed" rather than "pulled" out of regular farm household status. In even the more remote rural districts, the number of non-farm households (based on the 1960 definition) typically grew by over 3 percent per year during 1960–1966, several times the rate realized during 1966–1970. There was also a significant increase in the number of rural non-primary workers and the share this held of total non-agricultural employment growth.[15] The large fraction of off-farm migrants that apparently did not leave the rural sector, and the fact that adolescent off-farm migration rates were relatively low when compared to the rates among cohorts in their late twenties and early thirties may be taken as an indication that much off-farm migration during this period was the result of the displacement from agriculture of young adults having access to little or no land. Quite likely this population was particularly "ripe" for urbanward migration when city-centered economic growth took hold after 1965. Others may have found it possible to reintegrate into farming. Certainly a pattern of this sort is consistent with the relatively high rural-urban migration rates observed during the second period among cohorts age 35–49, and with the net in-migration to farm households in the cohort age 40–44. It also helps explain the flood of in-migrants to Seoul during the late 1960s, and the substantial under-representation of former farmers among in-migrants to Seoul squatter settlements from rural areas.

The degree to which net rural-urban migration rates exceeded off-farm migration rates among older cohorts during 1966–1970 suggests that, even aside from the urbanward movement of widowed rural women, older farm leavers in general were moving to the city rather than remaining in the rural sector, while net in-migration to farming perhaps largely reflected the re-entry

FIGURE 29 Off-Farm and Rural-Urban Average Annual
Net Migration Rates, 1966–1970

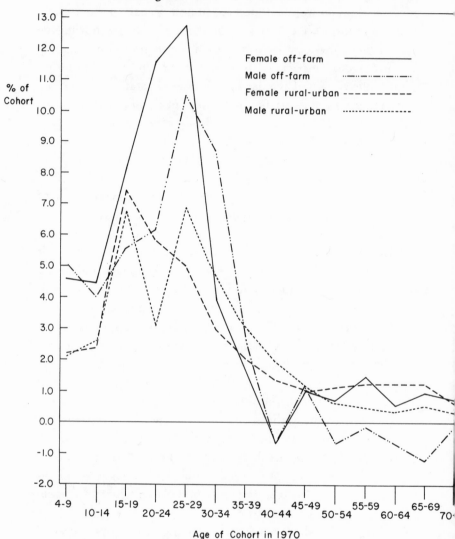

Age of Cohort in 1970

Note: There are slight differences between the rates of off-farm migration presented in this figure and those presented for the same period in Figure 27 because that figure used migration rates based on the 1960 census definition of farm households in order to achieve comparability with the migration rates estimated for the 1960–1966 period. In Figure 29 above, estimates based on the slightly broader 1966 census of farm households are employed.

to farming by aging members of the rural non-farm population.

INTRA-PROVINCIAL VARIATIONS
IN THE RATES OF FARM
AND NON-FARM HOUSEHOLD GROWTH

The closer a rural area is to a city, the greater the exposure of
the population to the opportunities, experiences, and ideas that
alter values and broaden the horizons of economic and social
aspirations. More concretely, being near a regional city raises
the aspirations for secondary education and lowers the effective
cost of such education, since secondary schools tend to be
located in urban centers. In addition, the lower cost of visiting
the city or of migrating there on a trial basis after leaving school
makes more accessible the kinds of experiences that engender
a permanent transfer of commitment from agriculture to the
non-agricultural sectors later in life. On the other hand,
agricultural areas just beyond the urban fringe have clear
advantages over more remote regions for the development of
specialized commercial agriculture. Not only do farmers in the
former areas face lower transportation costs in getting their
produce to the local urban market, but they are also in a position
to enjoy fuller and more immediate market information. Even if
the demand of the local urban market is limited, inter-urban
transportation linkages bring the farmers on the outskirts of a
smaller city into close proximity to the urban markets of the
broader region or the nation as a whole. The effective price of
certain production inputs (fertilizers, improved seeds, pesticides,
herbicides, tools, and so forth) is also likely to be lower on the
outskirts of the city, since these, like modern consumer goods,
are frequently distributed from urban centers.[16] The immediate
rural hinterland of the city also offers greater opportunities
for farm family members to engage in non-agricultural occupations on a part-time or full-time basis. In remote rural districts,

the range of choice may be limited to continuing in subsistence or quasi-subsistence farming[17] or migrating out of the region.

To examine the relationship between access to urban areas and the growth of farm and non-farm households during the late 1960s, the kun (counties) of each province were grouped into "near-urban," "peri-urban," and "outlying" provincial sub-regions. The classification is based solely on geographical distance from nearby urban areas (whether or not within the same province), and the density of transportation linkages to these urban areas. Global access to the national urban sector as a whole was not considered, and there are large qualitative differences between similarly classified sub-regions of different provinces.

The peri-urban region in each province includes those kun from which annexations to urban areas were made in either 1963 or 1973. The peri-urban regions thus include areas directly influenced by the expansion of the urban fringe. The near-urban regions encompass other kun in the immediate hinterland of major cities or accessible to groups of smaller cities. The remaining kun comprise the designated outlying region of each province. Figure 30 maps the urban areas and the rural sub-regions of each province; the rate of change in total, farm, and non-farm households in each sub-region between 1966 and 1970 is shown in Table 138.[18] The proportion of all households that were farm households in 1970 is also indicated in that table.

Although the rate of decline in the number of farm households was greatest in the urban areas of each province (and roughly twice as fast in the urban sector of Kyŏnggi province as in the urban areas of other provinces), the rate of decline of farm households in peri-urban areas was less than in near-urban areas in six of the seven provinces where both these rural sub-regions are defined. In the seventh case, Kyŏnggi province, the rate of decline in the two regions was identical. In half the former six cases (that is, in North Chŏlla, North Kyŏngsang, and South Kyŏngsang provinces), the outlying region also showed a more

FIGURE 30 Provincial Sub-Regions

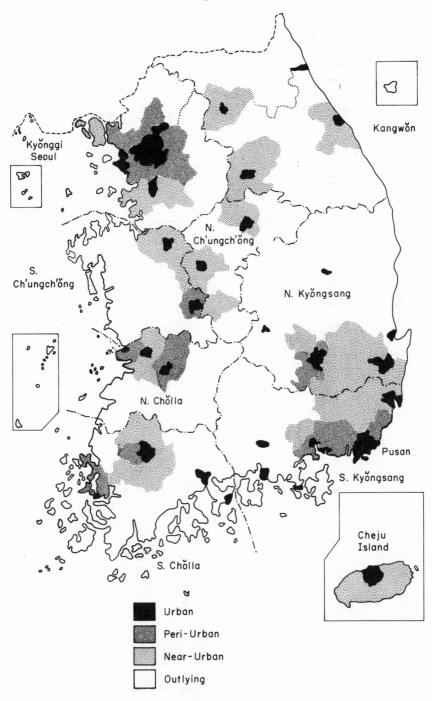

Kangwŏn

Kyŏnggi
Seoul

N.
Ch'ungch'ŏng

S.
Ch'ungch'ŏng

N. Kyŏngsang

N. Chŏlla

Pusan

S. Kyŏngsang

Cheju
Island

S. Chŏlla

Urban

Peri-Urban

Near-Urban

Outlying

TABLE 138 Percentage Change in Households: Provincial Sub-Regions, 1966–1970

	Province								
	Kyŏnggi & Seoul	Kangwŏn	North Ch'ungch'ŏng	South Ch'ungch'ŏng	North Chŏlla	South Chŏlla	North Kyŏngsang	South Kyŏngsang & Pusan	Cheju Island
All Households									
Whole Province	34.8	4.0	0.4	3.5	-1.1	3.8	6.8	13.8	11.8
Urban	47.1	18.6	13.7	27.7	12.0	22.6	27.4	33.0	27.2
Peri-urban	30.8	–	–	11.4	-0.9	2.0	3.2	4.8	–
Near-urban	0.8	-1.5	-3.8	-0.6	-3.4	-4.0	0.5	-1.0	6.7
Outlying	-3.0/-0.4	-4.5/4.0	-0.9	-1.3	-4.7	-0.3	-1.6	-4.5	–
Farm Households									
Whole Province	-8.0	-7.2	-4.1	-3.7	-5.6	-2.6	-3.9	-5.8	-5.8
Urban	-33.1	-16.3	-17.5	-18.2	-16.8	-16.8	-11.5	-18.1	-15.5
Peri-urban	-3.9	–	–	-4.1	-2.0	-2.2	-2.6	-4.3	–
Near-urban	-3.9	-7.4	-2.8	-4.4	-4.2	-3.2	-3.4	-4.7	-3.9
Outlying	-11.2/-3.9	-10.3/-4.8	-3.9	-2.9	-6.2	-1.6	-3.6	-4.7	–
Non-Farm Households									
Whole Province	46.6	13.9	9.3	16.5	7.3	16.5	20.7	30.2	69.1
Urban	50.1	24.2	21.8	34.0	16.9	29.4	31.6	37.6	70.1
Peri-urban	68.9	–	–	46.7	1.9	18.0	21.7	27.4	–
Near-urban	16.6	9.0	-7.6	7.9	-1.1	-6.5	10.5	10.0	67.8
Outlying	3.4/6.1	1.3/13.0	7.1	3.4	0.4	5.0	4.1	-3.4	–

TABLE 138 (continued)

				Province					
	Kyŏnggi & Seoul	Kangwŏn	North Ch'ungch'ŏng	South Ch'ungch'ŏng	North Chŏlla	South Chŏlla	North Kyŏngsang	South Kyŏngsang & Pusan	Cheju Island
Non-Farm Households as a Percentage of Total Households, 1970									
Whole Province	85.3	58.0	35.6	40.2	38.0	37.6	49.3	62.3	35.5
Urban	98.4	90.1	85.1	92.3	89.1	90.0	93.2	94.9	66.6
Peri-urban	61.6	–	–	40.1	29.8	24.1	28.0	35.0	–
Near-urban	26.4	39.8	18.8	38.9	26.6	23.8	30.5	28.2	23.2
Outlying	59.7/36.8	52.5/53.6	20.6	26.4	24.2	21.3	27.0	21.3	–

Notes: The provincial sub-regions employed in this table are shown in Figure 30.
The two figures given for the outlying regions of Kyŏnggi province and Kangwŏn province reflect a further division of the outlying regions of these provinces into border-zone kun and non-border-zone kun (data in the above tables presented in this same order).
Farm Household is defined to include private farm households regardless of the amount of land worked by the household. The households of landless agricultural laborers are included among *Non-Farm Households* within these tables (see text). The number of total farm households in each sub-region was calculated from the National Population Census reports for 1966 and 1970. The number of farm households in corresponding years was calculated from MAF, *Yearbook of Agriculture and Forestry Statistics 1967*, and MAF, *Agricultural Census 1970*. Non-farm households were calculated residually.

rapid decline in the number of farm households than the peri-urban region.

The outlying border-zone kun of Kyŏnggi and Kangwŏn provinces are best treated as a special case, characterized as they are by a heavy concentration of military encampments with a substantial number of military families, civilian employees and a large service population in the adjacent settlements.[19] In Figure 30, the southern boundary of this outlying border-zone region in these two provinces is indicated by a dotted line. The border-zone regions of Kyŏnggi and Kangwŏn provinces show much greater rates of farm household decline—exceeding 10 percent during 1966–1970—than other rural sub-regions. A con-tributing factor in this decline may have been the settlement of refugees from adjacent areas of North Korea in this region during the 1945–1954 period, and the displacement and losses experienced by the settled population of the area during the war years. Such a population is likely to be more mobile than a farm population characterized by long and uninterrupted settlement. The border zones of both Kyŏnggi and Kangwŏn provinces showed a substantial decline in the total number of households and much lower rates of increase for non-farm households than in the non-border-zone outlying counties of the same provinces.

It seems quite clear that increased off-farm migration, in response to a quantum increase in the expansion of the non-agricultural sector, did not occur simultaneously throughout the rural sector and in direct relationship to objective differences between long-term economic expectations in local agriculture and non-agricultural prospects elsewhere. Rather, it was the near hinterlands of the developing centers that responded first. With continued growth of the non-agricultural sector, rapid off-farm migration has gradually taken hold in regions more remote from the center(s) of growth. As a behavioral pattern, off-farm migration appears to have diffused in a manner similar to the diffusion of information about modern urban life, technological and organizational innovations, new consumption patterns, new social and economic attitudes, and so on. As with these others, the diffusion of off-farm migration as an accepted behavioral alternative appears to be simultaneously filtered downward through the hierarchy of urban places, and broadcast outward from these points of contact between the agricultural and non-agricultural sectors.[20] An interpretation of this pattern implied by the economic literature on inter-regional migration is that the non-instantaneous diffusion of information about non-agricultural opportunities and the costs of migrating inhibit an immediate response in the more remote rural areas, even though these may be among the poorest rural districts. But the contention that the diffusion of new attitudes, tastes, and information about urban life is a principal stimulus to rural out-migration is flawed in one critical respect. The "information" that most often triggers migration tends to be highly specific: knowledge about living conditions, employment possibilities, and often even the promise of a job in a particular locale obtained through inter-personal contacts. A significant outflow of population can only occur after a period of more restricted out-migration has created a relatively broad and strong chain of inter-personal ties between out-migrants in particular cities and potential migrants in specific villages. Clifford J. Jansen, summarizing the findings of D. J. Bogue and others, states that "there is a series of stages in the development of any migration stream. From initial invasion it develops into a phase of settlement which at its peak becomes routine, institutionalized."[21] Ezra Vogel has drawn attention to the role of kinship and community ties in facilitating the movement of rural Japanese youths to the city and their integration into the non-agricultural sector,[22] and surveys conducted in Seoul during the late 1960s typically indicate that over half of migrant heads of households reported receiving substantive assistance from relatives or acquaintances when they first arrived. Perhaps 70 percent or more of in-migrant household heads had contacts of some sort in Seoul prior to arriving.[23] Indeed, fragmentary evidence suggests one-third or more of household heads coming from rural areas may have been migrating to a waiting job.[24]

Since the inter-personal channels for rural-urban migration only develop gradually as the number of initial migrants to a specific destination increases and these forerunners become sufficiently established to provide assistance to others, it is the lag in the development of these migration channels rather than the speed of the diffusion of information in a more general sense that controls the spatial diffusion of rural out-migration.

Econometric models of inter-regional migration which include among the independent variables a measure of the number and distribution of former migrants living in the alternative areas of destination typically find this to be the most important determinant of the direction of current migration.[25] Even under relatively tight labor market conditions, a large community of former migrants in a destination region may be able to secure many more jobs for potential migrants "at home" than can be provided by a smaller number of "friends and relatives" in areas where economic conditions are more favorable. Hence a deterioration of the economy in destinations to which most previous migrants from an area have gone is unlikely to result in a rapid change in the direction of out-migration but rather in a decline in the overall volume of out-migration until new channels of migration to more promising destinations become established.

THE AGING OF FARM OPERATORS
AND CHANGING LANDHOLDING PATTERNS

Not unexpectedly, the increasing rate of off-farm migration in the post-Korean War period has brought about pronounced changes in the pattern of birth cohort commitment to the agricultural sector. In the 1955, 1960, 1966, and 1970 census data there is a strong positive correlation between the age of a male cohort and the percentage of that cohort in farming and related occupations. For all age groups this proportion has dropped during each inter-censal period but, when this pattern is simultaneously considered by age group and birth cohort, as in Figure 31, it is clear that the decline in the commitment to farming has principally occurred through a sharp decline in the proportion of each successive cohort initially entering farming and to a much lesser extent through a decline over time in the proportion of farmers within a given birth cohort. This pattern is consistent with the contrast between very high off-farm migration rates among young adults and much more modest rates among those over 30. For example, according to the 1955 census of population, 55.3 percent of males age 25–29 (corresponding to the birth cohort of 1925–1929) were farmers or related workers. By 1970, the proportion of farmers and related workers among males age 25–29 had fallen to 24.7 percent, while for the 1925–1929 birth cohort the proportion had dropped to 37.4 percent. For cohorts already past 30 in 1955, the extent of the subsequent decline was much less.

The difference between birth cohorts in the proportion of farmers provides an indication of whether the economic conditions favored or disfavored movement into the non-agricultural sector at the time each successive birth cohort passed through the early working years. Thus, the relatively narrow difference in the proportion of farmers between the 1925–1929 and 1930–1934 birth cohorts suggests that, in the late 1950s, the relative absorptive capacity of the non-agricultural sector and the incentives it offered to potential off-farm migrants were weaker than during the previous decade, in spite of the post-Liberation dislocation and the wartime upheaval. Indeed, in 1955 the proportion of farmers among males age 25–29 exceeded that among males age 30–34. This suggests that the events of the first post-Liberation decade affected the work lives of different cohorts unevenly, increasing the opportunities outside agriculture for some birth cohorts with the departure of Japanese who had been employed in the secondary and tertiary sectors, while diminishing the opportunities for other cohorts through the disruption of education and career formation.

FIGURE 31 Male Participation in Agriculture, 1955–1970

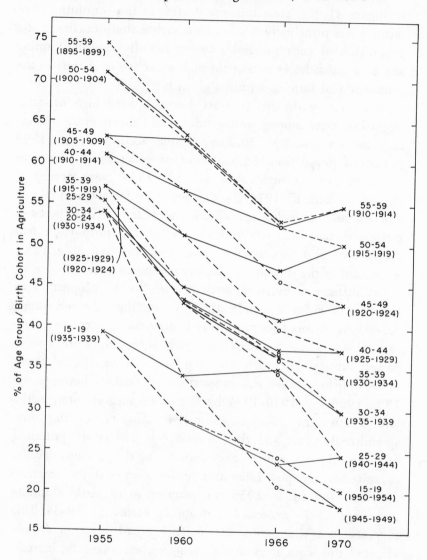

Note: The solid line connects the proportion of farmers and related workers among males of the indicated birth cohort in the 1960, 1966, and 1970 Population Censuses, while the broken line links the proportion of farmers in the same age group. Because of errors in the age reporting in the 1960 census, the values for 1960 were interpolated. Interpolation was also required to obtain the 1966 birth cohort values, since the age groupings for which occupation data are reported do not correspond to the same birth cohort groupings as in 1960 or 1970. For this reason, two sets of data points are given for 1966, one for age groups (o) and one for birth cohorts (x).

This apparent increase between 1966 and 1970 in the proportion of farmers among cohorts born before 1930 is noteworthy in light of the absence of any similar pattern in the two previous inter-censal periods. Quite probably, the rise in the rate of participation in agriculture among cohorts born during 1910–1914 and 1915–1919 was at least in part the result of delayed retirement from farming brought about by the absence of an immediate successor. The increase in the commitment to farming among cohorts born between 1920 and 1929 suggests that net return migration to farm households among middle-aged males was accompanied in many cases by a return to farming as an occupation.

It should come as no surprise that the precipitous decline in the commitment to agricultural careers among young male cohorts has resulted in a rapid shrinking of the young adult work force in farm households. The extent of the changes wrought on the demographic structure of the agricultural village is most vividly conveyed by a comparison of the age-sex pyramids of the farm household population in 1960, 1966, and 1970 (see Figure 32). The pronounced wasp-waist shape of the 1970 pyramid reflects the widespread departure of farm youth. It is this fall in the number of younger members of the potential agricultural work force rather than the reduction in the absolute size of the farm household working-age population as a whole that has necessitated higher labor force participation rates among older men and middle-aged and older women in farm households.

It might be assumed that the withdrawal of young cohorts from agriculture has resulted in an across-the-board rise in the age of farm operators. This is indeed the case in the aggregate, but closer examination indicates that, between 1960 and 1970, the median age of operator actually declined among farm households cultivating between 7 and 30 tanbo (see Table 139; 1 tanbo = 1/10 chŏngbo or 0.245 acres). Peasant farms are only in the rarest of circumstances passed unchanged from father to son, generation after generation. Under partible inheritance, a surplus of claimant heirs over current operators necessitates a division of family holdings which insures, in the context of a relatively fixed and intensely exploited land base, that the majority of farmers will enter operator status with a holding smaller than that of the household in which they were raised. As a result of the termination of some family lines and the withdrawal of others from farming, however, land periodically becomes available for purchase or rental and, by this process, an inexorable, geometric fragmentation of ancestral lands is avoided and a "life-cycle" pattern of land acquisition and division emerges. One implication of such a cycle is a positive correlation between age of operator and the size of land holdings up to the age—approximately 60—at which an operator's minor heirs marry and establish new households with some portion of the family holdings. This correlation holds for Korea in both 1960 and 1970.[26] Table 140 indicates that the median area of land cultivated by operators under age 25 was 5.7 tanbo in 1960 (6.2 tanbo in 1970), and that this rose to 7.5 tanbo for operators age 45–54 (8.0 tanbo in 1970), followed by a decline to 6.9 tanbo (6.7 tanbo in 1970) for operators age 65 or older.[27] Table 140 also indicates that in 1960—prior to extensive off-farm migration—older operators clearly made up a disproportionate share of the operators of larger holdings, while younger operators were heavily overrepresented on smaller farms.

The changing relationship between age of operators and farm size revealed in Table 140 provides indirect evidence that young male adult off-farm migration has been heaviest from those households with the most limited holdings wherein the prospects in agriculture are extremely poor, even if the principal heir can anticipate an undivided inheritance of family holdings. At the same time, however, the equally poor prospects in agriculture for minor heirs, even in relatively well-to-do farm households, means that farms of all sizes have contributed to the outflow. Although the size and quality of the anticipated inheritance does not alone determine who leaves and who stays, it can probably be assumed that, on the whole, young adult males choosing to

FIGURE 32 Farm Household Population Cultivating 1 Tanbo or More, 1960, 1966, 1970

(1,000s)

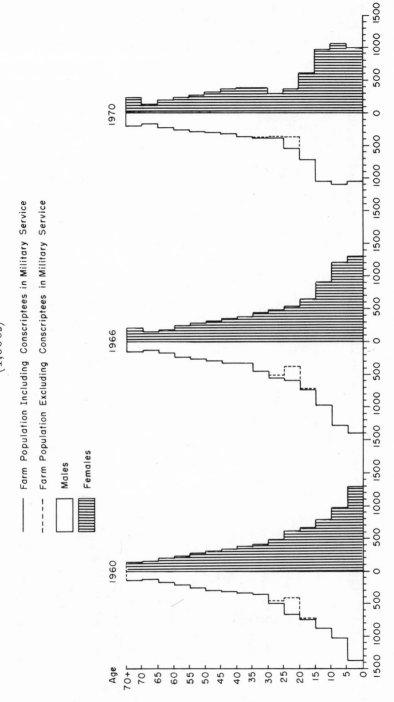

Source: KASS-POP Adjusted Farm Population.

TABLE 139 Changes in Farm Size Distribution and Age of Operator on Household Farms, 1960–1970

			tanbo			
	1-2	2-3	3-5	5-7	7-10	10-15
Percent of All Farms						
1960 (%)	7.1	9.1	18.7	17.5	18.5	17.2
1970 (%)	7.3	8.0	16.6	16.3	18.3	18.7
Percent of Farm Land						
1960 (%)	1.3	2.7	9.0	12.7	18.9	25.3
1970 (%)	1.1	2.2	7.3	10.8	17.1	25.2
Median Age of Operator						
1960	42.9	42.3	42.8	44.2	45.8	47.3
1970	45.1	44.3	44.2	44.7	45.5	46.6
Change in Number of Farms						
1960–1970 (%)	5.0	-10.7	-9.3	-4.6	1.3	11.8
Change in Farm Land						
1960–1970 (%)	0.7	-11.2	-9.2	-4.3	1.5	12.0
Average Size of Holding						
1960	1.47	2.44	3.93	5.91	8.33	12.04
1970	1.40	2.42	3.93	5.92	8.35	12.06
Average Size of Household						
1960	4.79	4.91	5.27	5.78	6.35	7.03
1970	4.71	4.90	5.22	5.60	5.99	6.43

TABLE 139 (continued)

	tanbo				
	15–20	*20–25*	*25–30*	*30+*	*Numerical Totals*
Percent of All Farms					
1960 (%)	7.0	2.9	1.2	0.8	2,329,128 farms
1970 (%)	8.1	3.6	1.5	1.6	2,384,842 farms
Percent of Farm Land					
1960 (%)	14.6	7.7	3.0	3.8	18,988,685 tanbo
1970 (%)	15.4	8.9	4.7	7.3	21,344,717 tanbo
Median Age of Operator					
1960	48.2	48.6	48.8	48.9	45.1 years (overall)
1970	47.5	48.3	48.6	49.2	45.8 years (overall)
Change in Number of Farms 1960–1970 (%)	18.6	29.6	31.7	90.9	2.4
Change in Farm Land 1960–1970 (%)	18.8	29.6	31.5	118.8	12.4
Average Size of Holding					
1960	17.00	22.01	27.11	36.47	8.15 tanbo
1970	17.02	22.00	27.05	41.67	8.94 tanbo
Average Size of Household					
1960	7.70	8.15	8.50	8.81	6.11 persons
1970	6.84	7.10	7.25	7.37	5.81 persons

Source: Based on agricultural census data for 1960 and 1970.

Note: All data refer to private farm households directly managing at least 1 tanbo of cultivated land.

TABLE 140 Farm Size Distribution by Age of Operator and Changes in the Number by Farm Size, 1960–1970

	Farm Size Class (tanbo)							
	1-5		5-10		10-20		>20	
	%	%	%	%	%	%	%	%
1960								
% of Operators								
Age ≦24	41.9	(5.3)	38.6	(4.7)	17.2	(3.1)	2.3	(2.0)
25–34	44.7	(23.9)	35.0	(18.2)	17.5	(13.6)	2.8	(10.7)
35–44	35.7	(27.1)	36.5	(27.0)	23.1	(25.4)	4.6	(25.1)
45–54	29.9	(22.4)	35.9	(26.3)	28.1	(30.6)	6.1	(32.8)
55–64	30.0	(14.7)	35.9	(17.2)	28.0	(19.9)	6.1	(30.1)
≧65	33.8	(6.6)	34.8	(6.7)	25.6	(7.3)	5.7	(8.0)
Total	35.0	(100.0)	36.0	(100.0)	24.1	(100.0)	4.9	(100.0)
1970								
% of Operators								
Age ≦24	37.8	(3.7)	37.6	(3.4)	21.1	(2.4)	3.6	(1.6)
25–34	36.5	(19.9)	35.7	(18.0)	23.2	(15.1)	4.6	(12.0)
35–44	32.3	(27.8)	35.2	(28.0)	26.4	(27.1)	6.1	(25.0)
45–54	28.3	(23.0)	33.9	(25.4)	29.6	(28.6)	8.2	(31.5)
55–64	29.1	(16.9)	33.7	(18.1)	29.0	(20.1)	8.1	(22.3)
≧65	36.6	(8.6)	32.5	(7.1)	24.2	(6.8)	6.7	(7.5)
Total	31.9	(100.0)	34.6	(100.0)	26.8	(100.0)	6.7	(100.0)
% Change in Number of Operators (Farm Males), 1960–1970								
Age ≦24	-35.0	(-61.1)	-29.7	(-16.7)	-11.7	(-7.5)	13.4	(11.9)
25–34	-22.2	(-22.9)	-2.6	(-19.5)	26.0	(2.9)	58.1	(31.2)
35–44	-4.0	(-3.9)	2.4	(2.8)	21.2	(8.5)	40.4	(26.3)
45–54	-4.4	(-3.6)	-4.8	(-0.7)	6.3	(12.9)	35.0	(18.3)
55–64	7.4	(1.4)	3.7	(0.5)	14.4	(31.5)	46.8	(103.1)
≧65	21.2	(17.4)	4.7	(16.4)	5.8	(4.4)	31.6	(0.5)
Total	-6.7	(-31.6)	-1.6	(-7.7)	13.8	(5.0)	40.6	(24.5)

TABLE 140 (continued)

	Numerical Totals (persons)	%	Median Size (tanbo)	Gini Coefficient[a]
1960				
% of Operators				
Age ≤24	102,667	(4.4)	5.7	.3498
25–34	435,693	(18.7)	5.5	.3718
35–44	617,680	(26.5)	6.6	.3698
45–54	612,397	(26.3)	7.5	.3619
55–64	400,208	(17.2)	7.5	.3621
≧65	160,045	(6.9)	6.9	.3758
Total	2,328,690	(100.0)[b]	6.7	.3703
1970				
% of Operators				
Age ≤24	74,067	(3.1)	6.2	.3698
25–34	415,525	(17.4)	6.5	.3791
35–44	656,084	(26.8)	7.2	.3814
45–54	617,665	(25.9)	8.0	.3820
55–64	442,081	(18.5)	7.8	.3849
≧65	179,420	(7.5)	6.7	.4081
Total	2,384,842	(100.0)	7.3	.3855

% Change in Number of Operators (Farm Males), 1960–1970

	All Farms	
Age ≤24	-27.9	(-31.9)
25–34	-4.6	(-11.9)
35–44	6.2	(3.9)
45–54	0.9	(4.1)
55–64	10.5	(15.1)
≧65	12.1	(10.8)
Total	2.4	(-10.4)

TABLE 140 (continued)

Sources: Values for 1960 were calculated from data on farm size and age of operator given in MAF, *Agricultural Census 1960* (Volume 8, Tables 1–4), and on the age and sex distribution of the farm population by farm size given by Kyu-sul Lee *Population Pressure and its Effect on Korean Farm Households* (Seoul, 1967), Tables 1–3. The latter study is based on a sample retabulation of the 1960 Population Census and 1960 Agricultural Census data. Since the total males in each farm size class reported by Lee do not correspond to the total reported in the 1960 Agricultural Census (differing by as much as 9% in the case of farms of 20 tanbo and over), the numbers of farm males by holding size and age on which the third panel is based were estimated by applying Lee's age distribution to the Agricultural Census totals. Values for 1970 are based on MAF, *Agricultural Census 1970* (Volume 10, Tables 1–7).

Notes: All values refer only to farm households managing more than 1 tanbo of cultivated land. The values in parentheses in the upper two panels indicate the age distribution of the operators in each farm size class by the percentage in each age group. The parenthesized values in the third panel indicate the percentage change in the number of farm males of the indicated age in farm households of each holding-size class. (In the first row the change in the number of males age 15–24 is presented, rather than the change in the total number of males under age 25.)

[a]The calculation of median farm size and the distribution Gini Coefficients are based on the ten-class division for farms of over 1 tanbo used in Table 139.

[b]Includes 438 households for which age of operator is not reported.

remain in farming face better initial prospects as independent operators than those who have chosen to leave—indeed, the departure of other claimants to family holdings directly contributes to the improvement of those prospects.

During the 1960s, the departure of heirs to very small holdings and of minor claimants to larger holdings was not merely sufficient to prevent a deterioration in the size of holdings worked by young operators but of such magnitude that the average acreage cultivated by this group increased measurably. For example, between 1960 and 1970, the total number of operators age 25–34 declined by 4.6 percent, while the proportion of operators in this age group holding farms with less than 5 tanbo of cultivated land decreased from 44.7 to 36.5 percent. At the same time, the number of farm males[28] age 25–34 (including operators and non-operators) declined by 11.9 percent. The decline in the numbers of both operators and farm males in this age group was greatest in the smallest farm-size class. In the 1–5 tanbo class, the number of operators in this age group declined nearly as rapidly as the number of farm males. In all larger-size classes the number of operators declined at a much slower rate or increased at a more rapid rate than the number of farm males, and the difference in these rates of change increased with farm size. At the upper end of the farm-size spectrum, a very different picture emerges. Not only did the proportion of operators age 24–34 with holdings in the 10–20 tanbo range increase from 17.5 percent in 1960 to 23.5 percent in 1970— compared to an increase in the share of such farms among all farms of 1 tanbo or more from 24.1 to 26.8 percent—but the proportion of operators 25–34 among all operators of farms in this size class rose from 13.6 to 15.1 percent in spite of the absolute and relative decline in the total number of young operators. A similar pattern can also be observed in the over-20 tanbo farm-size class. A natural corollary to these changes is that the ratio of operators to farm males for this age group rose substantially in all farm-size classes, with the greatest absolute and relative increases occurring in the 5–10 tanbo farm-size class.

Among operators 65 and older, a very different pattern is evident. Whereas in 1960 operators in the oldest age group had been overrepresented only in the two upper-size classes, the 1960s saw an increase in the proportion of such operators on farms of less than 5 tanbo, with the result that, by the end of the decade, operators over 65 were overrepresented in the smallest farm-size class, and underrepresented in the two middle classes. The larger proportion of older operators on small holdings is most likely a consequence of the departure from the agricultural sector of heirs to such farms, leaving the holdings under the management of aging parents unable to "retire" from operator status due to the absence of successors. A contributing factor may also be the limited labor resources of farm households with aging heads and no younger adult members, which might lead such households to reduce the amount of land directly cultivated through the sale or renting out of some parcels or by returning previously rented land.

The recorded increase in the number and proportion of larger holdings between 1960 and 1970 resulted in the distribution of land becoming somewhat more unequal, as reflected in a rise of the Gini Coefficient from 0.3702 to 0.3855. According to Agricultural Census data, the proportions of total cultivated acreage held by the upper one-tenth of farm households and by each of the subsequent three-tenth groupings of farm households in 1960 and 1970 were as shown in Table 141.

TABLE 141 Cultivated Acreage Held by Farm Households,
1960 and 1970

		0–10th	10th–40th	40th–70th	70th–100th
Percentage of Cultivated Land	1960	26.1%	40.4%	22.8%	10.6%
	1970	27.2%	40.4%	22.6%	9.8%
Average Size of Holdings (tanbo)	1960	21.3	11.0	6.2	2.9
	1970	24.3	12.1	6.7	2.9

A comparison of the Gini Coefficient of the pattern of land distribution for specific operator age groups in 1960 and 1970 (see Table 140) indicates that the land distribution within each age group became more unequal; the greatest shift, not un-expectedly, occurred among operators over age 65, with operators 55–64 and under 25 showing the second and third greatest increases in the Gini Coefficient. As a consequence of these shifts, the overall correlation between the extent of the inequality in the landholding pattern and the age of operator was altered. In 1960 this relationship took the form of a wave function, with the greatest inequality exhibited among operators over 65 and between 25–34, with a relatively more equal distribution among operators 45–64, and the greatest equality occurring among operators in the youngest age class. By 1970 this pattern had changed to a monotonically increasing degree of intra-cohort inequality with increasing age of operator.

SOCIO-ECONOMIC SELECTIVITY
IN VILLAGE OUT-MIGRATION

Direct empirical evidence bearing on the relationship between pre-migration socio-economic status and the rate of off-farm migration generally is provided by two village surveys conducted by Chong-ju Yun in 1970 and 1971 and a unique study by Man-gap Yi in which six villages which had been studied in 1958 were resurveyed in 1969.

In Chong-ju Yun's first study, conducted in Pongsan township[29] of Yesan county in South Ch'ungch'ŏng province, information was gathered from current residents on the characteristics of households and individuals who left the village during the preceding five years. Calculations made from Yun's data indicate that departed households for which the size of holdings is known include proportionately more households with less than 1,000 p'yŏng (3.3 tanbo) than were observed among households remaining in the villages. There was also a substantially higher proportion of tenants and of non-farm households (based in this case on the non-farm occupation of the head) among households that had left. While the highest departure ratio is observed among tenant-farm households, specifically those with less than 500 p'yŏng of land, when one looks beyond the very smallest holding class, there appears to be little difference in the departure ratio between tenant-farm households and owner-operator households within the same farm-size class. Non-farm households also show a high departure rate, in spite of a probable downward bias due to the inclusion of newly formed non-farm households in the denominator.

The types of occupations reported for absent members of families in Pongsan township suggest that young adult out-migrants were positively selected in terms of educational attainment,[30] but Chong-ju Yun presents no information on the age-specific educational levels that would provide more definitive evidence. Generally, however, studies of migration in Korea have shown the existence of a strong positive correlation between level of education and age-specific mobility rate for the population as a whole.[31] Chong-ju Yun's second rural survey, conducted during the summer of 1971 in Ian township, Sangju county, North Kyŏngsang province, provides detailed age-specific information on the educational attainment levels of residents and absent household members. Table 142, derived from her data, indicates that over half the male household members age 20–24 were living away from the villages at the time of the survey, with an absentee rate of 39 percent for males 15–19 and 29 percent for those 25–29—although the exclusion from the denominator of members who have been away for more than five years means that the cumulative departure rate through age 29 will have been substantially higher than this last figure would suggest. Although both the standardized educational attainment rates for migrants and residents and the age-specific absentee rates indicate lower-than-average mobility among males with no education, and higher-than-average departure rates among males with more than a high school education, the difference in the absentee rates between those with 1–6, 7–9, and 10–12 years of schooling is relatively slight. Ian township is a remote and relatively poor rural area, and the degree of educational selectivity in out-migration may have been less than elsewhere; but a more important implication of the Ian township data is that exposure to the world beyond the village is widely distributed among farm youths of all levels of educational attainment, and this experience is not solely limited to that gained in the course of military service. Paralleling Chong-ju Yun's findings in villages beyond the metropolitan penumbra, Man-gap Yi's study of six villages situated to the south of Seoul in Kyŏnggi province also indicates that the highest rates of departure are found among non-farming households, the households of agricultural laborers, households with small holdings, households of more recent arrival in the village, and among younger and better-educated household members and heads.[32]

For over two decades the basic skills required for successful integration into modern non-farming work lives have been made

TABLE 142 Absence Ratio by Age of Cohort and
Educational Attainment in Ian Township, 1971

*I. Percentage of Household Members Absent by Age and Years of
Education*[a]

		Men				
Age	Total	0 Years	1–6 Years	7–9 Years	10–12 Years	13+ Years
10–14	4.7	0.0	4.1	5.9	33.3	–
15–19	39.0	28.6	43.8	35.0	37.9	100.0
20–24	56.2	50.0	53.5	57.6	54.2	77.3
25–29	29.0	0.0	25.0	30.0	31.9	66.7
30–34	3.2	0.0	2.5	7.5	5.6	0.0
		Women				
10–14	6.3	25.0	5.4	6.8	100.0	–
15–19	35.8	35.3	35.5	36.4	29.4	100.0
20–24	23.6	0.0	18.1	40.0	42.9	28.6
25–29	1.1	0.0	1.6	0.0	0.0	–
30–34	0.5	0.0	0.8	0.0	0.0	–

*II. Standardized Educational Attainment Levels for Residents and
Absentees in the Principal Out-Migration Ages*[b]

		Years of Education				
	Total	0 Years	1–6 Years	7–9 Years	10–12 Years	13+ Years
		Men Age 15–29 (%)				
Absentees	100.0	1.5	39.8	30.3	20.0	8.4
Residents	100.0	4.1	42.3	32.0	19.9	2.0
		Women Age 10–24 (%)				
Absentees	100.0	4.6	61.6	25.5	6.3	2.0
Residents	100.0	3.9	72.6	18.5	4.1	0.9

Source: Calculated from data presented in Chong-ju Yun, *A Study of Rural Population*, pp. 109, 214.

Notes: Absentees are non-resident household members that have left the household
during the preceding five years.

[a]Calculated as follows: $\dfrac{\text{Absentees}}{\text{Absentees} + \text{Residents}} \times 100.0$.

[b]Calculated with use of identical weights for each of the component five-year age
groups.

available to the children of even very poor farm households through an effective system of compulsory primary school education.[33] In more prosperous farm households, resources are applied to assist sons in the move to the non-agricultural sector by providing secondary and higher education. An increasing number of farm girls are also being sent beyond primary school, probably in part to improve their chances for a favorable marriage. That farm households may be consciously viewing expenses for schooling as a substitute for a patrimony in land, with the intention of launching sons into non-agricultural careers, is suggested by the low orientation toward farming as an occupation for their children among farm parents,[34] and among students in agricultural high schools, virtually all of whom come from farm backgrounds.[35] The reluctance of farm households to employ in "dirty" field work sons who have been sent beyond the primary grades has been seen as a display of conspicuous consumption,[36] but may also be intended to reinforce the parents' expectations that such sons move out of agriculture.

When well-to-do farm households maintain an identification with the traditional elite class, the *yangban*, the incentive to educate sons may spring from status considerations as well. In the wake of the land reform and rapid industrialization, wealth based on rural land ownership is both proscribed by law and lacking in the prestige it formerly enjoyed. Where family status is a concern, it must now be sought in urban-centered professional, commerical, or political careers. One Western anthropologist with extensive field experience in Korea has noted that, whereas farm households without *yangban* identifications invest in fields, *yangban* households invest in educating their sons.[37]

THE ECONOMIC INTEGRATION
OF OFF-FARM MIGRANTS[38]

Because of the apparent ease of entry into traditional or bazaar-sector activities in the urban economies of less developed countries, it has frequently been argued (or simply assumed) that rural-urban migrants typically enter such occupations as a first step in assimulating to the urban economy. There is increasing evidence, however, that the concentration of migrants in casual occupations characterized by high underemployment has been seriously overstated.[39] Data for Korea also argue against such an assumption, and suggest that recent in-migrants to the urban sector, particularly among those in the most mobile ages, have been absorbed into modern sector employments in proportions that fall only slightly, if at all, behind those of non-migrants of the same age and education.

The KDI Migration Data indicate that male in-migrants to Seoul under 35 and female in-migrants of all ages had somewhat higher labor force participation rates in 1970 than their non-migrant peers,[40] although if adjusted for differences in marital status and education the gap would probably be diminished. At the same time, male in-migrants under 30 and female in-migrants under 45 had rates of open unemployment which were far less than the rates exhibited by non-migrants.[41] At older ages, however, migrants showed somewhat higher unemployment rates than non-migrants. An analysis of the occupational structure of migrants and non-migrants in Seoul based on the KDI Migration Data and using single-digit occupational categories[42] indicates that, when age is held constant, the most pronounced differences between the occupational distribution of male in-migrants and non-migrants in 1970 was the underrepresentation of the former in professional, administrative, and clerical positions, substantial overrepresentation in service occupations among migrants under age 25, and a general but modest over-representation of migrants of all ages among production workers and laborers. Female in-migrants showed even more pronounced underrepresentation in professional, administrative, and clerical positions, heavy overrepresentation in services, and substantial underrepresentation in sales.

A major part of the difference between migrant and non-migrant economic characteristics is probably a consequence of differences in levels of education, marital status, or other factors unrelated to migrant status or place of origin per se. Where migrant status itself plays a role, this would appear to be less a consequence of labor market discrimination for or against migrants than an outgrowth of the constraining circumstances surrounding their economic decisions. For example, an over-representation of migrants in lower-status and poorer-paying jobs (and that this is the case is by no means clear) may simply be due to the fact that few migrants can afford to remain in the urban sector while exercising a reservation price against certain kinds of work. Furthermore, young in-migrants from rural areas may have lower aspirations with respect to the salary and prestige of initial urban employments, and young off-farm migrants in particular may view their initial non-agricultural employments more in the light of experimentation or as an adventure in independence than their urban peers for whom successful integration into the non-agricultural sector is an absolute necessity and for whom getting the "right" first job may take on greater importance.

The very marked increase of female workers in urban man-ufacturing and modern-sector service activities in recent years has meant a rapid rise in the real wages of female workers out-side agriculture and a correspondingly rapid increase in the expected real earnings of young female off-farm migrants. Yukio Masui has argued that increased wage-earning opportunities for married women in the post-war period in Japan contributed to a shift from the seasonal or short-term (*dekasegi*) male-dominated off-farm migration to a pattern of family off-farm migration by reducing the supply price of non-successor males in family off-farm migration to the level of the supply price of male labor in individual off-farm migration.[43] In Korea as well, the expansion of female employment opportunities in the non-agricultural sector may also have contributed to the growth of family off-farm migration, although for somewhat different reasons than in Japan. In Korea a very small proportion of married women are economically active during the child-rearing ages, and there remains widespread discrimination against the employment of young married women, particularly in the modern manufacturing and service sectors. However, the very rapid expansion of employment opportunities for young unmarried women in the modern sector has probably worked to keep female earnings relatively buoyant in petty trading, family enterprises, and domestic service, even in the wake of rapid in-migration to urban areas. It is in these latter activities that the labor of married and older women is chiefly utilized (and probably underreported). The custom of partible inheritance also provides the successor-heir with an incentive to assist his brothers in integrating into the non-farm sector. On average, the Korean farmer enjoys fewer opportunities for non-agricultural employment within commuting distance of the farm, and fewer opportunities to supplement farm incomes through seasonal inter-regional agricultural labor migration than were available to the pre-war Japanese farm operator, and this also serves to reduce the opportunity costs of family off-farm migration.

It was suggested earlier in this chapter that the increase in return migration to farm households among older age groups in the late 1960s may have been in part a response to increasing inter-cohort competition in the non-agricultural labor market. It appears that recent changes in the industrial structure have generally favored the integration of young work cohorts into the modern sector, with migrant or native status being a relatively unimportant factor. At the same time, older urban workers with limited education and few accumulated industrial skills may have been increasingly locked out of better occupations. Surveys of squatter colonies and low-income districts conducted in Seoul during the late 1960s indicated that a large proportion of the middle-aged and older male heads of household in these areas were characterized by high unemployment, and unskilled or marginal occupations, while younger male household heads in the same areas exhibited more stable work lives and economically more rewarding occupations.[44]

Younger workers as a group enjoy a general educational advantage over older workers, which facilitates movement into more remunerative career paths, and also have in a longer future work life an incentive to accept lower wages and poorer conditions over the short run in jobs believed to provide training and experience contributing to permanent integration into the modern sector.[45] In addition to such direct economic incentives, young workers with few family obligations are in a better position to forego present earnings for higher future income, and may have families who are willing to subsidize them during their early work years. The willingness of young workers to "invest" in themselves through lower wages, their greater trainability and possible greater tolerance of long working hours, uncompensated layoffs, irregular wage payments, and so on, undoubtedly favor a preference among employers for young, new entrants to the labor force. This preference is particularly striking in light export industries: in 1974, 86 percent of all employees in the Masan Free Export Zone were under 30, and a tour through the Kuro Industrial Export Estate in Seoul leaves the impression that at least 90 percent of the on-line production workers are under 25. A strong perference for young workers among modern-sector employers carries some serious implications. In this case, modern-sector employment expansion will largely be filled through the recruitment of young workers, many of whom will be drawn from farm households. Moreover, in times of contraction or slower expansion, the employer will try to shift the burden of unemployment to older unskilled workers, even while continuing to recruit new labor force entrants. Many of the older workers bearing the brunt of the contraction do not have the option of returning to the agricultural sector, even if they were drawn from the farm household population initially, since the organization of the agricultural sector precludes the ready reintegration of persons who have largely cut their ties to the farming community. At the same time, the continued recruitment of young workers prevents off-farm migration from serving as an effective and responsive adjudicator of imbalances between the farm and non-farm labor markets.

SYNOPSIS AND CONCLUSIONS

In Korea, as in most other nations, modern industrial growth has been accompanied by a relative and subsequently an absolute decline in the population directly dependent on agriculture for a livelihood. Off-farm migration as an inseparable part of the transition from a rural agrarian society to one that is predominantly urban and industrial is more than simply the underlying mechanism of this transition. The structure and dynamics of the off-farm migration process play an important role in determining the relationship between economic growth and social change. For this reason, off-farm migration deserves examination apart from its function as a supplier of labor to the non-agricultural sectors.

The decline of the farm population has frequently coincided with the rapid growth of the largest city or cities. In view of this coincidence, it has often been assumed that the streams of in-migrants to the centers of modern growth consist largely of former agriculturalists, and indeed the poorest of the rural peasantry. In fact, the road between field and factory appears to be much more circuitous. As James White notes in a recent review of the literature on rural-urban migration in prewar Japan,[46] inter-prefectural migrants to the major urban centers included a disproportionately large number of migrants from regional cities and out-migrants from rural non-farm families. In this, Japan was not unlike most European countries during the early stages of industrialization. Although rural-urban migration in Korea during the 1960s occurred within a more compact spatial system, both aggregate data and city and village surveys indicate a similar pattern. Migrants to Seoul have included a large number of inter-city movers, while former farmers appear to be underrepresented even among the poorer migrants from rural areas. Likewise, in farm villages, the highest rates of household departure are found among non-farm households, households of agricultural laborers, and farm households cultivating less than 0.5 chŏngbo ($\cong$0.5 hectare). It may be noted that large fractions of the latter two groups are typically engaged in non-agricultural activities as well as farming. However, indications of positive educational selectivity in individual off-farm migration and changes in the demographic structure of larger as well as smaller farm-size categories show that all farm classes have contributed to the outflow from agriculture.

Off-farm migration is not a once-and-for-all, one-way movement of population, although movement into and out of agriculture is almost entirely confined to the farm-born. Yet, while many of the current group of young adults in agriculture have undoubtedly had some non-agricultural work experience, mobility between the sectors drops off sharply after age 30–35, and there is no reason to believe that individuals lacking ready access to land or years of accumulated experience in farming can successfully re-enter farming from the non-agricultural sector in later life. Seasonal movement between agriculture and non-agricultural employments is not a prevalent pattern in Korea. Seasonal migration has not been considered within the body of this chapter, but it can be noted that it chiefly entails the movement of men into construction and the urban "informal sector" during the spring, summer, and fall months. Conversations with rural officials suggest, moreover, that there is no large-scale return of out-migrants during the periods of peak labor demand, although this may be more common in peri-urban districts. In general, seasonal migration appears to supplant work in agriculture rather than complement it, and in many instances may preface household off-farm migration.

The intensity of off-farm and rural-urban migration during the 1960s was undoubtedly influenced by the extensive out-migration from the southern provinces during the colonial period, which had served to dislodge much of the rural population from the soil of the ancestral village. Further disruptions of agrarian life accompanied the economic and social upheavals of the post-Liberation period, the Korean War, and post-war reconstruction. Yet, if the potential for migration from agriculture was high, the capacity of the urban sector to absorb migrants remained weak until the middle 1960s.

The farm population continued to increase through 1967 and, as a result of rural population pressure and despite the urban labor absorption difficulties, the volume of net off-farm migration during 1960–1966 was substantial, averaging 243,000 persons per year, not including newborn children of migrant women. This was equivalent to an average annual net off-farm migration rate of 1.8 percent, with the rates for the most mobile age groups reaching 3–4 percent per year. Net rural-urban migration rates were much lower, especially among young adults, suggesting that a large fraction of off-farm migration during this period consisted of newly formed households that were squeezed out of agriculture but continued to remain in the rural sector and perhaps in the same village.

The average annual volume of off-farm migration more than doubled between 1960–1966 and 1966–1970, exceeding 568,000 persons per year in the second period. Whereas there had been slightly more male than female net migration during 1960–1966, there was a substantial excess of female off-farm migration during 1966–1970, reflecting the expansion of demand for female workers in the labor-intensive manufacturing industries. The overall net off-farm migration rate reached 4.2 percent per year in the second half of the decade, with annual rates of 10–12 percent exhibited by the most mobile cohorts. Rural-urban migration rates continued to be significantly below off-farm migration rates for young adult cohorts, if not for adolescents, indicating that off-farm migration in association with new family formation often did not simultaneously entail urbanward movement. In both periods there was net out-migration from the farm population among older female cohorts, probably a reflection of a high rate of departure among widows. Older male cohorts generally showed modest rates of net in-migration. Since this was accompanied in the second period by relatively high rates of rural out-migration among both older men and women, it may be suggested that some non-farm households displaced from agriculture in the first period were able to reintegrate into farming during the second while others were moving to the city. Fully two-thirds of net off-farm migration was accounted for by the cohorts between age 15 and 35 during both 1960–1966 and 1966–1970, and most of the remainder were under age 15. In both periods, net migration from farm households accounted for 60 percent of the increase in the non-farm working-age population.

All provinces showed higher rates of off-farm migration during 1966–1970 than during 1960–1966, but, surprisingly, the provinces with relatively higher migration rates in the first period had relatively lower off-farm migration rates subsequently. While the reasons for this pattern can only be imperfectly speculated upon, the source of the aggregate trend may be traced to changes in the relative intensity of migration among younger and middle-aged male working-age *birth* cohorts. Prima facie evidence suggests this phenomenon cannot be directly related to changes in the relative level of farm incomes among provinces. Rather, higher off-farm migration rates during the first period appear to have reduced the potential for additional off-farm migration among members of the affected cohort, while at the same time the early outflow contributed to the formation of a large stock of potential return migrants, some of whom presumably re-entered farming in the second period, reducing the net rate of outflow.

Improvements in non-agricultural labor market conditions suggest that off-farm migration among adolescents and young adults became more positively selective in the late 1960s with respect to education and other human capital assets. As a result, earlier migrants may have found it more difficult to compete with younger migrants in the non-farm labor market. Certainly the competitive disadvantage of older cohorts vis-à-vis younger ones is suggested by the increasing concentration of older production and construction workers in the less skilled occupational categories in these industries.

Although migrants from farm and non-farm backgrounds cannot be separated within the aggregate data, occupational differences between young in-migrants to Seoul and their non-migrant peers were not particularly great in 1970 and can largely be explained in terms of differences in educational attainment. The heavy concentration of migrants observed in production and service occupations is very largely a function of their age, and there is some indication that young out-migrants from rural villages in particular are more likely to move into sales than into services, although production activities absorb the largest fraction of economically active young rural out-migrants of both sexes.

In determining the rate of off-farm migration from a given area, treating information and risk merely as modifying factors in the calculation of the economic advantages of migrating obscures the fact that, quite frequently, the central element of the decision to migrate is knowledge of and access to work opportunities in areas to which relatives or friends have previously gone. Migration follows upon former population movements, and even the intensity and direction of migration for employment is in large measure dependent on the location and number of former out-migrants. With the establishment of migration channels between urban centers and increasingly remote rural districts, successive phases of migration intensity, selectivity, and associated changes in demographic structure and landholding patterns spread through the rural sector in a manner that may be characterized as hierarchical diffusion. The relationship between the diffusion of rural-urban migration channels and off-farm migration appears to be indirect as well as direct: it is the non-farm population of the rural sector and the provincial cities that is particularly drawn to the more distant centers of urban growth, and their departure makes it possible for new off-farm migrants to secure a niche in the rural or provincial city non-agricultural economy, while perhaps even continuing to provide labor to agriculture on a part-time or seasonal basis. In short, a substantial fraction of off-farm migration appears initially to entail short-range stepwise replacement migration. Stepwise migration may be a more important avenue for off-farm migration in Korea than is generally realized. A large fraction of net in-migration to provincial cities from their rural hinterlands during the 1960s went to replace the net outflow from regional cities to the metropolitan centers of Seoul and Pusan, and it can be argued that the share of rural out-migration going to nearby provincial cities was disproportionately large, given the typically great difference in size between the provincial urban sector and the major metropolitan destination.

At the level of national aggregates, the rapid increase in off-farm migration since the middle 1960s has resulted in a dramatic transformation of the demographic structure of the farm household population. The massive departure of farm youths is clearly apparent in the age-sex distribution of the 1970 farm population. The further decline of the farm population to 13.2 million persons or 38 percent of the total population in 1970 was undoubtedly accompanied by an intensification of this pattern. The decreasing commitment of farm youths to agriculture has also affected the pattern of landholding. Although the proportion of farm households cultivating 1.0–1.5 chŏngbo or more increased only marginally between 1960 and 1970, and the total distribution of landholdings became slightly more unequal, beneath this aggregate pattern there was a more substantial increase in the average holding size among young operators, particularly for the group age 25–34, which worked to narrow the difference in average holding size between younger and older farm operators. At the same time, the proportion of farm operators over 65 on very small holdings increased. It may be presumed that this also is a reflection of the widespread departure of would-be heirs to small family farms, as well as an indication that aging operators without heirs in residence are forced to reduce the scale of their farming activities.

It has been suggested that the dynamics of off-farm migration are deeply rooted in the social fabric. As a result, the intensity of off-farm migration and of the accompanying structural changes varies greatly between rural districts and lacks close correspondence to inter-local differences in the productivity of labor in agriculture. Extensive off-farm migration also tends to weaken the sense of long-term commitment to agriculture among those remaining in the area. For these several reasons, rapid off-farm migration may contribute to serious regional imbalances in the pattern of rural development, since there is no correspondingly fluid redistribution of factors of production within the agricultural sector to compensate for the uneven pace of population withdrawal. The retention of land by off-farm migrants or casual part-time farmers inhibits the movement of land into the hands of those more committed to its full utilization. Other serious social and economic constraints also hamper the inter-regional mobility of agricultural workers in an environment of village-centered family farming. As two recent studies of rural development in Japan seem to suggest, an unbalanced pattern of off-farm migration may contribute to the emergence of a pattern of land utilization that is seriously dysfunctional for the society as a whole.[47]

The Korean economy has experienced unprecedented growth since 1960, but the concentration of modern-sector growth in the two metropolitan regions, and especially around Seoul, has limited the possibilities for farm households to pursue a measured and gradual integration into the non-agricultural sector. The pattern of population redistribution and investment allocation that has emerged may be a contributing factor in the perpetuation of significant pockets of rural underdevelopment and imbalances between regions in the effectiveness with which Korea's scarce agricultural resources are utilized. Fortunately, the government's current concern with rechanneling new development away from Seoul[48] may presage a more concerted pursuit of policies to avert further aggravation of regional disparities and the realization of a more balanced improvement of rural welfare.

THIRTEEN

Conclusion

The drama in Korean economic growth over the past decade and a half was to be found in the rapid growth of cities and modern industries and in the extraordinary rate of expansion of manufactured exports. But when this great economic transformation began, well over half the Korean people still lived in the countryside; nearly 40 percent still lived there as late as 1975. What happened to these people and what bearing do their actions have on any overall appraisal of South Korea's economic performance over the three decades 1945–1975?

The central point of the previous chapters is that, while the Korean countryside never progressed at rates of 10 percent a year and more, agricultural production and the incomes of the great majority of South Korean farmers did rise and by a substantial degree. Over the entire three-decade period, agricultural value added grew at roughly 3 percent a year. Since the size of the rural population in 1975 was only slightly higher than what it was in 1945 (and the same as in 1955), this increase in agricultural output represented a rise in rural per capita production and not simply the impact of an increasing rural population and labor force trying desperately to keep food production growing ahead of the number of mouths to be fed.

The presumption, therefore, is that this rise in production led to a rise in rural per capita income and hence an improvement in the farmers' standard of living. In fact, the rise in the rural standard of living of most farmers was significantly higher than the increase in per capita production. The main reason, of course, was land reform. During the Japanese colonial period, the rate of rural tenancy had risen from a high to a still higher level, and landlords, as a result, had been the main beneficiaries of the agricultural growth of that period. By being beneficiaries under the hated Japanese rule, however, landlords became politically vulnerable once that rule ended, and it was this fact more than any other, even the Korean War, that made a thoroughgoing transfer of land on near confiscatory terms possible. By the mid-1950s, 85 percent of South Korea's land was being farmed by its owners, as contrasted with less than 40 percent in the 1930s, and this transfer of ownership made possible a 20 to 30 percent increase in income for everyone except former landlords.

The other principal determinant of farm income was the price paid for produce sold off the farm. During the 1950s and much of the 1960s, the government, whose purchases dominated the grain and several other agricultural markets, held prices paid to farmers down as part of an effort to maintain general price stability. Reinforcing this emphasis on price stability were large imports of PL 480 grain from the United States. But by 1969 the government reversed this policy and raised grain prices paid to farmers. Because the government did not allow a comparable rise in urban grain prices, the result was a deficit financed by the government by means that contributed to a substantial rise in the money supply. This same policy, however, also led to a real rise in farm incomes of about 10 percent.

By 1975, therefore, farm income per capita had more than doubled over the level of the 1930s and had increased by an even greater amount compared to the late 1940s. As pointed out in Chapter 11, deliberate efforts to redistribute income to farmers accounted for roughly 40 percent of this rise from the 1930s to 1975, and efforts to increase production accounted for the remainder.

Higher farm purchase prices also stimulated output, but there were many other measures that contributed to the rise in agricultural output as well. These measures can be discussed on two levels. At one level are the direct inputs into the agricultural production function—land, labor, and various kinds of capital—together with increases in the productivity of those inputs. At the second level are the market conditions and government policies, including the institutions created to implement those policies, that help or inhibit those increases in inputs and productivity.

Since there has been little net increase in the amount of arable land during the past three decades, and there has been no rise in the rural labor force from the mid-1950s on, it follows that increases in these inputs cannot account for much of the increase in farm output. Korean farmers, however, have made substantial efforts to adjust their methods to a land-short, and in later years an increasingly labor-short, factor endowment. The key to the successful adaptation to Korea's poor endowment of arable land has been the shift away from grain, a land-intensive crop, to various cash crops, sericulture, and the like, where the value of output per unit of land is much higher than for grain. To carry out this adaptation, Korea had to be in a position to import increasing amounts of grain to fill the rising gap between supply and demand and, in addition, to provide a market either at home or abroad for the expanding output of non-grain farm products.

Labor supply did not really become a problem until the mid-1960s, when migration to the cities and other off-farm opportunities led to an absolute decline in the number of workers remaining in agriculture. Prior to the mid-1960s, there were shortages of labor during peak seasons, but the removal of a worker from the farm on the average reduced output less than consumption, leaving those who remained with more consumption goods per capita than had been the case before. By the late 1960s and particularly in the 1970s, however, farmers were under great pressure (incentive) to mechanize key activities in order to compensate for the loss of labor. Between 1964 and 1975, as a result, power threshers in use rose 9-fold and power tillers nearly 80-fold.

The fixed capital stock did rise substantially over this three-decade period, notably through a steady expansion in the amount and reliability of irrigation. But the key increases affecting agricultural production were the rises in such current inputs as improved seeds and chemical fertilizer. Both improved plant varieties and chemical fertilizer were in use in Korea in the 1930s, much earlier than in most currently less-developed economies. After 1945 and despite the Korean War, Korean farmers were able to obtain even larger amounts of chemical fertilizer than before through the American aid program. And by the early 1960s, Korea had begun to produce its own chemical fertilizer with production levels in the mid-1970s that saturated the domestic market and left a small surplus for export.

Finally, there were the many sources of increased factor productivity. Some of these sources of productivity growth, such as the various improvements in plant varieties, were embodied in particular inputs. Others, such as the rising educational level of farmers and the improvements in marketing and transport, were not so embodied and, instead, probably improved farm management and hence the productivity of all inputs. Taken together, these various sources of productivity growth accounted for roughly half the rise in agricultural production.

Some of these increases in inputs and their productivity can be attributed to particular efforts of the government, but others cannot. Many of the influences that made for a successful agricultural performance were the result of a gross national product that was expanding at 10 percent a year during the latter half of this three-decade period and exports that were rising even faster. Certainly the shift in emphasis toward cash crops, for example, would have been impossible if there had not been a rapidly expanding urban market. And the mechanization of agriculture would have proceeded much more slowly if more of the children of farm families had stayed home on the farm when they grew up instead of leaving for cities and their rapidly growing employment opportunities. Nor, as another study in this series demonstrates, was the rapid expansion of the educational system as much a result of government action as it has been in most other countries. Instead, private demand, the product of rising incomes and a Confucian heritage, accounted for an unusually large share of the resources in the education system.

The Korean government, however, did play some role in the farm output and income even in the early years, and that role has increased through time. Government price policies have already been mentioned, and the government also played a major role in the land reform. Other areas where a government role is essential if anything much is to happen include agricultural research and extension of the results of that research to farmers and large-scale investments in irrigation and land development schemes. Farmers, even in rich countries, lack the resources to invest in research on their own, and they are even less able to contemplate the construction of a hundred-million-dollar dam. Nor are private firms in a poor country such as Korea likely to find investments in these areas attractive, either because returns are apt to be low or, more likely, because such returns as there are will be difficult to capture in the form of private-firm profits.

Government investments in agriculture and in research and

extension in Korea go back far into the period of Japanese colonial period and were never completely abandoned. But it was not until the 1960s that serious efforts to develop agriculture got under way and, even in the 1960s, these efforts were quite modest. In a very real sense, the 1960s was a period of institution-building more than a period of major government efforts to assist farmers. More and more money was pumped into the credit operations of the National Agricultural Cooperative in an eventually successful effort to make them a major source of rural finance. The agricultural research effort in Suwŏn was upgraded in quality and expanded in scope during the period. In fact it would not be much of an exaggeration to say that the research and extension activities of the Office of Rural Development were recreated almost from scratch in the early 1960s after being virtually abandoned in the 1945–1960 period.

When one speaks of investment in agriculture in the 1960s and of the creation of agricultural support institutions, one is speaking about the role of United States aid and not just of activities of the Korean government acting on its own. In many areas, such as upland development and tideland reclamation, aid funds accounted for most of the activity. Aid was also central to the government's fertilizer importation policy as well as to the construction of the first domestic chemical fertilizer plants. Foreign advisers were scattered through the various agricultural organizations, and many Koreans were sent abroad for training under programs such as that involving the University of Minnesota with various activities in Suwŏn. Some of the funds were wasted and many of the advisers were not terribly effective, but, by the end of the 1960s, Korea had a wide variety of functioning rural support institutions, even if those institutions were starved for funds and staffed by not always highly qualified personnel. U.S. aid clearly deserves a significant share of the credit for this accomplishment, just as it must also share credit or blame for the low-price policy for grain that existed alongside this institution-building.

It was not until after the bad harvests of the late 1960s, and even more the election of 1971, that the Korean government began to place a high priority on rural development but, when priorities did change, the rural support organizations, together with the newly revitalized rural governmental offices of the Ministry of Home Affairs, were capable of responding. It will be some time before a complete appraisal of the New Community Movement and the other rural development activities of the early 1970s is possible. But there is no question that the early 1970s witnessed a great upsurge in government initiated and encouraged rural activities. In part, these activities simply reflected the fact that the government in the mid-1970s was finally allocating substantial sums to rural investment. But more was involved than money; for really the first time since independence was regained in 1945, rural government personnel were not just engaged in political control and the inefficient and corrupt allocation of key inputs such as fertilizer, but were actively involved in aiding and leading rural development efforts. Not all these efforts were well advised or welcomed by the farmers, but many of them were, and that was a change.

Are there any lessons in Korea's rural development experience over the past thirty years with relevance for other developing countries? Clearly there are many limitations on the transferability of any one country's experience to another, and Korea is no exception to this more general rule. Korea, for example, inherited considerable experience with modern agricultural techniques from the Japanese colonial period. And Korea's Confucian heritage had much to do with the high levels of education found even in rural areas. But many other aspects of the Korean rural heritage were anything but unique—poverty, small fragmented holdings, and an unmechanized technology, to name only a few.

One major lesson of Korea's rural development experience, suggested in the Introduction, is that rural development is a lot easier to accomplish if it is taking place in the context of a rapidly growing economy. Put differently, the fast pace of industrial development and Korea's increasingly strong balance of payments position were vital parts of the nation's rural development effort. This lesson is one frequently forgotten in many of the current discussions of rural development.

A second lesson, however, is that, if a nation waits until population growth has crowded as many people onto as limited a land area as that in Korea, raising grain output is a slow and expensive process, however modern the technology. Korea, in fact, like Japan, has opted, albeit unwillingly, for an emphasis on non-grain crops at home and for increasing reliance on grain from abroad. Self-sufficiency might be possible, but the price would be a lower and more slowly rising standard of living.

Korea is also one of several examples of the enormous impact that deliberate efforts to redistribute income, if actually carried out, can have on the quality of rural life. If Korea's countryside were still riddled by high levels of tenancy and if grain prices were still being manipulated for the benefit of urban workers and industrial profits, a book on Korea's agricultural experience might not look all that different in key respects from one on, say, Indonesia's.

Korea's rural development experience is also full of examples of specific techniques that have either helped or hindered farmers, some of them described in previous chapters and many more not. Although these experiences may hold valuable lessons for others, there is little room in a study like this for the kind of specificity required to judge the usefulness of these experiences.

Finally, can one say anything about the transferability of Korea's rural institution-building experience and of the role of outside assistance in that effort? Certainly massive amounts of American assistance did not create institutions that are carbon copies of American models, although there was more than a little effort to do precisely that. But Korean government and private efforts, together with United States aid, did create a set of institutions that do seem to work reasonably well. Perhaps the specific form of the institutions created is less significant than the fact that in a still poor Asian economy they exist and that they do work.

Appendixes

APPENDIX A

Agricultural Data Collection Methods

STATISTICAL SYSTEM

The statistical system of Korea being a decentralized one, each ministry has its own statistical unit to collect data relevant to its own administrative purpose. The Bureau of Statistics, under the Economic Planning Board, is the largest agency of the government responsible for the collection, compilation, analysis, interpretation, and publication of general-purpose statistics. It also serves as the central coordinating agency for inter-ministerial statistical activities. As a consultative body, the Statistical Council is organized within the Bureau of Statistics in order to give advice on the establishment of statistical standards and other statistical problems. The council consists of 32 members, representing policy-makers, academics, and other users of government statistics.

As for agricultural statistics, the Ministry of Agriculture and Fisheries is in charge of the compilation of Crop Statistics, Basic Agricultural Statistics, Livestock Statistics, Farm Household Survey, Cost of Production Survey for Major Food Crops, Food-Grain Consumption Survey, and the

Agricultural Census. To conduct these surveys, the Bureau of Agricultural Statistics acts as the central organ, with local Offices of Agricultural Statistics located in each of 9 provinces. Under the Provincial Offices there are 139 Kun (County) Branch Offices of Agricultural Statistics, one in each kun. The major functions of the Bureau of Agricultural Statisitcs are planning of agricultural statistical surveys, supervision of the local offices, collection, compilation, analysis, and publication of the results. The local Offices of Agricultural Statistics are responsible for conducting various surveys, and the branch office in each kun is directly engaged in the enumeration, preliminary compilation, and reporting.

In addition, other organizations such as the National Agricultural Cooperative Federation (NACF), Office of Rural Development (ORD), and National Agricultural Economic Research Institute (NAERI) also collect data on limited items. The Research Department of the National Agricultural Cooperative Federation conducts the survey on rural prices, wages and various charges. The Office of Rural Development mainly conducts experimental surveys such as the soil survey, and the National Agricultural Economic Research Institute conducts surveys on farm management, distribution of farm products, and the cost of production for non-food items.

In recent years, the Government of Korea has been laying great emphasis on the development of statistics, particularly on agricultural statistics in view of the high dependence of the country on agriculture. In 1972, the government raised the status of the Statistics Division to bureau level. The main idea behind this upgrading was to give due recognition to agricultural statistics as an essential part of agricultural development planning.

CROP PRODUCTION ESTIMATES

There are two factors involved in grain production estimates: 1) area planted to specified crops; and 2) yield per unit area. The two multiplied together provide the production estimate for the crop.

AREA ESTIMATES

The discussion of crop area estimates here is limited to identification of possible sources of non-sampling error in the survey procedures.

Following a cadastral survey of 1918, there was no further official survey of crop area conducted in Korea. In the administrative or official estimates of grain production, the planted area was estimated through interviews with or reports from individual farmers who reported what was planted in each field. Each field and its official area were registered in the myŏn (township) office (the smallest administrative unit in Korea).

Inasmuch as the enumerators were located in the myŏn office and engaged in other administrative work, the area statistics were likely to be affected by administrative decision and subject to bias. In the early period, it seems likely that this bias was downward because of a long history of under-reporting to reduce taxes and forced sales of grain to the government. For example, government rice purchase quotas for an individual village were often based on production performance and, the higher the production, the higher the purchase quota.

A sample survey of field areas was jointly conducted in March 1965 by the Ministry of Agriculture and Forestry (later Fisheries) and the USAID/ Korea in order to appraise the accuracy of grain production estimates. In this survey, 600 sample map areas were selected by random sampling. The result of actual measurement showed that the area of paddy was 5.1 percent greater than the area registered as paddy at the time, as shown in Table A.1. For upland, the actual measurement showed that the area was 12.9 percent greater than the registered area.

TABLE A.1: Registered versus Measured Area
(p'yŏng)[a]

	Registered Area (A)	Measured Area (B)	B/A
Paddy	7,252,914	7,621,792	105.1
Upland	5,165,355	5,829,632	112.9
Total	12,418,269	13,451,424	108.3

Source: The Ministry of Agriculture and Forestry (Fisheries).

Note: [a]One p'yŏng = 36 square feet = 1/3,000 hectares

Since it was a practice of myŏn enumerators to include some of the non-registered paddy area in the official estimates, the most that can be said is that the extent of under-reporting was a maximum of 5.1 percent for paddy and 12.9 percent for upland.

Against this tendency toward under-reporting in the past, the recent administrative estimates are believed (although there are no data on which to base this belief) to be biased upward. This judgment is instead based on the following facts:

1) Because fertilizer distribution came under strict government control and the distribution quota is based on the planted acreage, farmers as well as myŏn enumerators have strong incentives to over-report the field area in order to receive more fertilizer. For the same reason, there is a

possibility that a part of cultivated land lost to urbanization and industrial sites may not be immediately reflected in the administrative estimates.

2) The compulsory quota system in the government rice purchase program has been substantially eased. Instead, the government procures the required amount of rice from farmers on a voluntary basis.

YIELD ESTIMATES

The yield estimates currently used in the official production statistics are based on the crop-cutting method. The crop-cutting estimates are a set of yield estimates derived from the production-cost survey. The production-cost survey is based on a sample of 1,200 grain farmers who are assisted by 80 enumerators in keeping daily records of their production activities.

Tables A.2 and A.3 present comparisons between two sets of yield data for three major grain crops—rice, common barley, and naked barley—for the years 1961 through 1972. With the same acreage base (the official area estimates) for both sets of yield data, the official production estimates currently used by the government are generally higher than those obtained from the production-cost statistics.

Since the official yield estimates are based on the random crop-cutting survey under the direct supervision of the central government (Statistics Bureau of the Ministry of Agriculture and Fisheries), they are less likely to be affected by administrative elements at the local level. Confusion still remains, however, in determining which series provide more accurate estimates.

DISCREPANCY BETWEEN
PRODUCTION AND CONSUMPTION DATA

Other data available for comparative purposes are the aggregate consumption estimates obtained from the household-consumption survey. Table A.5 provides a comparison between the aggregate supply and the aggregate consumption for both rice and barley, the aggregate supply defined as domestic production plus imports minus exports minus net increase in the ending stock.

Differences in the two series were found, the consumption data being lower than the supply data. The average rate of discrepancy between the two series for each grain for the period 1964 through 1975 is approximately 7 percent for rice and 22 percent for barley. If the relative accuracy of the statistics on imports, exports, and inventory changes is assumed, production estimates are definitely higher than consumption estimates.

Overall, the production estimates that are officially used in Korea seem to be biased upward, less for rice, more for barley.

TABLE A.2 Official Production Estimates versus Estimates from Production-Cost Survey (Rice), 1961–1975

Year	Official Estimates			Estimates from Production-Cost Survey		A/B
	Area 1,000 ha	Yield kg/ha	Production (A) 1,000 M/T	Yield kg/ha	Production (B) 1,000 M/T	%
1961	1,132.9	3,050	3,458.8	2,560	2,900.2	119.1
1962	1,143.1	2,630	3,011.1	2,230	2,549.1	117.9
1963	1,158.2	3,240	3,751.7	2,980	3,451.4	108.7
1964	1,191.3	3,310	3,940.9	2,836	3,378.5	116.7
1965	1,208.9	2,870	3,464.4	2,543	3,074.2	112.9
1966	1,209.3	3,200	3,870.5	2,797	3,382.4	114.4
1967	1,214.3	2,940	3,571.9	2,751	3,382.4	106.9
1968	1,136.3	2,790	3,165.9	2,596	2,949.8	107.3
1969	1,208.0	3,360	4,057.1	2,841	3,431.9	118.3
1970	1,193.4	3,270	3,906.8	2,589	3,089.7	126.3
1971	1,187.8	3,350	3,975.3	2,952	3,506.4	113.4
1972	1,187.6	3,310	3,933.4	2,905	3,450.0	114.0
1973	1,181.7	3,560	4,211.6	2,966	3,504.9	120.2
1974	1,204.4	3,690	4,444.9	3,233	3,893.8	114.2
1975	1,218.0	3,830	4,669.1	3,193	3,889.1	120.1
Average						(115.3)

Sources: MAF, Yearbook of Agriculture and Forestry Statistics (Grain Statistics), 1961–1975. MAF, Cost of Production Survey, 1967–75.

TABLE A.3 Official Production Estimates versus Estimates from Production-Cost Survey (Common Barley), 1961–1975

Year	Area 1,000 ha	Official Estimates Yield kg/ha	Production (A) 1,000 M/T	Estimates from Production-Cost Survey Yield kg/ha	Production (B) 1,000 M/T	A/B %
1961	476.5	1,890	898.2	1,264	602.3	149.5
1962	474.9	1,620	767.7	1,150	546.1	140.9
1963	487.4	1,200	582.7	1,030	502.0	116.5
1964	518.9	1,730	895.5	1,250	648.6	138.4
1965	545.0	1,750	951.1	1,297	706.9	134.9
1966	498.4	1,960	975.3	1,626	810.4	120.5
1967	502.2	1,850	930.9	1,434	720.2	129.0
1968	478.9	1,760	840.9	1,380	660.9	127.5
1969	446.3	2,050	916.4	1,575	702.9	130.1
1970	423.5	1,930	819.0	1,599	677.2	120.7
1971	382.0	1,940	742.2	1,608	614.3	120.6
1972	363.4	2,070	751.1	1,588	577.1	130.4
1973	342.2	1,940	662.4	1,533	524.6	126.3
1974	375.0	1,770	665.7	1,343	503.6	132.2
1975	325.1	2,160	704.9	1,577	516.7	136.4
Average						(130.3)

Sources: Same as Table A.2.

TABLE A.4 Official Production Estimates versus Estimates from Production-Cost Survey (Naked Barley), 1961–1975

	Official Estimates			Estimates from Production-Cost Survey		
Year	Area 1,000 ha	Yield kg/ha	Production (A) 1,000 M/T	Yield kg/ha	Production (B) 1,000 M/T	A/B %
1961	332.6	1,740	580.1	1,871	622.3	93.0
1962	363.3	1,680	610.5	1,600	581.3	105.0
1963	407.6	820	335.3	1,210	493.2	67.8
1964	423.0	1,460	619.1	1,590	672.6	91.8
1965	386.5	1,760	855.9	1,871	910.2	94.1
1966	470.9	2,210	1,042.9	1,938	912.6	114.0
1967	476.6	2,070	985.1	1,710	815.0	121.1
1968	506.8	2,450	1,242.8	1,993	1,010.1	122.9
1969	503.1	2,290	1,150.1	1,919	965.4	119.3
1970	488.4	2,360	1,155.0	2,096	1,023.7	112.6
1971	457.4	2,440	1,115.3	2,033	929.9	120.0
1972	512.0	2,370	1,213.4	2,010	1,029.1	117.9
1973	473.5	2,360	1,115.4	1,908	903.4	123.5
1974	484.2	2,150	1,038.9	1,646	797.0	130.4
1975	385.9	2,580	995.0	2,048	790.3	125.9
Average						(110.6)

Sources: Same as Table A.2.

TABLE A.5 Discrepancy Between Supply and Consumption Data, 1964–1975
(1,000 M/T)

Rice Year[a]	Rice			Barley		
	Supply[b] (A)	Consumption (B)	B/A (%)	Supply (A)	Consumption (B)	B/A (%)
1964	3,736	3,252	87.0	1,606	1,344	83.6
1965	3,926	3,374	85.9	1,655	1,358	82.0
1966	3,594	3,451	96.0	1,874	1,559	83.1
1967	3,954	3,835	96.9	2,085	1,513	72.5
1968	3,822	3,910	102.3	2,106	1,511	71.7
1969	3,946	3,817	96.7	2,142	1,599	74.6
1970	4,394	4,175	95.0	1,880	1,454	77.3
1971	4,776	4,234	88.6	2,098	1,411	67.2
1972	4,356	4,298	98.6	2,173	1,343	61.8
1973	4,272	3,998	93.6	2,076	1,789	86.2
1974	4,641	4,331	93.3	2,089	1,758	84.2
1975	4,775	4,442	93.0	2,106	1,862	88.4
Average						(77.7)

Source: MAF, *Yearbook of Agriculture and Forestry Statistics* (Grain Statistics), 1964–1975.

Notes: [a]Rice-year begins on November 1 in the previous year and ends on October 31 in the current year

[b]Supply = Domestic Production + Imports – Export – Net Increase in the Ending Inventory

FARM HOUSEHOLD SURVEY

The purpose of this survey is to understand the rural economy and to measure the various changes in the agricultural structure with a view to improving farm management practices, studying the level and pattern of rural consumption, providing data needed for farm policy making, estimating agricultural income, and collecting other data which will furnish basic materials for research dealing with agriculture and its contribution to national income. The survey also provides data for estimating cost of production for important food crops such as rice, barley, naked barley, and wheat.

This survey was initially conducted by the Research Department of the Bank of Korea, beginning in 1952 with a sample of 500 grain farmers. Since July 1961, the survey has been carried out by the Bureau of Statistics, MAF. The sample size was expanded to 1,200 farm households spread over 80 enumeration districts throughout the country.

Beginning in 1974, the sample has been redesigned based on the results of the 1970 Agricultural Census, doubling the sample size to 160 enumeration districts and 2,500 farm households. Information on over 500 items is collected through the survey. The various survey items could be classified broadly into two categories—"dynamic" survey items such as farm revenue, farm expenditure, and so on, and "static" survey items such as farm assets, farm liabilities, and so on. The principal items under the two categories are listed below.

1) DYNAMIC SURVEY ITEMS

Farm revenue and expenses
Non-farm revenue and expenses
Living expenses and consumption of farm products for home use
Farm production and utilization of farmland
Inputs of farm labor, fertilizer, etc.
Changes in farm assets
Sales of farm products
Purchased physical amount of other inputs
Production costs of major crops

2) STATIC SURVEY ITEMS

Composition of farm population, degree of education, and type of employment
Inventories of assets
Inventories of farm products and inputs
Liabilities
Cash and quasi-cash of household

The survey design employed is a three-stage stratified sampling: si (city), ŭp (town), and myŏn (township) constitute the primary sampling unit; the enumeration districts established in the 1970 Population Census form the secondary sampling unit; a farm household is the tertiary sampling unit.

The data on survey items are collected completely by the cost-accounting method in which a whole-time enumerator is permanently stationed in the survey district for keeping a selected group of holdings under observation and is in a position to see the operations being carried out in these holdings and is in close personal contact with the farmer.

In spite of the fact that there are limitations to these sample data, this survey is the best source of data available for understanding the Korean farmers as economic entities and is frequently used for obtaining national estimates of certain economic aspects of Korean agriculture.

BASIC AGRICULTURAL STATISTICS

The main purpose of this survey is to determine the status of basic factors of agriculture, such as cultivated land, rural households, and population, so as to provide necessary statistical data to policy-makers, research institutes, and administrators in the field of agriculture. Data are compiled on the basis of reports submitted by agricultural officers through proper administrative channels.

Report numbers are set as follows in accordance with provisions governing control of administrative reports as embodied in Cabinet Decree No. 258, December 1961:

Report No.

MAF – 28	Report on number of farm households and population by nationality
MAF – 29	Report on number of farm households by type of farming
MAF – 30	Report on number of farm households by size of farms
MAF – 31	Report on rural population by age group
MAF – 32	Report on acreage of cultivated land

COST OF PRODUCTION OF MAJOR CROPS

Estimates of components of costs of major crops, such as rice, barley, and wheat area, are obtained for the country as a whole from a sub-sample of 2,500 farm households selected for the Farm Household Survey. The data are compiled every year and published in the *Report on the Results of Farm Household Economic Survey* and *Cost of Production Survey* Published by the Ministry of Agriculture and Fisheries.

RURAL PRICE, WAGE SURVEY

Price statistics of agricultural products received and paid by farmers and rural wages are compiled by the National Agricultural Cooperative Federation (NACF).

The purpose of the survey is to understand farmers' exchange situtaion, and measure fluctuation in price of farm products as well as in the purchase of farm supplies. The survey provides data to study the parity index in the rural areas for improving agricultural economic activities and government farm-policy decisions. The survey is classified into two major categories, one for prices received by farmers consisting of 74 articles of agricultural and livestock products, and another for prices paid by farmers consisting of 54 farm supply articles, 101 household goods, and 10 farm wages and charges. This survey is conducted once a month (around the 15th of each month).

The survey covers 69 kun in the rural areas which account for the major portion of the total amount of agricultural product transactions as well as of agricultural production of such items as rice, common barley, naked barley, soybeans, sweet potatoes, cattle, and pigs. One market in each selected kun is chosen in accordance with quantity of sale of agricultural products.

From the survey data, indexes of prices received and paid by farmers are also prepared monthly and annually (calendar year).

LIVESTOCK STATISTICS

Currently, the information on livestock statistics is collected through administrative channels on the basis of data obtained from each farm. The number and products of important livestock are collected and compiled annually. Livestock statistics also are collected during the agricultural census.

Based on the figures of i (villages), totals for myŏn (townships), kun (counties), provinces and the whole country, are built up. These statistics lack objectivity and, therefore, have their own limitations.

Recently, the responsibility for collecting and disseminating statistics pertaining to the number of livestock has been transferred from the Livestock Bureau to the Research and Statistics Bureau, MAF. There is now a plan to introduce a sample survey for estimating the number of important livestock such as cattle, pigs, chickens, and so on, and livestock products such as milk, poultry, and so on. The survey would also provide information on animal husbandry practices.

GRAIN CONSUMPTION SURVEY

The daily consumption data on cereals, pulses, and potatoes for both rural and urban areas are collected through a nationwide survey conducted by the Research and Statistics Bureau, MAF. A sample of 1,800 households (overall sampling fraction 1/3,000) has been selected through a three-stage stratified self-weighting sampling design with myŏn/rural tong, i/pan, and segment forming respectively the first, second, and third stage units. One segment is selected at random from each i/pan and all households falling within it are enumerated. The data are collected by the food-accounting technique, in which a whole-time enumerator is permanently stationed in the sampled i/pan for keeping a group of households in the selected segment under observation. One enumerator is responsible for about 30 households. The data are tabulated to obtain daily per household and per capita consumption by month and by grain and published annually.

AGRICULTURAL CENSUSES

THE 1960 CENSUS OF AGRICULTURE

The Republic of Korea conducted an agricultural census for the first time in February 1961. It was carried out on the basis of complete enumeration of all agricultural holdings.

The field enumeration of the census was carried out for a period of about ten days commencing from February 1, 1961 by a staff consisting of 4,528 supervisors and 25,682 enumerators under the close supervision of 40 Central and 43 Provincial Office personnel of the Ministry of Agriculture and Forestry. The total number of holdings enumerated was 2.3 million. The publication of census results was completed by December 1964.

Sampling method was used for obtaining the advance estimates of census results as also for the post-enumeration check survey. For advance estimates, the sampling design employed was a two-stage stratified sample with a uniform sampling fraction in each stratum. The sampling fraction was 6/1,000. The final stage of selection was that of enumeration districts (ED). A total of 404 EDs was selected. The census schedules for all agricultural holdings within the selected EDs were tabulated (this included schedules for both private and quasi-farm households). The total number of agricultural holdings included in the sample was 14,226. The report on advance estimates of the 1960 census of agriculture was published in June 1962.

The post-enumeration survey designed to test, among other things, the completeness of coverage, was conducted on March 1, 1961. The sample used for the survey was the same as that used for the advance estimates, except that, in the post-enumeration survey, a third stage of sampling within the selected EDs was introduced. Together with the post-enumeration survey, actual measurement of the fields was also done on an experimental basis. Field areas were calculated by triangulation and plane-table methods. The results showed that the plane-table method gave better results than the triangulation method.

THE 1970 CENSUS OF AGRICULTURE

The Republic of Korea conducted the second decennial census as of December 1970 within the framework of the 1970 World Census of Agriculture. The census of population conducted in October 1970 provided the necessary basis for preparation of enumeration districts (EDs) for the agricultural census. During December 1970, a fishery census, through separate questionnaires, was also conducted. The legal enforcement of the agricultural census was based on Statistical Laws as also on the Agricultural Census Regulation.

The second census, like the first one, was conducted on the basis of complete enumeration of all farms including cooperatives, schools, religious institutions, and so on, having agricultural activities.

APPENDIX B

Agricultural Output and Input Data

The main sources for the data in this appendix are:

1) *Yearbook of Agriculture and Forestry Statistics* (hereafter *YBAFS*), 1953, 1956, 1958, 1965 and subsequent editions published by the Ministry of Agriculture and Forestry (now Ministry of Agriculture and Fisheries).

2) Ministry of Agriculture and Forestry (Fisheries), *Report on the Results of Farm Household Economic Survey* and *Cost of Production Survey,* 1962 through 1975.

3) National Agricultural Cooperative Federation, *Agricultural Yearbook,* various years.

4) The Korean Agriculture Bank, *Agricultural Yearbook 1960.*

5) National Agricultural Cooperative Federation, *Summary of Rural Prices 1974.*

6) Sŏng-hwan Pan, *Han'guk nongŏp ŭi sŏngjang, 1918–1971,* Korea Development Institute, 1974.

OUTPUT

Statistics on output in this study are based on data in *YBAFS*. However, a major change in data collection procedure occurred with the shift to a sample survey method from an administrative reporting system for rice output in 1965 and for barley, wheat, and potatoes in 1966. The sample survey resulted in much higher output estimates than the output estimate based on the administrative reporting system. The percentage increase in output reported after the shift to the sample survey as compared to the output estimates reported when the data were based on the administrative reports for the above crops are as follows:

Crops	Percent
Rice	32.27
Barley	94.66
Naked barley	32.37
Wheat	126.34
Rye	109.86
Sweet potatoes	77.98
White potatoes	33.29

Accordingly, the output series in the earlier years should be corrected, based on the above ratios, so that output series are consistent. Official revisions of the output levels were made back to 1955 and reported in the 1966 yearbook for rice and in the 1967 yearbook for other crops. The adjustments in the output data were extended, in this study, to the period from 1945 to 1954.

Output of grain is reported in sŏk, a unit of volume prior to 1955. The statistics on output in volume were converted into metric tons for the period from 1945 to 1954. The conversion rates of sŏk per metric ton are as follows:

Crops	Conversion Rate (sŏk per M/T)
Rice	6.9444
Barley	9.09
Naked barley	7.09
Wheat	8.85
Rye	10.70
Millet	7.04
Barnyard millet	7.246

Crops	Conversion Rate
Glutinous millet	6.944
Sorghum	7.09
Corn	7.407
Buckwheat	8.33
Soybeans	7.40
Red beans	6.70
Green beans	6.70
Kidney beans	6.90
Peas	6.90
Peanuts	7.60
Other beans	6.95

The quantity of fruits and vegetables are converted from kwan (= 3.75 kg.) to metric tons.

ESTIMATION OF VALUE
OF AGRICULTURAL PRODUCTION

VALUE OF CROP PRODUCTION

Value of production of each crop was obtained by multiplying the quantity of each crop produced in each year by the prices received by the farmers in 1970. Values of crops for which price information in 1970 is not available were estimated by multiplying estimated price to quantity produced. The unit prices were estimated by dividing the value of the commodity in question in 1970 by the quantity produced in that year. Information on the value of the commodities was obtainable from *YBAFS*.

Sources of farm price information are: 1) National Agricultural Cooperative Federation, *Summary of Rural Prices 1974,* and 2) Ministry of Agriculture and Forestry *YBAFS 1971* and *1972,* and subsequent editions.

VALUE OF MONOPOLY CROPS PRODUCTION

Monopoly crops consist of tobacco and Korean ginseng. Value products of tobacco were estimated by multiplying the 1970 constant price by quantity produced in each year. The value of ginseng production was first estimated at 1965 constant prices and then converted to 1970 constant prices by multiplying by a factor of 4.838, which is the ratio of the price in 1970 over that of 1965. For 1961 through 1974, the value of ginseng in

YBAFS was deflated by the price index. For 1945 through 1960, index of quantity of ginseng manufactured (base of 1961) was multiplied by the value in 1961.

VALUE OF LIVESTOCK PRODUCTION

The value of various elements of livestock production except Korean native cattle for various periods was estimated as follows:

1961-1974: Value of livestock production at current prices reported in MAF's *YBAFS* was deflated by price indexes (1965 = 100) for individual kinds of livestock, except dairy cattle and sheep, for which price indexes are not available. The price index for all livestock (1965 = 100) was applied to dairy cattle and sheep.

1945-1960: Value of livestock production for this period is not available. Therefore indexes of each kind of livestock, with numbers of each livestock in 1961 as base, were constructed. The value of individual livestock was estimated by multiplying this index by the value of individual livestock at the 1965 price.

The estimated value of each kind of livestock was multiplied by the following factors to obtain value product at 1970 constant prices. The factors are a ratio of the prices in 1970 over those in 1965.

Livestock	Factors
Dairy Cattle	1.791
Pigs	1.822
Goats	3.045
Sheep	2.127
Rabbits	1.598
Chickens	1.583
Ducks	1.591
Horses	1.262

The value of product of Korean native cattle was estimated differently from other livestock. It was estimated as a sum of newly born cattle and increase of value of existing cattle. In the estimation, a consideration has been given to both sex and age groups. (For details see Sŏng-hwan Pan, *Han'guk nongŏp ŭi sŏngjang, 1918-1971,* 1974, pp. 121, 122).

VALUES OF LIVESTOCK PRODUCTS
AND SILKWORM COCOON

The values of livestock products for various periods were estimated by the following methods:

1961–1974: Values of livestock products at current prices reported in *YBAFS* were deflated by the price indexes (1965 = 100) of individual livestock products.

1955–1960: Values were estimated based on the production index of each product with base of the three-year average of 1964–1966 (that is, 1964–1966 = 100). However, the estimated value of honey with the use of the production index in 1961 was 18.6 percent less than the reported value. Therefore the production index of honey received upward adjustment by 18.6 percent from 1945 through 1960. The production index of livestock products were multiplied by the value of each kind of livestock in 1965.

1945–1954: First, indexes of numbers of each livestock, taking 1955 as base year, were constructed. Second, this index was multiplied by the production index of 1955 with base year of 1964–1966 to estimate production index with base year of 1964–1966 (that is, 1964–1966 = 100). Value products of each commodity were obtained by multiplying these indexes by the value in 1965.

The value products at 1965 prices were converted to the value products at 1970 prices by multiplying the rate of price changes between two years. The rates are as follows:

Commodities	*Rates*
Eggs	1.368
Eggs of ducks	1.269
Milk	2.127
Honey	1.444
Goat's milk	2.127

The value of silkworm cocoon was estimated by multiplying 1970 prices of medium grades of spring and fall cocoon to the quantity produced in each year.

ESTIMATION OF INPUTS

Inputs consist of land, labor, fixed capital, and current inputs.

1) Land: Table B.2a shows the area of cultivated land of various categories and crop area. Sources are: *YBAFS 1953, 1956, 1957, 1962, 1967, 1969,* and *1975.* Areas reported in chŏngbo were converted to hectares by multiplying by a factor of 0.99174.

2) Labor: Labor input is measured in a flow concept in terms of male

equivalent units. Data are from Sung Hwan Ban's *Growth of Korean Agriculture*, pp. 246–247. (This long English summary to Song-hwan Pan's *Han'guk nongŏp ŭi sŏngjang, 1918–1971* will hereafter be referred to as *GKA*). The same measurement technique was extended to the period from 1972 to 1974. Labor input in this study is labor actually used for agricultural production.

3) Current Inputs: Current inputs include expenses of fertilizer consumption, other chemicals to control insects and disease, purchase of minor farming tools and materials, purchased seed, and feed.

Expenditure on fertilizer consumption was estimated by multiplying the quantity consumed in element by the prices of each element. The prices of fertilizer in terms of elements were estimated for major brands of fertilizer for both the years 1965 and 1970.

Expenditure on other chemicals was obtained by deflating previous estimates (Ban, *GKA*) with a price index of pesticides (1970 = 100) for the period from 1955 to 1971. For the periods 1945–1954 and 1972–1974 a quantity index of pesticide consumption (1955 = 100) was multiplied by the expenditure in 1955.

Expenditure on minor farming tools was estimated based on the data from *GKA*, that is, expenditure in each year at 1965 prices is multiplied by a factor 1.6835, which is the ratio of prices of farm implements in 1970 to that in 1965 for the period from 1945 through 1971. For the period from 1972 to 1974, expenditure was first estimated at current prices by multiplying average farm expenditure on minor farming tools by the number of farm households in each year. These figures were deflated by the price index of farm implements (1970 = 100) to obtain expenditures at 1970 constant prices.

Material expenditure in 1970 prices was estimated by multiplying a factor 1.9493 to the expenditure series at 1965 constant prices in *GKA* for 1945 through 1971. The factor is the price index in 1970 (1965 = 100). For the period from 1972 to 1974 the expenditure at current prices on minor farming tools above was deflated by the price index of farm supplies. However, figures in 1955, 1957, 1960, 1962, and 1963 in the original series at 1965 constant prices were adjusted because of abnormally high or low values. The average figure of 1954 and 1956 was used for the 1955 value; the average figure of 1956 and 1958 for the 1957 value; the average figure of 1959 and 1961 for the 1960 value; and the average figure of 1961 and 1964 for the values of 1962 and 1963 were applied respectively.

Expenditure on purchased seed was estimated for radish, Chinese cabbage, cabbage, watermelon, cucumber, sweet melon, and onion. It is assumed that farmers purchase seed for the above vegetables. Total seed requirements for each vegetable were estimated by multiplying quantity required per hectare to crop area. The resulting quantity is multiplied by 1970 prices to obtain expenditure on purchased seed. The seed requirement per hectare and prices are as follows:

Commodity	Seed per Hectare, in Liters	1970 Price per Liter, in Wŏn
Radish	12	1,500
Chinese cabbage	5	2,389
Cabbage	2	2,389
Watermelon	2.65	7,189
Cucumber	2.85	1,172
Sweet melon	1	6,478
Onion	7	5,111

Expenditure for purchased feed at 1970 constant prices for 1955–1971 was estimated by deflating data series at current prices in *GKA* by the price index of feed. For 1972 through 1974, expenditure was estimated by multiplying a factor of 22,451.89 by the quantity consumed. The factor is the price of feed per metric ton at 1970 prices, obtained by dividing the total expenditure of feed by quantity consumed in 1970. For 1945 through 1954, total expenditure for feed consumption was estimated by multiplying a factor 0.3038 by the value of livestock and its products. The factor is an average ratio of value of feed consumption to value of livestock and its products between 1955 and 1964. Expenditure on purchased feed for this period was calculated by multiplying the 1970 constant prices of grain by-products by the quantity of by-products of imported grains. The remainder was assumed to be supplied by domestic farmers.

FIXED CAPITAL

Fixed capital consisted of depreciation charges on perennial trees, farm machinery and equipment, and farm buildings, irrigation fees, and cattle service prices. For 1945 through 1971 time series data in *GKA* at 1965 constant prices (p. 246) are converted to 1970 constant prices by multiplying the following factors to values in 1965:

Items	Factors
Perennial trees	2.1645
Cattle service prices	2.492

Items	Factors
Machinery & equipment	1.6835
Buildings	1.783
Irrigation fees	1.949

The factors are ratios of the price indexes of respective items in 1970 (1965 = 100). Deflators and sources are: price index paid by farmers for farm implements and machinery in *YBAFS* and *Summary of Rural Prices* for farm machinery and equipment, price index of fruits for depreciation on perennial trees from same data source as above, price index paid by farmers for farming supplies and materials from same source as above, residential house deflator in National Income Statistics for the depreciation of buildings, and index of rental fees of draft cattle in *YBAFS* for cattle service prices. The same estimating procedures as in *GKA* were followed for the same period from 1972 through 1974.

FACTOR SHARES

Factor shares for 1945 through 1971 are taken from *GKA* with a slight adjustment. Average factor shares for 1951 and 1953 are adapted as factor shares in 1952. Average factor share of fixed capital in 1954 and 1956 was taken as the factor share of capital in 1955. The difference caused by this adjustment was allotted to factor shares of land and labor in proportion to respective factor shares prior to the adjustment. The average factor share of current inputs in 1955 and 1958 was applied to factor shares in 1955 and 1957. The difference caused by the adjustment was allotted to factor shares of land and labor as in fixed capital above.

The ratios of values of current inputs and fixed capital to total output at 1970 constant prices are calculated to estimate factor shares of these inputs from 1972 through 1974. These ratios are multiplied by 1.7999 for factor shares of current inputs and 2.0044 for fixed capital. The multipliers are ratios of factor shares estimated as a ratio of current inputs or fixed capital to total output. The residual was allocated to factor shares of land and labor in the proportion of 0.436 for land and 0.344 for labor.

TABLE B.1 Value of Agricultural Production by Group, 1945–1974
(1970 constant prices Sample Survey)
(million wŏn)

Year	Rice	Barley & wheat	Miscel- laneous Grains	Pulses	Potatoes	Total Food-Grains, Pulses and Potatoes
1945	186,802	23,986	3,061	10,737	15,484	239,872
1946	175,184	39,220	3,175	11,540	11,497	240,617
1947	201,341	34,820	2,392	10,694	12,103	261,352
1948	225,125	36,743	3,190	12,065	13,237	290,363
1949	214,193	47,974	4,467	16,131	14,524	297,291
1950	212,356	49,515	2,881	11,227	12,619	288,600
1951	164,991	28,776	4,048	10,239	11,087	219,143
1952	134,961	40,543	6,627	11,108	8,064	201,303
1953	205,503	45,989	3,492	12,707	19,003	286,695
1954	218,104	59,197	3,478	14,155	15,274	310,210
1955	225,848	49,635	3,733	13,286	16,695	309,199
1956	186,051	52,304	3,325	13,609	16,948	272,238
1957	229,118	46,383	3,307	13,605	16,649	309,064
1958	241,258	56,041	4,453	13,799	17,160	332,712
1959	240,388	64,743	3,490	12,489	16,665	337,778
1960	232,527	64,983	3,261	11,923	18,172	330,867
1961	264,278	70,084	3,903	15,053	21,169	374,490
1962	230,113	65,364	3,991	14,376	23,480	337,326
1963	286,832	45,419	4,335	14,444	26,852	377,885
1964	301,826	72,226	4,878	15,129	48,012	442,073
1965	267,223	83,412	4,560	16,162	53,240	424,600
1966	299,139	92,448	4,086	15,700	50,469	461,844
1967	275,006	87,866	3,984	18,866	33,560	419,284
1968	243,883	95,112	6,019	23,132	39,929	408,077
1969	312,203	95,379	4,950	22,068	40,722	475,323
1970	300,664	91,053	4,320	22,407	41,012	459,458
1971	305,119	85,105	3,749	21,239	37,312	452,526
1972	302,032	86,832	3,248	20,992	34,870	447,977
1973	321,452	76,777	3,562	22,567	32,029	456,389
1974	339,253	73,093	3,300	29,336	28,767	473,752

TABLE B.1 (continued)

Year	Fruit	Vege-tables	Special Crops	Leaf-Tobacco & Ginseng	Sub-total	By-products of Crops[a]	Total
1945	2,263	25,284	17,068	7,956	292,445	16,669	309,115
1946	3,289	29,294	15,252	9,373	297,828	16,976	314,805
1947	2,256	32,943	10,121	11,072	317,751	18,111	335,863
1948	4,538	50,068	14,349	12,105	371,425	21,171	392,596
1949	4,986	50,960	20,728	16,668	390,655	22,267	412,922
1950	3,640	35,817	15,791	7,220	351,070	20,011	371,081
1951	3,701	69,207	19,008	2,841	314,902	17,949	332,851
1952	5,718	53,809	12,870	4,155	277,857	15,837	293,695
1953	5,852	55,135	12,663	3,544	364,039	20,750	384,790
1954	5,878	52,135	12,975	6,964	388,164	22,125	410,290
1955	6,311	49,299	15,539	6,203	386,553	22,033	408,588
1956	6,312	47,380	13,507	6,256	345,696	19,704	365,400
1957	6,834	54,027	8,340	7,131	385,398	21,967	407,365
1958	8,207	56,213	9,299	7,321	413,755	23,584	437,339
1959	8,927	47,929	10,242	7,612	412,490	23,511	436,002
1960	8,955	46,556	7,980	7,599	401,958	22,911	424,870
1961	8,071	62,324	9,615	9,326	463,828	26,438	490,266
1962	10,638	61,436	7,637	10,797	427,836	24,386	452,223
1963	9,580	56,678	5,797	9,971	459,913	26,215	486,128
1964	12,318	73,140	6,642	14,924	549,099	31,298	580,398
1965	16,444	76,482	6,720	16,099	540,347	30,799	571,147
1966	17,628	97,784	9,211	20,464	606,933	34,595	641,528
1967	18,987	102,256	10,659	18,425	569,613	32,467	602,081
1968	21,404	122,485	10,228	21,246	583,442	33,256	616,698
1969	22,802	114,573	11,752	18,535	642,987	36,650	679,637
1970	23,437	110,186	11,240	18,604	622,927	35,506	658,433
1971	22,839	136,518	11,759	22,257	645,902	36,816	682,718
1972	28,817	131,379	10,646	37,695	656,515	37,421	693,937
1973	30,749	132,165	11,445	35,853	666,603	37,996	704,600
1974	38,487	137,891	12,021	36,550	698,704	39,826	738,530

Note: [a]Value of by-product of crops was considered to be 5.7 percent of value of all crops before adding by-product. This figure was obtained from *YBAFS*.

TABLE B.1 (continued)

Year	Crops Total	Livestock Production			Silkworm Cocoon	Agricultural Production Total
		Livestock	Livestock Products	Sub-total		
1945	309,115	19,669	884	20,554	2,608	332,277
1946	314,805	22,025	994	23,020	2,816	340,642
1947	335,863	26,169	1,128	27,297	3,045	366,207
1948	392,596	29,708	1,269	30,978	3,096	426,670
1949	412,922	28,145	1,356	29,502	3,435	445,860
1950	371,081	20,375	492	20,868	2,621	394,570
1951	332,851	16,899	764	17,664	2,201	352,716
1952	293,695	24,963	1,048	26,011	2,827	322,535
1953	384,790	27,836	1,172	29,008	2,820	416,619
1954	410,290	40,600	2,677	43,278	2,759	456,328
1955	408,586	53,467	4,298	57,766	3,146	469,499
1956	365,400	53,984	7,158	61,142	2,855	429,399
1957	407,365	56,485	7,396	63,881	2,770	474,018
1958	437,339	59,598	7,710	67,308	2,724	507,372
1959	436,002	65,578	8,492	74,070	2,637	512,710
1960	424,870	65,554	10,353	75,907	2,215	502,993
1961	490,266	60,893	10,368	71,262	2,356	563,885
1962	452,223	57,192	11,298	68,490	2,653	523,367
1963	486,128	69,093	14,016	83,110	2,966	672,205
1964	580,396	70,981	13,353	84,334	2,811	667,544
1965	571,147	70,822	14,773	85,596	3,735	660,479
1966	641,528	72,622	17,119	89,741	4,617	735,887
1967	602,081	77,988	21,592	99,581	5,241	709,904
1968	616,698	72,179	21,516	93,696	7,980	718,374
1969	679,637	91,855	31,556	123,412	9,964	813,014
1970	658,433	85,860	33,814	119,675	10,290	788,400
1971	682,718	81,956	29,761	111,718	11,861	806,299
1972	693,937	82,090	38,881	120,971	12,873	827,782
1973	704,600	92,349	36,045	128,395	14,882	847,877
1974	738,530	98,039	42,319	140,358	17,858	896,747

TABLE B.2a Area of Cultivated Land[a], 1945–1974
(1,000 hectares)

Year	Total	Area of Cultivated Land (paddy field)			Of which Rearranged Paddy Field
		Sub-total	One-Crop	Two-Crop	
1945	2,067	–	–	–	–
1946	1,901	–	–	–	–
1947	1,854	–	–	–	–
1948	2,028	–	–	–	–
1949	2,053	1,226	–	–	–
1950	1,954	1,148	–	–	–
1951	1,942	1,149	818	331	–
1952	1,942	1,153	818	334	–
1953	1,939	1,152	804	348	–
1954	1,950	1,160	809	351	
1955	1,994	1,187	774	412	–
1956	1,991	1,188	797	391	–
1957	1,998	1,192	807	385	–
1958	2,012	1,199	800	398	–
1959	2,016	1,202	805	397	–
1960	2,024	1,206	807	399	–
1961	2,032	1,210	767	443	–
1962	2,062	1,223	757	465	–
1963	2,079	1,228	744	484	–
1964	2,171	1,261	738	522	–
1965	2,256	1,286	704	581	–
1966	2,293	1,287	696	590	–
1967	2,311	1,290	678	612	–
1968	2,318	1,289	653	636	88
1969	2,311	1,283	641	641	100
	(2,118)	(1,194)	(841)	(353)	(123)
1970	2,297	1,272	634	638	111
1971	2,271	1,264	635	628	127
1972	2,242	1,259	642	616	150
1973	2,241	1,262	659	603	173
1974	2,238	1,268	541	727	235

Note: [a]Chŏngbo are converted to hectares (1 chŏngbo = 0.99174 hectare)

TABLE B.2b Area of Upland, Crop Area, and Labor Input
(Area in 1,000 hectares)
(Labor input in 1,000 man-equivalent unit)

| Year | Area of Cultivated Land (Upland) | | | Crop Area | Labor Actually Used |
	Sub-total	Of which Orchard Field	Of which Mulberry Field		
1945	–	14	42	2,932	1,691
1946	–	15	34	2,714	1,588
1947	–	14	28	2,765	1,627
1948	–	18	29	2,848	1,656
1949	827	20	30	2,859	1,676
1950	805	20	30	2,779	1,646
1951	792	19	30	2,560	1,502
1952	789	18	28	2,667	1,497
1953	787	18	29	2,805	1,656
1954	789	18	29	2,855	1,678
1955	807	19	32	2,920	1,629
1956	803	20	34	2,977	1,719
1957	805	20	35	3,037	1,831
1958	812	22	36	2,956	1,928
1959	813	23	35	2,967	2,154
1960	818	22	36	2,966	2,234
1961	821	22	23	3,061	2,349
1962	839	22	27	3,104	2,095
1963	851	23	30	3,154	2,239
1964	909	28	41	3,341	2,315
1965	970	42	50	3,560	2,335
1966	1,005	44	61	3,456	2,341
1967	1,021	47	67	3,515	2,328
1968	1,029	50	93	3,526	2,168
1969	1,028	55	98	3,549	2,096
	(923)	(–)	(–)		
1970	1,024	59	84	3,479	2,010
1971	1,006	54	80	3,301	2,031
1972	982	58	77	3,268	1,934
1973	978	65	79	3,213	1,916
1974	969	74	87	3,291	1,524

Output and Input Data

TABLE B.3 Current Inputs, 1945–1974
(1970 constant prices, 1,000,000 wŏn)

Year	Fertilizer	Other Chemicals	Material Expenditure	Farming tools	Purchased Seed	Purchased Feed	Total
1945	121	73	2,245	1,140	869	482	4,933
1946	2,623	117	2,289	1,161	873	539	7,605
1947	6,505	11	2,450	1,243	1,085	840	12,136
1948	6,788	103	2,812	1,427	1,072	634	12,818
1949	11,081	60	2,928	1,486	1,155	75	16,786
1950	1,011	82	2,727	1,384	825	44	6,074
1951	4,134	42	2,307	1,171	1,008	301	8,963
1952	8,464	44	2,094	1,611	1,063	698	13,976
1953	6,840	42	2,821	1,649	997	1,574	13,926
1954	9,758	92	3,007	1,622	1,114	361	15,956
1955	11,038	222	3,179	902	1,133	268	16,745
1956	12,896	328	3,351	1,746	1,152	1,529	21,005
1957	12,504	416	3,108	1,837	1,235	2,474	21,576
1958	14,218	419	2,864	1,943	1,100	2,681	23,227
1959	13,192	414	2,702	1,461	1,128	605	19,504
1960	16,774	544	2,820	1,618	1,178	1,337	24,274
1961	17,676	333	2,939	1,783	1,200	1,827	25,762
1962	17,383	624	2,404	1,436	1,196	1,666	24,711
1963	17,090	1,050	2,404	2,206	1,156	4,246	28,155
1964	18,757	2,004	1,870	2,286	1,297	2,799	29,015
1965	20,793	1,336	2,023	1,852	1,467	1,815	29,289
1966	22,456	1,515	2,123	1,749	1,470	1,428	30,744
1967	25,711	2,671	1,979	2,069	1,683	3,876	37,791
1968	25,660	3,420	2,617	2,096	1,809	7,619	43,224
1969	28,620	4,259	3,620	1,616	2,150	9,561	49,829
1970	30,658	6,652	4,971	1,802	2,430	12,778	59,293
1971	32,065	6,948	5,469	2,203	2,372	16,785	65,845
1972	34,257	7,699	5,403	2,338	2,271	12,545	64,516
1973	40,496	8,441	7,420	2,086	2,317	13,847	74,610
1974	43,132	14,518	9,325	2,629	2,505	16,899	89,011

TABLE B.4 Fixed Capital, 1945–1974
(1970 constant prices, 1,000,000 wŏn)

Year	Perennial Trees	Cattle Service Prices	Machinery Equipment	Building	Irrigation	Fixed Capital Total
1945	531	14,092	383	1,271	3,902	20,181
1946	558	14,342	390	1,295	3,908	20,496
1947	526	15,066	418	1,387	3,914	21,313
1948	677	15,945	480	1,592	3,920	22,615
1949	733	16,263	500	1,657	3,926	23,081
1950	734	9,757	465	1,544	3,636	16,136
1951	721	13,568	394	1,306	3,639	19.629
1952	671	15,440	542	1,797	3,653	22,105
1953	664	15,596	555	1,840	3,649	22,306
1954	682	17,609	545	1,809	3,681	24,328
1955	703	19,812	548	2,165	3,776	27,006
1956	707	19,442	587	1,955	2,740	25,433
1957	709	19,071	620	1,932	2,532	24,865
1958	767	18,700	652	2,483	2,535	25,140
1959	794	20,103	848	2,874	1,840	26,463
1960	784	18,883	864	3,438	2,539	26,511
1961	698	16,472	725	2,772	3,494	24,183
1962	704	15,606	1,072	2,664	4,916	24,965
1963	700	15,869	1,152	3,549	4,598	25,871
1964	804	14,984	1,478	3,243	5,792	26,304
1965	1,134	16,866	1,671	2,202	5,281	26,974
1966	1,185	16,636	1,810	2,331	6,489	28,453
1967	1,233	16,726	1,887	2,589	6,646	29,983
1968	1,341	15,726	2,405	2,980	6,286	28,740
1969	1,381	14,269	2,546	3,350	7,362	28,910
1970	1,434	13,960	2,437	3,057	7,465	28,354
1971	1,497	13,925	2,867	3,065	7,257	28,613
1972	1,659	13,123	3,571	3,619	7,327	29,301
1973	1,784	12,130	4,441	5,020	7,332	30,707
1974	2,003	10,081	5,182	4,786	7,465	29,520

TABLE B.5 Gross Agricultural Production, Output, and Value Added, and Farm-
Supplied and Purchased Intermediate Goods
(1970 constant prices, 1,000,000 wŏn)

Year	Gross Agricultural Production	Farm Supplied Seed & Feed	Gross Agricultural Output	Total Current Inputs	Gross Value Added
1945	332,277	13,902	318,375	4,933	313,442
1946	340,642	13,919	326,722	7,605	319,116
1947	366,207	15,110	351,096	12,136	338,959
1948	426,670	16,756	409,914	12,818	397,095
1949	445,860	17,012	428,847	16,786	412,060
1950	394,570	13,982	380,588	6,074	374,513
1951	352,716	12,070	340,646	8,963	331,682
1952	322,535	14,147	308,387	13,976	294,410
1953	416,619	15,309	401,309	13,926	387,383
1954	456,328	20,885	435,442	15,956	419,486
1955	469,499	23,230	446,268	16,745	429,523
1956	429,399	27,639	401,759	21,005	380,753
1957	474,018	27,077	446,941	21,576	425,364
1958	507,372	27,603	479,769	23,227	456,542
1959	512,710	30,905	481,804	19,504	462,300
1960	502,993	28,492	474,501	24,274	450,227
1961	563,885	29,441	534,443	25,762	508,681
1962	523,367	27,915	495,451	24,711	470,740
1963	572,205	27,220	544,985	28,155	516,830
1964	667,544	35,916	631,628	29,015	602,612
1965	660,479	36,580	623,898	29,289	594,809
1966	735,887	38,560	697,326	30,744	666,582
1967	706,904	39,447	667,456	37,791	629,665
1968	718,374	43,397	674,976	43,224	631,752
1969	813,014	51,092	761,921	49,829	712,092
1970	788,400	53,183	735,216	59,293	675,922
1971	806,299	57,270	749,028	65,845	683,183
1972	827,782	78,862	748,919	64,516	684,402
1973	847,877	81,124	768,753	74,610	692,143
1974	896,747	83,187	813,560	89,011	724,548

Appendix B

TABLE B.6 Indexes of Agricultural Outputs, Inputs, and Productivities, 1946–1973

(1946 = 100, 3-year moving average)

Year	Total Production (1)	Total Output (2)	Total Inputs (3)	Total Productivity in Terms of Total Production (4)=(1)/(3)	Total Productivity in Terms of Total Output (5)=(2)/(3)	Labor Productivity in Terms of Total Output (6)	Land Productivity in Terms of Total Output (7)
1946	100.0	100.0	100.0	100.0	100.0	100.0	100.0
1947	109.1	109.2	102.8	106.1	106.2	110.0	109.2
1948	119.2	119.4	108.7	109.7	109.8	118.1	118.0
1949	121.9	122.4	109.7	111.1	111.6	120.6	121.6
1950	114.8	115.5	109.4	104.9	105.6	117.4	115.5
1951	103.0	103.4	109.4	94.1	94.5	109.2	105.4
1952	105.1	105.4	113.6	92.5	92.8	111.1	107.7
1953	115.1	115.0	118.2	97.4	97.3	116.8	117.4
1954	129.8	128.8	121.5	106.3	106.0	127.3	130.2
1955	130.4	128.8	124.9	104.4	103.1	125.7	129.1
1956	132.1	130.0	128.1	103.1	101.5	123.1	129.2
1957	135.8	133.4	132.1	102.8	101.0	119.4	132.2
1958	143.8	141.4	136.0	105.7	104.0	117.3	139.6
1959	146.6	144.2	140.6	104.3	102.6	112.0	141.7
1960	152.0	149.6	144.9	104.9	103.2	108.9	146.5
1961	153.0	151.0	145.9	104.9	103.5	110.9	146.7
1962	159.7	158.1	147.6	108.2	107.1	116.0	152.3
1963	169.7	167.9	150.2	113.0	111.8	123.9	158.1
1964	182.9	180.7	155.7	117.5	116.1	128.7	165.2
1965	198.6	196.0	159.6	124.4	122.8	137.5	173.5
1966	202.4	199.6	163.2	124.0	122.3	139.8	173.1
1967	208.0	204.8	165.1	126.0	124.0	146.9	176.0
1968	215.4	211.2	166.5	129.4	126.8	157.1	181.0
1969	223.2	218.0	166.8	133.8	130.7	170.5	187.5
1970	231.7	225.5	168.6	137.4	133.7	180.3	195.2
1971	233.1	224.2	168.6	138.3	133.0	184.1	196.0
1972	238.9	227.3	169.7	140.8	133.9	189.6	200.3
1973	247.6	233.8	168.1	147.3	139.1	213.5	206.9

TABLE B.7 Factor Shares, 1945–1974

Year	Land	Labor	Fixed Capital	Current Inputs
1945	46.90	37.00	10.32	5.78
1946	46.69	36.84	10.27	6.20
1947	44.28	34.93	9.97	10.81
1948	45.56	35.95	9.23	9.27
1949	44.46	35.08	9.05	11.41
1950	50.26	39.67	6.98	3.08
1951	46.49	36.68	9.80	7.03
1952	46.03	36.36	9.44	8.17
1953	45.59	35.97	9.10	9.33
1954	45.50	35.90	9.25	9.35
1955	44.67	35.25	10.85	9.24
1956	43.07	33.98	12.44	10.51
1957	43.31	34.18	12.00	10.51
1958	43.00	33.93	11.29	11.78
1959	44.70	35.27	10.87	9.17
1960	44.51	35.12	11.00	9.38
1961	45.08	35.57	8.83	10.52
1962	43.94	34.67	9.01	12.38
1963	43.10	34.00	8.39	14.51
1964	44.10	34.80	8.01	13.09
1965	43.92	34.05	7.11	14.31
1966	46.57	36.74	5.94	10.75
1967	46.39	36.60	6.41	10.60
1968	45.12	35.60	6.88	12.39
1969	44.18	34.86	7.53	13.44
1970	43.83	34.58	7.67	13.91
1971	43.70	34.48	7.30	14.53
1972	42.84	33.81	7.84	15.51
1973	41.62	32.84	8.03	17.51
1974	40.83	32.21	7.27	19.69

TABLE B.8 Value of Estimated Farm-Supplied Seed and Feed, 1945–1974
(1970 constant prices, 1,000s wǒn)

Year	Seed	Feed	Total
1945	8,139,788	5,762,260	13,902,048
1946	7,466,017	6,453,924	13,919,941
1947	7,658,351	7,452,380	15,110,731
1948	7,979,393	8,777,117	16,756,510
1949	8,125,174	8,887,777	17,012,951
1950	7,686,710	6,295,699	13,982,409
1951	7,005,321	5,065,263	12,070,584
1952	6,943,509	7,203,705	14,147,214
1953	8,070,698	7,238,373	15,309,071
1954	8,098,677	12,786,954	20,885,631
1955	8,234,246	14,996,739	23,230,985
1956	8,386,343	19,253,226	27,639,569
1957	8,689,221	18,388,208	27,077,429
1958	8,527,172	19,076,496	27,605,670
1959	8,526,307	22,379,593	30,905,900
1960	8,829,003	19,663,041	28,492,044
1961	9,055,883	20,385,699	29,441,582
1962	9,452,212	18,462,948	27,915,160
1963	9,795,214	17,425,237	27,220,451
1964	10,862,210	25,054,253	35,916,463
1965	11,832,737	24,747,733	36,580,470
1966	11,690,260	26,870,088	38,560,348
1967	11,912,513	27,535,432	39,447,945
1968	11,800,074	31,597,525	43,397,599
1969	11,866,656	39,226,114	51,092,770
1970	11,651,059	41,532,239	53,183,298
1971	11,021,293	46,249,481	57,270,774
1972	10,881,168	67,981,473	78,862,641
1973	10,696,738	70,427,819	81,124,557
1974	10,650,343	72,537,512	83,187,855

TABLE B.9 Value of Estimated Purchased Seed, 1945–1974 (1970 constant prices, 1,000s wŏn)

Year	Radish	Chinese Cabbage	Cabbage	Watermelon	Cucumber	Sweet Melon	Onion	Total
Price per liter	1,500	2,389	2,389	7,189	1,172	6,478	5,111	
1945	515,574	282,881	1,797	29,511	12,880	21,773	5,188	869,604
1946	505,836	291,159	1,390	31,718	14,162	26,566	3,041	873,872
1947	606,366	393,301	1,362	35,377	17,275	29,352	2,612	1,085,647
1948	596,196	381,559	1,840	31,761	16,988	29,384	14,955	1,072,653
1949	624,924	432,397	1,988	32,214	17,873	28,127	17,495	1,155,018
1950	440,136	316,650	1,390	22,538	12,512	19,687	12,236	825,149
1951	612,054	300,783	2,852	14,191	7,305	23,528	40,249	1,006,962
1952	631,602	334,484	2,958	22,653	5,514	27,784	38,925	1,063,920
1953	542,340	341,627	3,703	21,071	12,225	26,501	49,980	997,447
1954	614,700	379,194	4,028	21,315	10,261	25,588	58,960	1,114,048
1955	614,502	390,351	4,444	19,626	11,210	19,965	73,629	1,133,727
1956	615,186	388,547	4,315	20,402	10,084	19,609	94,237	1,152,380
1957	656,424	433,341	5,528	20,273	11,044	18,676	90,909	1,235,195
1958	589,590	365,027	5,342	25,090	11,123	22,705	81,178	1,100,055
1959	597,186	371,860	6,092	27,146	10,201	20,768	94,809	1,128,062
1960	615,852	378,477	6,063	48,598	10,568	24,934	94,022	1,178,512
1961	639,126	389,299	4,988	44,809	9,299	26,534	86,580	1,200,633
1962	644,346	402,618	4,324	46,506	10,672	28,464	59,569	1,196,499
1963	628,272	391,270	4,510	50,977	11,220	30,382	40,249	1,156,880
1964	686,448	416,940	5,055	73,076	13,401	37,708	65,150	1,297,779
1965	748,422	466,321	7,248	93,428	15,499	39,140	97,349	1,467,407
1966	759,924	467,432	7,802	76,908	15,977	40,060	102,608	1,470,711
1967	803,370	589,056	7,243	93,594	15,886	43,066	131,552	1,683,787
1968	888,120	605,719	8,653	110,632	19,023	43,001	134,629	1,809,777
1969	979,902	803,397	11,912	122,364	23,161	48,870	160,997	2,150,603
1970	1,195,686	852,049	13,627	140,006	27,603	57,874	143,323	2,430,168
1971	1,128,402	888,278	9,327	139,718	25,750	52,206	128,654	2,372,335
1972	1,101,888	764,205	8,854	139,912	26,604	48,404	181,783	2,271,650
1973	1,121,004	801,545	8,854	152,924	26,682	52,368	153,698	2,317,075
1974	1,211,544	818,555	9,675	166,806	30,038	62,364	206,219	2,505,201

TABLE B.10 Value of Chemical Fertilizer Consumption by Elements, 1945–1974
(1970 constant prices, 1,000s wŏn)

Year	Nitrogen	Phosphorus	Potash	Total
1945	81,706	31,411	8,561	121,678
1946	2,064,891	369,842	188,814	2,623,547
1947	5,573,587	904,749	26,970	6,505,306
1948	4,472,461	2,050,090	246,218	6,768,769
1949	8,729,276	1,657,266	695,031	11,081,573
1950	960,344	51,037	–	1,011,381
1951	3,315,878	818,736	–	4,134,614
1952	8,241,407	50,073	172,558	8,464,038
1953	5,963,304	811,104	65,792	6,840,200
1954	7,596,178	2,115,218	47,507	9,758,903
1955	9,636,070	1,183,378	218,901	11,038,349
1956	10,440,172	2,255,414	200,814	12,896,400
1957	9,469,171	2,873,523	161,992	12,504,686
1958	11,294,469	2,799,530	124,185	14,218,284
1959	10,643,254	2,400,516	148,879	13,192,649
1960	14,283,983	2,315,174	175,428	16,774,585
1961	13,872,096	3,388,006	416,647	17,676,749
1962	13,242,590	3,672,842	468,063	17,383,495
1963	12,613,084	3,957,637	519,479	17,090,200
1964	11,390,977	6,440,307	925,933	18,757,217
1965	14,336,414	5,178,758	1,278,817	20,793,989
1966	15,758,444	5,233,570	1,454,443	22,456,457
1967	18,259,299	5,565,963	1,885,738	25,711,000
1968	18,809,467	5,089,516	1,761,207	25,660,190
1969	21,066,190	5,483,221	2,071,410	28,620,821
1970	23,390,212	5,215,034	2,053,620	30,658,866
1971	22,848,662	6,920,863	2,295,878	32,065,403
1972	24,510,877	7,168,920	2,577,528	34,257,325
1973	27,053,571	9,736,765	3,706,402	40,496,738
1974	29,563,110	9,724,226	3,845,037	43,132,373
1970's prices[a]	₩65,786	₩41,937	₩24,743	

Note: [a]Prices per metric ton in element.

TABLE B.11 Pesticide Consumption, Quantity and Value, 1945–1974
(quantity in 1000 kg., value in 1,000,000s wŏn at 1970 constant prices)

Year	Fungicides	Insecticides	Herbicides	Other Pesticides	Total	Value
1945	–	–	–	–	–	73
1946	1,815	258	–	58	2,130	117
1947	22	174	–	4	201	11
1948	1,788	67	–	8	1,864	103
1949	786	277	–	22	1,086	60
1950	496	488	–	500	1,485	82
1951	381	385	–	3	770	42
1952	514	280	–	11	806	44
1953	334	380	–	62	776	42
1954	944	715	–	11	1,671	92
1955	2,006	1,924	1	87	4,020	222
1956	1,889	2,789	21	171	4,872	328
1957	3,755	2,971	33	3	6,764	416
1958	2,666	2,386	41	1	5,096	410
1959	2,544	3,003	17	1	5,567	414
1960	3,452	2,392	9	2	5,857	544
1961	3,421	2,120	9	4	5,556	333
1962	4,604	2,788	9	18	7,420	624
1963	15,099	3,587	52	33	18,772	1,050
1964	18,357	4,885	39	72	23,355	2,004
1965	9,432	3,185	25	84	12,728	1,336
1966	7,786	4,513	130	118	12,549	1,515
1967	1,934	7,675	274	104	9,988	2,671
1968	2,089	7,283	470	139	9,982	3,420
1969	8,517	7,512	1,225	274	17,530	4,259
1970	10,925	8,862	4,957	277	25,023	6,652
1971	6,916	13,936	8,280	826	29,960	6,948
1972	4,708	17,924	10,152	411	33,197	7,699
1973	6,345	19,716	9,666	669	36,397	8,441
1974	6,857	35,909	19,428	407	62,602	14,518

TABLE B.12 Expenditure on Minor Farming Tools and Farm Supplies
(1965 and 1970 constant prices, 1,000,000 wŏn)

Year	Farming Tools at 1965 prices	Farming Tools at 1970 prices	Material Expenditure 1965 prices	Material Expenditure 1970 prices
1945	677	1,140	1,152	2,246
1946	690	1,161	1,174	2,289
1947	738	1,243	1,257	2,450
1948	848	1,427	1,442	2,812
1949	882	1,486	1,502	2,928
1950	822	1,384	1,399	2,727
1951	695	1,171	1,183	2,307
1952	957	1,611	1,074	2,094
1953	980	1,649	1,447	2,821
1954	963	1,622	1,542	3,007
1955	536	902	1,631	3,179
1956	1,037	1,746	1,719	3,351
1957	1,091	1,837	1,594	3,108
1958	1,154	1,943	1,469	2,864
1959	868	1,461	1,386	2,702
1960	961	1,618	1,447	2,820
1961	1,059	1,783	1,507	2,939
1962	853	1,438	1,233	2,404
1963	1,310	2,206	1,233	2,404
1964	1,358	2,286	959	1,870
1965	1,100	1,852	1,037	2,023
1966	1,039	1,749	1,089	2,123
1967	1,229	2,069	1,015	1,979
1968	1,245	2,096	1,342	2,617
1969	960	1,616	1,857	3,620
1970	1,070	1,802	2,550	4,971
1971	1,309	2,203	2,806	5,469
1972	–	2,338	–	5,403
1973	–	2,086	–	7,420
1974	–	2,629	–	9,325

APPENDIX C

Revised Farm Household Income Estimates[1]

The Farm Household Surveys use a concept for household income that includes the change in value of the inventory of agricultural products and the change in value of animals and trees held by the household. There are several problems with these inventory figures. First, as is clear from Table C.1, the figures for the end of one year and the beginning of the next do not jibe as they should. Second, and more important, much of the increase in value of these inventories simply represents the rise in prices of grain, the main component of these inventories, between the beginning and end of the year. Hence, the more rapid the inflation in a given year, the higher the agricultural income. Such an income concept is clearly flawed, particularly when one realizes that much of this inventory (90 percent) is grain consumed by the farm household itself. Total elimination of the inventory component, however, has the disadvantage of reducing farm income below what was actually produced in years when the farmer built up his inventories in real terms (and vice versa when he ran down those inventories).

Thus the concept of inventories used in the revised farm household income estimates makes two corrections from the official survey estimates:

1) The beginning-of-the-year inventory figure has been changed to be the same as that for the end of the previous year.

2) The beginning-of-the-year figure has been inflated by the grain price index for the year (the grain price paid to farmers). Deflating the end-of-the-year figure would get similar but slightly lower results.

The revised inventory estimates appear in Tables C.2 and C.3. Lack of data made it impossible to make revisions prior to 1967. For these earlier years, inventories have been eliminated entirely.

The "increase in animals and trees" figure has been entirely eliminated from agricultural income. In most years, this figure is less than 5 percent of farm income in its unrevised form, and much of that 5 percent or less presumably represents price increases and other discrepancies. Lack of data available to the author made a more refined estimate impossible.

TABLE C.1 Agricultural Products Inventory, 1964–1975

	Beginning of the Year	*End of the Year*
1964	1,218 liters	1,854 liters
1965	1,528 "	1,800 "
1966	54,553 wŏn	64,609 wŏn
1967	64,466 "	74,022 "
1968	70,173 "	94,027 "
1969	89,226 "	110,407 "
1970	106,023 "	123,960 "
1971	127,014 "	187,705 "
1972	186,807 "	248,719 "
1973	234,304 "	293,433 "
1974	230,004 "	371,428 "
1975	361,241 "	472,136 "

Source: MAF, *Report on the Results of Farm Household Economic Survey,* various years.

TABLE C.2 Grain Prices and Inventories, 1967–1975

	Index of Grain Prices Paid to Farmers		*Revised Inventory Figures*	
	1970 = 100	*Previous Year = 100*	*Beginning of Year*	*End of Year (deflated)*
1967	64.5	112.2	72,472	74,022
1968	73.1	113.3	83,889	94,027
1969	88.9	121.6	114,346	110,407
1970	100.0	112.5	124,197	123,960
1971	124.9	124.9	154,826	187,705
1972	158.3	126.7	237,897	248,719
1973	169.0	106.8	265,532	293,433
1974	240.5	142.3	417,585	371,428
1975	303.1	126.0	468,111	472,136

Sources: MAF, *Yearbook of Agriculture and Forestry,* various years, and Table C.1.

TABLE C.3 Farm Household Income and Inventories, 1958–1975

	Farm Household Income *(Excluding Inventories)*	*Increase (+) or Decrease (–) of Inventory* *(End of Year Prices)*	*Farm Household Income* *(Including Revised Inventories)*
1958	42,910	–	42,910
1959	40,198	–	40,198
1960	44,750	–	44,750
1961	n.a.	–	–
1962	59,286	–	59,286
1963	82,799	–	82,799
1964	107,913	–	107,913
1965	109,839	–	109,839
1966	118,349	–	118,349
1967	137,168	1,550	138,718
1968	150,958	10,138	161,096
1969	190,791	–3,939	186,852
1970	230,407	–237	230,170
1971	284,357	32,879	317,236
1972	348,298	10,822	359,120
1973	398,855	27,901	426,756
1974	511,951	–46,157	465,794
1975	718,691	4,025	722,716

Source: Derived from the MAF Farm Household Surveys and Table C.2.

Notes

ONE   *Introduction*

1. Isabella Bird Bishop, *Korea and Her Neighbors,* (Seoul, 1970, originally published in 1898), p. 100.

2. The major sources of information for the period of Japanese control are the statistical handbooks and other publications of the Japanese Government General in Korea. Other major works that deal at length with agriculture in this period include Sang-Chul Suh, *Growth and Structural Changes in the Korean Economy, 1910–1940,* (Cambridge, Mass., 1978) and Andrew J. Grajdanzev, *Modern Korea* (New York, 1944).

3. Sang-Chul Suh, pp. 82–90.

4. This statement is based on the work of Roger Sedjo and is discussed at greater length in Chapter 4.

TWO   *Agriculture's Role in Korean Economic Development*

1. In addition to the formal labor surplus models of Fei and Ranis, the neoclassical versions of Jorgenson, Williamson, and others, there are a great number of studies dealing with the role of agriculture in economic development. See, for example, Bruce F. Johnston and Peter Kilby, *Agriculture and Structural Transformation* (London, 1975) and John W. Mellor, *The Economics of Agricultural Development* (Ithaca, 1966). For discussions that deal with the role of agriculture in Korean development, see Paul W. Kuznets, *Economic Growth and Structure in the Republic of Korea* (New Haven, 1977), Chapter 5, and Irma Adelman and Sherman Robinson, *Income Distribution Policy in Developing Countries: A Case Study of Korea* (Stanford, 1978).

2. This kind of counterfactual statement, of course, requires a number of assumptions other than those appearing in the text. In this case, one

must assume that the population that shifted into the urban areas would have remained on the farm but contributed nothing to farm income or food demand or alternatively emigrated abroad. Such assumptions are unrealistic, but more realistic assumptions would complicate the calculation and add little to clarity.

3. These estimates were obtained by applying the percentage shares of agricultural trade in Table 11 to total export figures in constant 1970 wŏn in BOK, *Economic Statistics Yearbook 1976*, pp. 296–297. This procedure introduces some error since the percentages were derived from dollar price figures, but the error should not be great.

THREE   *The Growth of Agricultural Output and Productivity*

1. Statistical series and the results of the analysis in this study differ somewhat from Sŏng-hwan Pan's previous studies, *Han'guk nongŏp ŭi sŏngjang 1955–1971*, KDI, 1974 and "Growth Rates of Korean Agriculture, 1918–1971," in *Agricultural Growth in Japan, Taiwan, Korea and the Philippines*, edited by Y. Hayami, V. Ruttan and H. Southworth (Hawaii University Press, forthcoming). The main difference has occurred due to the measurement procedures. In this study, time series data on outputs and inputs are constructed multiplying the quantity of individual product and factor by base year's price through the study period. However, in previous studies time series data were being constructed first at current price and then deflated by price received or paid by farms for a certain commodity group.

2. Gross agricultural output is estimated by subtracting the value of farm-supplied feed and seed from total agricultural production. Gross value added in agriculture is estimated by subtracting the value of current input from gross agricultural output.

3. Sŏng-hwan Pan, *Han'guk nongŏp ŭi sŏngjang*.

4. Han'guk Ŭnhaeng, *Chosŏn kyŏngje yŏnbo 1948* (Annual economic review of Korea, 1948), pp. 111–131.

5. Sŏng-hwan Pan, *Han'guk nongŏp ŭi sŏngjang*. Years in which shares of fixed capital and working capital are abnormally high or low received an adjustment. In working capital, average factor share of 1951 and 1953 was used for 1952's factor share, and average factor share of 1955 and 1958 for factor share of 1956 and 1957. In fixed capital, the average of 1951 and 1953 for 1952's factor share, and average of 1954 and 1956 for 1955's factor share. The difference between the original series and the new series due to the adjustment was proportionally allocated to land and labor.

FOUR   *The Sources of Agricultural Growth*

1. Yong Sam Cho, *"Disguised Unemployment" in Underdeveloped Areas With Special Reference to South Korean Agriculture* (Los Angeles and Berkeley, 1963), p. 76.

2. This statement is based on the monthly working hours per farm per month as estimated by the MAF *Report on the Results of Farm Household Economic Survey.*

3. The MAF *Survey* reports the number of hours worked per farm, including hired and exchange labor, but it does not report the number of labor hours worked per farm household. It includes only farm household members' work on the farm, not their work off the farm.

4. Under conditions of diminishing returns to scale, it is possible to conceive of circumstances when average product was rising while marginal product was falling, but in most circumstances both will move in the same direction.

5. See Roger A. Sedjo, "The Turning Point for the Korean Economy," in Sung-hwan Jo and Seong-Yeung Park, *Basic Documents and Selected Papers of Korea's Third Five-year Economic Development Plan* (Seoul, 1972), pp. 207–221; and Roger Sedjo, "Korean Historical Experience and the Labor Surplus Model," *Journal of Developing Areas,* June 1976, pp. 213–221.

6. The correlation coefficient (R) using data in Table 27 for female wages is 0.996 (urban textile wage = –809.7 + 1.213 X rural female wage). That for male wages (R) is 0.999 (urban manufacturing wages = –1663.31 + 1.1019 rural male wage). These calculations were made using data for 1964–1974. High correlation coefficients are, of course, to be expected when one is using time series data in current prices with a high rate of inflation.

7. United States Operations Mission to Korea, *Rural Development Program Evaluation Report, Korea, 1967,* p. D-3.

8. This discussion is based on that of Jin Hwan Park, *An Economic Analysis of Land Development Activities in Korea* (Seoul, 1969), pp. 18–21. The effort used to make these calculations involved PL 480 grain in the amount of 8,280 metric tons which was 36.5% of the investment cost on these 11,000 hectares.

9. Jin Hwan Park, p. 20.

10. USOM/K, *Rural Development Program,* p. 52.

11. These figures were taken from Korean Agricultural Sector Study, "An Analysis of New Land Development in Korea," Special Report No. 3, pp. 13–14.

12. All figures in this paragraph are from Jin Hwan Park, pp. 18–19.

13. Jin Hwan Park, pp. 31–34, 51.

14. Vincent Brandt, *A Korean Village Between Farm and Sea* (Cambridge, Mass., 1971), pp. 193–195.

15. This statement is based on an interview with an individual who has had extensive experience with small tideland reclamation projects.

16. This calculation was made using 1967 data and by assuming that grain yields on new upland after several years were 70% of grain yields on existing upland. Upland grain output accounted for slightly less than half of total grain output, and about 75% of existing upland acreage was sown to grain. (Jin Hwan Park, p. 89.)

17. Calculations of this sort depend on a variety of assumptions and accurate yield response functions. This calculation assumed most of the fertilizer was used on rice and that a 1965 TVA-estimated yield response function for Korea was applicable.

18. These estimates were made by Dong Hi Kim, "Economics of Fertilizer Use in Production of Foodgrains in Korea" (unpublished PhD dissertation, University of Hawaii, 1971), pp. 63, 83. The maximum useful levels of N-P-K used in these calculations were, on rice 14-6-8 kg. per tanbo, and on barley 8-4-2.5 kg. per tanbo.

19. USOM/K, *Rural Development Program*, p. 81.

20. For a further useful discussion of how these key, but difficult to quantify, inputs play a role in rising farm output, see Paul W. Kuznets, *Economic Growth*, pp. 130–145.

FIVE   *Regional Agricultural Production and Income*

1. They are not valid because they do not sum to unity as true weights should. True current-value weights would be w_{ij} for the ith crop in the jth year:

$$w_{ij} = \frac{P_{ij}Q_{ij}}{\sum_i P_{ij}Q_{ij}}.$$ It is clear that $w_{ij} = 1$, and the jth

year's terms of trade relative to the base year zero would be $T_j = \dfrac{\sum_i P_{ij}w_{ij}}{\sum_i P_{io}w_{io}}$.

The purchasing power ratios used here, T_j^*, imply "weights" w_{ij}^*, where $w_{ij}^* = Q_{ij}$, and it is clear that $\sum_i w_{ij} \neq 1$.

2. See Albert Keidel, "South Korean Regional Farm Products and Income, 1910–1975," unpublished PhD dissertation, Harvard University, 1978.

3. Ibid.
4. Ibid.

EIGHT   *Farm Price Policy*

1. Pal Yong Moon and B. S. Ryu, "An Analysis of the Effects of Government Buffer Stock Operation" (Seoul, 1974), pp. 35–42.

2. The law was enacted in 1966 with the aim of improving the rural economic situation by stabilizing the prices of farm products. Although the law provides that the financial resources of the fund shall be enlarged to not less than 20,000 million wŏn over five years following the establishment of the fund, the actual sum came to 17,400 million wŏn as of the end of 1975. The resources of the fund are supposed to be used for grain management operation, stockpiling of agricultural commodities, preparation for farm product exports. In fact, however, no money from the fund has been used for grain management operation, because this activity has been financed under the Grain Management Special Account of the government (GMSA).

3. This fund has been in operation since July 1975, with the aim of contributing to steady growth of the livestock industry by achieving long-term stability in the prices of compound feeds. The financial resources of the fund are scheduled to be increased to not less than 10,000 million wŏn. Funding sources are: 1) differences between the import prices of foreign feeds and the prices of such feeds charged to end users; 2) borrowings from banks; 3) profits on the fund's operation; and 4) government subsidies and loans. As of July 1976, the resources of the fund totaled 900 million wŏn.

NINE   *Local Government and Rural Development*

1. Sources of data for the analysis in this chapter were:
   1) Personal observation and conversations with farmers
   —1965–1966 approximately one year's residence in rural Korea (South Ch'ungch'ŏng province) with visits to other villages in South Ch'ungch'ŏng province (4), Kangwŏn province (3) and Kyŏnggi province (2);
   —1969 visits to villages in South Ch'ungch'ŏng province (3), Kyŏnggi province (1), and South Chŏlla province (3).
   —1972 visits to villages in Kangwŏn province (2), South Kyŏngsang province (2), and South Chŏlla province (1);

–1975 visits to villages in South Ch'ungch'ŏng province (2), and South Chŏlla province (4);

–1976 visits to villages in South Ch'ungch'ŏng province (2), Kyŏnggi province (2), and South Chŏlla province (2).

2) Interviews with officials of the Office of Rural Development (4), the National Agricultural Cooperative Federation (2 national, 4 local), the Ministry of Home Affairs at local and national levels (7), and the Ministry of Agriculture and Fisheries (1).

3) Documentary material on local government, mostly in the *Haengjŏng nonjip*, Seoul National University.

4) Conversations with other foreigners who have lived for extended periods recently in rural Korea (5).

N.B. 2 of the South Ch'ungch'ŏng province villages and 1 of the South Chŏlla province villages are same–i.e., visited again in different years.

2. Further discussion of these issues can be found in United States Operations Mission to Korea, *Rural Development Program Evaluation Report, Korea, 1967,* Chapters 7 and 8; Korean Agricultural Sector Study, *Agricultural Research and Guidance* (KASS Special Report No. 5, 1972).

3. For a further discussion of the New Community Movement, see Ronald Aqua, *Local Institutions and Rural Development in South Korea* (Ithaca, Rural Development Committee, 1974); and Sung Hwan Ban, "The New Community Movement in Korea" (Seoul, Korea Development Institute Interim Report No. 7502, 1975).

TEN   *Land Reform*

1. Jae Hong Cho, "Post-1945 Land Reforms and Their Consequences in South Korea," (unpublished PhD dissertation, Indiana University, 1964), pp. 100, 104.

2. For example, Ki Hyuk Pak, et al., *A Study of Land Tenure System in Korea* (Seoul, 1966); Jae Hong Cho; Theodore Reynolds Smith, *East Asian Agrarian Reform: Japan, Republic of Korea, Taiwan, and the Philippines* (Hartford, n.d.); and many others.

3. Ki Hyuk Pak, p. 127.

4. The market price used here refers to prices in a period when land reform legislation itself does not lower the price of land.

5. The availability of government loans at rates of interest that reflect the social cost of capital might make it possible for a few to buy land, but the numbers are not likely to be large.

6. According to Jae Hong Cho (p. 63), after 1946 farmers, including tenants, had to deliver grain at a set quota to the government at "legal" prices and landlords received rent at those same "legal" prices so that their rent was only about 7% of the output on tenanted land.

7. These are the calculations of Jae Hong Cho, pp. 88–89.

8. Theodore Smith, p. 71, quoting an estimate by R. H. Du Pasquier.

9. If the land directly sold by landlords to tenants was sold at an average price of 10 sŏk (roughly halfway between the 3.5 and 19 sŏk figures discussed in the text) and average annual main crop output on that land was 20 sŏk, then that land was worth 60 sŏk in the 1930s (3 times the yield) and payment was one-sixth of that original worth. Slightly different assumptions, of course, give different results.

10. An infrequent but not unheard of event. In one village, for example, a landlord dispossessed his tenants and converted 6 hectares of land into saltpans; Brandt, *A Korean Village,* p. 55.

11. Sŏng-hwan Pan, *Han'guk nongŏp ŭi sŏngjang,* p. 189. These growth rates are calculated from 5-year averages centering on the beginning and end years of the periods.

12. Rent was generally calculated as a percentage of the main crop (usually rice). Thus farmers had an incentive to increase inputs on other than the main crop. Of course, landlords and the Japanese government were mainly interested in rice and had more than a little power to offset tenant preferences based on relative rates of return to various crops.

13. Sŏng-hwan Pan, *Han'guk nongŏp ŭi sŏngjang,* p. 192.

14. Jae Hong Cho, p. 38. The analysis in these paragraphs is based on Cho's Chapters II and VI.

15. For a tenant whose income rose 33% (see above), for example, 80% of that increase would have to be saved to make up for the former landlord investment of 20% of total product.

16. The conditions under which loans of this sort would have been used for production as contrasted to consumption purposes are discussed elsewhere in this volume.

17. Ki Hyuk Pak, et al. (p. 175), for example, estimated production functions for rice and barley for owner-operator, owner-tenant, tenant-owner, and tenant farms. For rice the sum of the elasticities was 1.066, 0.976, 1.108, and 1.043 respectively.

18. The study of Ki Hyuk Pak, et al. (p. 473) also concludes that the 3-chŏngbo limit is not an inhibiting factor. The Korean Agricultural Sector Study, Special Report No. 3, pp. 23–24 also acknowledges that the 3-chŏngbo limit is above the optimum farm scale, and yet they go on to argue that it should be removed in order to increase non-farm investments in land. One would think that these latter could best be provided by the government rather than private investors if this is the case.

ELEVEN   *Rural Living Standards*

1. See discussion in Appendix C.

2. The best evidence that the 1930s Farm Household Surveys had an upward bias is the fact that average per household cultivated acreage was 1.5 hectares in 1933 and 1.6 in 1938 in the 8 southern provinces as contrasted to an average of 1.2 hectares in the national totals (in 1937), Chōsen Ginkō, *Chōsen keizai nenkan,* 1939, pp. 20, 23.

3. Agricultural output data connecting the 1930s with the post-1945 period for a comparable area is in Sŏng-hwan Pan, *Han'guk nongŏp ŭi sŏngjang,* pp. 243–244.

4. If farm household income before rent payments was about 300 yen (in 1934 prices) in both 1960 and the mid-1930s, then income after rent payments in the 1930s was about 225 yen and in 1960 about 295 yen. In 1975, farm household income deflated by prices paid *to* farmers would be roughly 500 yen and deflated by prices paid *by* farmers for living expenditures, about 560 yen, 550 after rent; 550 yen is 2.44 times or 325 yen above 225. Of that 325 yen increase, 135 yen could be attributed to income redistribution and 200 yen to increases in output.

5. These 1974 calculations were made by William I. Abraham, from Farm Household Survey data; "Observations on Korea's Income Distribution and The Adequacy of the Statistical Base," unpublished paper, 1976.

6. Specifically, unlike the other years' surveys, one has to guess in the 1963 survey at the total and average income of the 12 families (out of 652) in the top income group (over 200,000 wŏn).

7. This, at least, is the best guess of Cornelius Osgood, *The Koreans and Their Culture* (Tokyo, 1951), p. 134.

8. Rough estimates of the ratio of farm household incomes of the top and bottom 10% in the People's Republic of China can be derived from the study of C. R. Roll, "The Distribution of Rural Income in China: A Comparison of the 1930's and the 1950's," (unpublished PhD dissertation, Harvard University, 1974).

9. See Adelman and S. Robinson.

10. See, for example, the impressions of Isabella Bird Bishop, who traveled extensively in the Korean countryside in the 1890s, p. 79.

11. This estimate was derived from the data in Table 133 by assuming rates of literacy per family of 0%, 15%, 35%, and 75% in the 4 categories in the table.

12. Ok Ryun Moon and Jae Woong Hong, "Health Services Outcome Data: A Survey of Data and Research Findings on the Provision of Health Services in Korea" (unpublished paper, Seoul National University, December 1975), p. 86.

13. See Chapter 6.

14. Average farm household income in 1975 was over 500,000 wŏn in current prices, and the exchange rate was 500 wŏn to U.S. $1. Exchange rate conversions of this type probably understate the dollar value of Korean household income.

TWELVE   *Off-Farm Migration*

1. Simon Kuznets, "Economic Growth and the Contribution of Agriculture: Notes on Measurement," reprinted in C. Eicher and L. Witt, eds., *Agriculture in Economic Development* (New York, 1964), pp. 117–119.

2. A brief but still valuable discussion of the theoretical and empirical literature on migration as it relates to dualistic development is given by Donald W. Jones, *Migration and Urban Unemployment in Dualistic Economic Development* (Chicago, 1975), Chapters 1 and 2.

3. See, for instance, Bertrand S. Renaud, "Conflicts Between National Growth and Regional Income Inequality in a Rapidly Growing Economy: The Case of Korea," *Economic Development and Cultural Change* (April 1973) and Tae-yŏng Kim and Hyo-gu Yi, *Uri nara in'gu idong ŭi t'ŭkching, 1965–1970* (Seoul, 1976), Chapter 9. Se-yeul Kim, "The Economic and Social Determinants of Rural-Urban Migration in Korea: A Case Study of North Cholla Province" (unpublished PhD dissertation, University of Hawaii, 1973, pp. 85–110), has applied a Todaro-type model to rural-urban migration in North Chŏlla province, with mixed results.

4. Man-gap Lee, "Pushing or Pulling," in *Report of the International Conference on International Problems and Regional Development* (Seoul, 1970); Jung Ju Yoon, *A study of Fertility and Out-migration in a Rural Area* (Seoul, 1971), p. 156; and Jung Ju Yoon, *A Study of the Population of Seoul* (Seoul, 1975), p. 151.

5. The concept of the salvage value of labor as it applies to farm–nonfarm mobility has been formalized by Glen L. Johnson, "Theoretical Considerations," in *The Overproduction Trap in U.S. Agriculture* (Baltimore, 1972), pp. 29 ff. Marion Clawson "Aging Farmers and Agricultural Policy," *Journal of Farm Economics* (February 1963), p. 27, has also employed the term in this context.

6. John S. Nalson, *Mobility of Farm Families: A Study of Occupational and Residential Mobility in an Upland Area of England* (Manchester, 1968), provides a valuable book-length study of inheritance and the farm-search process in a semi-marginal agricultural region of England.

7. A more detailed account of pre-Liberation migration patterns is given by John Sloboda in Chapter 5, "Migration," in *Urbanization and*

Urban Problems in this series by Edwin S. Mills and Byung-Nak Song.

8. Tai Hwan Kwon, *Demography of Korea: Population Change and its Components, 1926–1966* (Seoul, 1977), p. 177 (Table 8.1) and p. 204 (Table 9.2).

9. All comparisons of off-farm migration between 1960–1966 and 1966–1970 rely on the narrower definition of farm households on which the 1966–1970 (I) estimates are based, since the more marginal farm household categories reflected in the 1966–1970 (II) estimates (see Table 136, Note 3) were not included among enumerated farm households in the 1960 census. Estimates of net migration presented in this chapter exclude net movement among children born during the period.

10. Because the 1960 Population Census was carried out as of December 1, while the 1966 and 1970 censuses were taken as of October 1, the first and second inter-censal periods are respectively 5.83 and 4.00 years in length. For this reason, the 1966 age groupings run one year ahead of those employed in 1970 in order to maintain constant birth cohorts. Further adjustment for the two-month difference between the census date in 1960 and that in the later enumerations is not practical, but the impact of this discrepancy on the estimates of off-farm migration during 1960–1966 is probably insignificant.

11. Unless otherwise noted, the figures cited in the text are based on estimates employing the 1960 farm household definition.

12. While information on marital status is not separately available for farm women, it may be assumed that the proportion married at any age was no less and probably greater than the proportion among all rural women. The contention that rural women migrating to the urban sector continued to marry somewhat earlier than their urban-born peers is suggested by the increase between 1966 and 1970 in the proportion of urban-dwelling women age 20–24 and age 25–29 who were currently married. Since this increase moves against the trend for the nation as a whole, it is generally attributed to an influx of earlier-marrying women from rural origins.

Percentage of Women Age 15–34 Currently Married, 1955–1970
(Rural/Urban/Total)

Age	1955 (R/U/T)	1960 (R/U/T)	1966 (R/U/T)	1970 (R/U/T)
15–19	16/ 9/14	9/ 3/ 7	5/ 2/ 4	4/ 2/ 3
20–24	80/62/75	72/48/65	58/33/48	49/37/42
25–29	91/86/90	96/88/93	94/83/90	93/84/88
30–34	91/88/90	89/86/92	91/87/94	96/92/95

Sources: Tai Hwan Kwon, *Demography*, Appendix Tables C; *1970 Population and Housing Census Report*, Vol. 1:12–1, Table 3.

13. Calculated from unpublished 1970 census data as compiled by KDI (hereafter referred to as the KDI Migration Data).

14. MAF, *Population and Housing Census Report, 1970,* Vol. 2:4–1, Table 3.

15. These statements are supported by calculations from census data for 1960 and 1966. A detailed examination of urban-centered and rural-centered non-agricultural employment growth by industry and worker category (i.e., self-employed, wage worker, family worker, etc.) is undertaken in the author's forthcoming dissertation.

16. Forrest R. Pitts, "A Factor Ecology of Taegu City," in *A City in Transition: Urbanization in Taegu, Korea* (Seoul, 1971) has shown the concentration of agriculture-oriented service activities around the wholesale produce market in the regional city of Taegu.

17. "Quasi-subsistence farming" as used here refers to farming in which the principal objective is production for direct household consumption and where market sales to obtain income for household cash requirements, however important to household welfare, remain a secondary objective of production decisions.

18. In Table 137 the definition of farm household follows the broader 1966 definition, which includes all direct cultivators regardless of holding size, but excludes households of pure agricultural laborers.

19. Although the military encampment population is excluded from the census population of the region, the officers living off base and their families make up a large fraction of the non-farm households in the towns of the border zone. Checkpoints along the routes to the south and the intentional funneling of traffic through these checkpoints inhibit movement in and out of the border zone and probably reduce the attractiveness of the zone to households economically unrelated to the military encampments.

20. Brian J. L. Berry, "Hierarchical Diffusion: The Basis of Development Filtering and Spread in a System of Growth Centers," in N. M. Hansen, ed., *Growth Centers in Regional Economic Development* (New York, 1972), provides a formal statement of hierarchical diffusion.

21. Clifford J. Jansen, "Migration: A Sociological Problem," in *Readings in the Sociology of Migration* (London, 1970).

22. Ezra F. Vogel, "Kinship Structure, Migration to the City, and Modernization," in *Aspects of Social Change in Modern Japan* (Princeton, 1967).

23. Jong Ju Yoon, *Population of Seoul,* p. 149.

24. Jong Ju Yoon, ibid., p. 152, Table 4–19. Evidence on this point is extremely limited but Yoon's data suggest that household heads migrating with or in advance of their families were much more likely to have a promise of employment or specific plans for starting a business than were single male migrants. See also Se-yeul Kim, p. 113.

25. Michael J. Greenwood, "An Analysis of the Determinants of Geographic Mobility in the United States," *The Review of Economics and Statistics* (May 1969).

26. Information on age-specific patterns of land ownership is unavailable, and the term "land holdings," like "farm size," refers to the total of both owned and rented land cultivated by the household. Overall, as farm size increases, the proportion of both paddy and dry field holdings that are in fact owned by the household also increases (through the 25–30-tanbo class), according to the *1970 Agricultural Census*. It is probably safe to assume a strong correlation between the amount of land cultivated and the amount of land owned by a household, with small holders generally owning less than they cultivate and large holders owning more than they cultivate.

27. As already noted, all inter-period comparisons in this section refer only to farm households and operators cultivating at least 1 tanbo of land.

28. There are, of course, female as well as male operators, but the proportion of women among young operators is assumed to be so small that it may be safely ignored.

29. A myŏn is a rural township, typically incorporating 10–20 administrative villages, each in turn including several natural villages. A typical myŏn might include a total population of 8,000–15,000 persons.

30. Jong Ju Yoon, *Fertility and Out-migration*, pp. 151 ff.

31. Tae-yŏng Kim and Hyo-gu Yi, pp. 68–72.

32. Man-gap Yi (Man-gap Lee), *Hanguk nongch'on sahoe ŭi kiyo wa pyŏnhwa* (Seoul, 1973), pp. 222–229.

33. According to the 1970 Population Census, 99% of primary school age children entered school at some point, and 98% of those entering completed the full 6 years.

34. Han'guk Nongch'on Sahoe Yŏn'guhoe, *Nongch'on sahoehak* (Seoul, 1965), p. 191, cited by Hŭi-sŏp Im, "Sŏul [Seoul]-si in'gu chipchung e mich'inŭn ŭisuk kjuo yoin." *Sŏul [Seoul] in'gu chipchung: kŭ munjechŏm kwa taech'aek* (Seoul, 1976), p. 109. See also Man-gap Yi, *Han'guk nongch'on*, p. 106.

35. Chae-bu Cho, "Nongch'on haksaeng tŭi ŭi ŭisik jujo e kwanhan chose yŏn'gu" (MA thesis, School of Education Ewha Womans University, 1976), cited by Hui-sop Im.

36. Brandt, p. 96.

37. Charles Goldberg, private communication.

38. Labor absorption and changes in industrial structure are treated at length elsewhere in this volume. The abbreviated discussion here is intended to provide a bridge between these topics and off-farm migration.

39. Lorene Y. L. Yap, "Internal Migration in Less Developed Countries:

A Survey of the Literature," (Washington, 1975), pp. 31 ff.; Dipak Mazumdar, "The Urban Informal Sector" (Washington, 1975), pp. 18–21, p. 57.

40. Tae-yŏng Kim and Hyo-gu Yi, p. 81. Seoul-based comparisons are more meaningful than those for the urban sector as a whole, since migration in the KDI Migration Data is defined as *inter-provincial* migration, and intra-provincial rural-urban migrants in provincial cities are included with the "non-migrant" population. Because of their special administrative status, all migration to Seoul or Pusan is by definition inter-provincial migration.

41. Tae-yŏng Kim and Hyo-gu Yi, p. 84. Among economically active males age 20–24, reported unemployment among migrants was 6.5% compared to 13.6% among non-migrants. Among economically active females of the same age, the unemployment rate was 3.0% among migrants and 6.8% among non-migrants.

42. Reported in full in "Occupational Characteristics of Migrants and Non-migrants," a chapter in John Sloboda's forthcoming dissertation.

43. Yukio Masui, "The Supply Price of Labor: Farm Family Workers," in *Agriculture and Economic Growth: Japan's Experience* (Princeton, 1970), pp. 231, 239–240.

44. The findings and implications of these studies are assessed in detail in "The Social and Economic Characteristics of Urban In-Migrants," a chapter in John Sloboda's forthcoming dissertation.

45. The theory of human capital investment as it relates to on-the-job training is treated in detail by Gary S. Becker, *Human Capital: A Theoretical and Empirical Analysis with Particular Reference to Education* (New York, 1964), Chapter 2.

46. James W. White, "Internal Migration in Prewar Japan," *Journal of Japanese Studies* (winter 1978), pp. 85 ff, 102 ff.

47. Mitsuru Shimpo, *Three Decades in Shiwa: Economic Development and Social Change in a Japanese Farming Community* (Vancouver, 1976), Chapter 5; Robert J. Smith, *Kurusu: The Price of Progress in a Japanese Village, 1951–1975* (Stanford, forthcoming), Chapter 7.

48. Son Ung Kim and Peter Donaldson, "Redistribution of Seoul's Population: Government Plans and Their Implementation," paper prepared for the annual meeting of the Association for Asian Studies, Chicago, 1978.

Appendix C

1. The discussion in this appendix owes a great deal to the work of Albert Keidel (in a private communication) and to William I. Abraham, "Observations on Korea's Income Distribution and the Adequacy of the Statistical Base" (unpublished paper, April 1976).

Bibliography
(Works Cited in Endnotes)

Abraham, William I. "Observations on Korea's Income Distribution and the Adequacy of the Statistical Base." Unpublished paper, 1976.

Adelman, Irma, and Sherman Robinson. *Income Distribution Policy in Developing Countries: A Case Study of Korea.* Stanford, Stanford University Press, 1978.

Aqua, Ronald. *Local Institutions and Rural Development in South Korea.* Ithaca, Rural Development Committee, 1974.

Ban, Sung Hwan (Pan, Sŏng-hwan). *Han-guk nongŏp ŭi sŏngjang, 1918–1971* (Growth of Korean agriculture, 1918–1971). Seoul, Korea Development Institute, 1974. Includes a long summary in English, referred to as *Growth of Korean Agriculture (GKA).*

——. "Growth Rates of Korean Agriculture, 1918–1971," in Y. Hayami, V. Ruttan, and H. Southworth, eds., *Agricultural Growth in Japan, Taiwan, Korea, and the Philippines.* Honolulu, University of Hawaii Press, forthcoming.

——. *Korean Agricultural Growth, 1955–1971.* Seoul, Korea Development Institute, 1974.

——. "The New Community Movement in Korea." Seoul, Korea Development Institute Interim Report No. 7502, 1975.

Bank of Korea (Han'guk Ŭnhaeng). *Chosŏn kyŏngje yŏnbo 1948* (Annual economic review of Korea 1948). Seoul.

——. *Economic Statistics Yearbook.* Various years. Seoul.

——. *Farm Household Survey.* Various years to 1960. Seoul.

——. *National Income in Korea, 1975.* Seoul, 1976.

——. *Price Survey, 1947–1970.* Seoul.

Barringer, Herbert R. "Migration and Social Structure," in M. G. Lee and H. R. Barringer, eds., *A City in Transition: Urbanization in Taegu, Korea.* Seoul, Hollym, 1971.

Becker, Gary S. *Human Capital: A Theoretical and Empirical Analysis with Particular Reference to Education.* New York, Columbia University Press, 1964.

Berry, Brian J. L. "Hierarchical Diffusion: the Basis of Development Filtering and Spread in a System of Growth Centers," in N. M. Hansen, ed., *Growth Centers in Regional Economic Development.* New York, Free Press, 1972.

Bishop, Isabella Bird. *Korea and Her Neighbors.* Seoul, Yonsei University Press, 1970. Originally published in 1898.

Brandt, Vincent. *A Korean Village Between Farm and Sea.* Cambridge, Harvard University Press, 1971.

Chang, Yunshik (Chang, Yun-sik). "Population in Early Modernization: Korea." Unpublished PhD dissertation, Princeton University, 1966.

Chang, Yun-sik (Chang, Yunshik). *Han'guk chŏllyŏk suyo mit kagyŏk ŭi punsŏk* (An analysis of demand and prices for Korean electrical power). Seoul, Korea Development Institute, 1977.

——, H. Y. Lee (Hae-yong Yi), E. Y. Yu (Ui-yong Yu), and T. H. Kwon (T'ae-hwan Kwon). *A Study of the Korean Population.* Seoul, Seoul National University, 1974.

Chayanov, A. V. *The Theory of the Peasant Economy.* Homewood, Illinois, Richard D. Irwin, 1966.

Cho, Chae-bu. "Nongch'on haksaeng tŭ ŭi ŭisik jujo e kwanhan chosa yŏn'gu" (A study of the cognitive structure of rural students). MA thesis, Ewha University, 1966.

Cho, Jae Hong. "Post-1945 Land Reforms and Their Consequences in South Korea." Unpublished PhD dissertation, Indiana University, 1964.

Cho, Yong Sam (Cho, Yong-sam). *"Disguised Unemployment" in Underdeveloped Areas with Special Reference to South Korea Agriculture.* Los Angeles and Berkeley, University of California Press, 1963.

Chōsen Ginkō (Bank of Korea). *Chōsen keizai nenkan* (Korean economic yearbook). 1939.

Chōsen Sōtokufu. (Government General in Korea). *Chōsen no nōgyō* (Agriculture in Korea). Seoul, 1936, 1940.

——. *Shōwa 8 nen nōgyō tōkei hyō* (1933 agricultural statistical tables). Seoul, 1937.

Clawson, Marion. "Aging Farmers and Agricultural Policy," *Journal of Farm Economics,* February 1963.

Easterlin, Richard A. *Population Labor Force and Long Swings in Economic Growth: The American Experience.* New York, Columbia University Press, 1968.

Economic Planning Board. *Annual Report on the Price Survey.* Various years. Seoul.

——. *Annual Report on the Family Income and Expenditure Survey.* Various years. Seoul.

——. *Korea Statistical Yearbook.* Various years. Seoul.

——. *1966 Population Census Report.* Seoul, 1968-1969.

——. *Population and Housing Census Report 1960* and *1970.* Seoul.

Food and Agriculture Organization, The United Nations. *Production Yearbook.* Various years. Rome.

Grajdanzev, Andrew J. *Modern Korea.* New York, John Day, 1944.

Greenwood, Michael J. "An Analysis of the Determinants of Geographic Mobility in the United States," *The Review of Economics and Statistics,* May 1969.

Han'guk Nongch'on Sahoe Yŏn'guhoe (Korean Rural Studies Society). *Nongch'on sahoehak* (Rural sociology). Seoul, 1965.

Harris, John R. and Michael P. Todaro. "Migration, Unemployment, and Development: A Two Sector Analysis," *American Economic Review,* March 1970.

Hasan, Parvez. *Korea: Problems and Issues in a Rapidly Growing Economy.* Baltimore, Johns Hopkins University Press, 1976.

Himeno, Minoru. *Chōsen keizai zuhyō* (Graphs of the Korean economy), Chōsen Tōkei Kyōkai (Korea Statistics Association) 1940.

Hong, Wontack (Hong Wŏn-t'aek). *Factor Supply and Factor Intensity of Trade in Korea.* Seoul, Korea Development Institute, 1976.

——. "Trade and Subsidy Policy and Employment Growth in Korea" Mimeographed report, Korea Development Institute, 1976.

Im, Hŭ-sŏp. "Sŏul [Seoul]-si in'gu chipchung e mich'inŭn ŭisik kjuo yoin" (Cognitive factors contributing to the concentration of population in Seoul), *Sŏul [Seoul] in'gu chipchung: kŭ munjcchŏm kwa taech'aek* (The concentration of population in Seoul: problems and counter-measures). Seoul, Korean Population Association, 1970.

Jansen, Clifford J. "Migration: A Sociological Problem," in *Readings in the Sociology of Migration.* London, Pergamon, 1970.

Johnson, Glen L. "Theoretical Considerations," in *The Overproduction Trap in U.S. Agriculture.* Baltimore, Johns Hopkins University Press, 1972.

Johnston, Bruce F. and Peter Kilby. *Agriculture and Structural Transformation.* London, Oxford University Press, 1975.

Jones, Donald W. *Migration and Urban Unemployment in Dualistic Economic Development.* University of Chicago, Department of Geography Research No. 165, 1975.

Keidel, Albert, "South Korean Regional Farm Products and Income, 1910-1975." Unpublished PhD dissertation, Harvard University, 1978.

Kim, Dong Hi. "Economics of Fertilizer Use in Production of Foodgrains in Korea." Unpublished PhD dissertation, University of Hawaii, 1971.

Kim, Kwang Suk (Kim Kwang-sŏk) and Michael Roemer. "Growth and Macro Change in the Structure of National Product, 1945-1975." Korea Development Institute Interim Report 7601, Seoul, 1976.

——. *Growth and Structural Transformation.* Studies in the Modernization of the Republic of Korea: 1945-1975. Cambridge, Council on East Asian Studies, Harvard University, 1979.

Kim, Sang-hyŏn. *Chae-Il Han'gugin* (Koreans in Japan). Seoul, Tan'gok Haksul Yŏn'guwon, 1969.

Kim, Se-yeul (Kim, Se-yŏl). "The Economic and Social Determinants of Rural-Urban Migration in Korea: A Case Study of North Cholla Province." Unpublished PhD dissertation, University of Hawaii, 1973.

Kim, Son Ung (Kim Sŏn-ung) and Peter Donaldson. "Redistribution of Seoul's Population: Government Plans and Their Implementation." Paper prepared for the Annual Meeting of the Association for Asian Studies, Chicago, 1978.

Kim, Tae-yŏng and Hyo-gu Yi. *Uri nara in'gu idong ŭi t'ŭkching, 1965–1970* (Characteristics of migration in Korea, 1965-1970). Seoul, Korea Development Institute, 1976.

Korean Agriculture Bank. *Agricultural Yearbook 1960.*

Korean Agricultural Sector Study (KASS). *Agricultural Research and Guidance.* Special Report N. 5. Seoul, Ministry of Agriculture and Forestry (Fisheries) and Michigan State University, 1972.

——. "An Analysis of New Land Development in Korea." Special Report No. 3. Mimeographed. Ministry of Agriculture and Forestry (Fisheries) and Michigan State University.

Korean Electric Association. *The Electric Yearbook 1974.* Seoul.

Korean Traders Association. *Statistical Yearbook of Foreign Trade.* Various years. Seoul.

Kuznets, Paul W. *Economic Growth and Structure in the Republic of Korea.* New Haven, Yale University Press, 1977.

Kuznets, Simon. "Quantitative Aspects of the Economic Growth of Nations: VIII Distribution of Income by Size," *Economic Development and Cultural Change,* January 1963.

——. "Economic Growth and the Contribution of Agriculture: Notes on Measurement," reprinted in C. Eicher and L. Witt, eds., *Agriculture in Economic Development.* New York, McGraw-Hill, 1964.

Kwon, Tai Hwan (Kwŏn T'ae-hwan). "Population Change and Its Components in Korea, 1925-66." PhD dissertation, Australian National University, 1972.

——. *Demography of Korea: Population Change and its Components, 1925-66.* Seoul, Seoul National University, 1977.

——, H. Y. Lee, Y. S. Chang, and E. Y. Yu. *The Population of Korea.* Seoul, Seoul National University, 1975.

Lee, Kyu-sul (Yi Kyu-sŏl). *Population Pressure and its Effect on Korean Farm Households.* Seoul, Economic Planning Board, 1967.

Lee, Man-gap (Yi Man-gap). "Pushing or Pulling," in *Report of the International Conference on Urban Problems and Regional Development.* Seoul, Yonsei University, 1970.

Yi Man-gap (Lee, Man-gap). *Han'guk nongch'on sahoe ŭi kiyo wa pyŏnhwa* (The social structure of the Korean village and its change). Seoul, Seoul National University, 1973.

Masui, Yukio. "The Supply Price of Labor: Farm Family Workers," in Kazushi Ohkawa, Bruce F. Johnston, and Hiromitsu Kaneda, eds., *Agriculture and Economic Growth: Japan's Experience.* Princeton, Princeton University Press, 1970.

Mazumdar, Dipak. "The Urban Informal Sector." World Bank Staff Paper No. 211. Washington, D.C., 1975.

Mellor, John W. *The Economics of Agricultural Development.* Ithaca, Cornell University Press, 1966.

Ministry of Agriculture and Forestry (Fisheries since 1973). *Agricultural Census, 1960.* Seoul, 1964.

——. *Agricultural Census, 1970.* Seoul, 1973.

——. "Agricultural Products Purchase Program." Seoul, 1975.

——. *Cost of Production Survey, 1962–1975.* Seoul.

——. *Grain Statistics Yearbook.* Various years. Seoul.

——. *Report on the Results of Farm Household Economic Survey.* Various years. Seoul.

——. *Yearbook of Agriculture and Forestry Statistics.* Various years. Seoul.

——, Agricultural Development Corporation. *Yearbook of Land and Water Development Statistics, 1975.* Seoul, 1975.

Ministry of Construction. *Statistical Yearbook.* Various years. Seoul.

Ministry of Finance. *Summary of Financial Implementation.* Various fiscal years. Seoul.

Moon, Ok Ryun (Mun Ong-nyun) and Jae Woong Hong (Chae-ung Hong). "Health Services Outcome Data: A Survey of Data and Research Findings on the Provision of Health Services in Korea." Unpublished paper, Seoul National University, December 1975.

Moon, Pal Yong (Mun P'ar-yong) and B. S. Ryu (Pyŏng-sŏ Yu). "An Analysis of the Effects of the Government Buffer Stock Operation." Seoul, Korea Development Institute. 1974.

Moon, Seung Gyu (Mun Sŭng-gyo). *Out-migration of Farm Families of Orientation in Two Rural Communities.* Seoul, Seoul National University, 1972.

Mun P'ar-yong (Moon, Pal Yong). *Kokka chŏngch'aek ŭi kyehoekhwa* (Planning good-grain policy). Seoul, Korea Development Institute, 1973.

Nalson, John S. *Mobility of Farm Families: A Study of Occupational and Residential Mobility in an Upland Area of England.* Manchester, Manchester University Press, 1968.

National Agricultural Cooperative Federation. *Agricultural Yearbook.* Various years. Seoul.

——. *Summary of Rural Prices.* Various years. Seoul.

——. *20-Year History of Agriculture.* Seoul, 1965.

Nongŏp Hyŏptong Chohap Chunganghoe, Chosabu (National Agricultural Cooperative Federation, Research Department). *Nongch'on mulka chongnam, 1959–1970* (Survey of agricultural prices, 1959–1970). Seoul, 1971.

Osgood, Cornelius. *The Koreans and Their Culture.* Tokyo, Charles Tuttle, 1951.

Pai, Gregory G. Y. "Rural to Urban Migration and Squatter Settlements with Special Reference to Seoul, Korea." Unpublished manuscript, Yonsei University, 1969.

Pak, Ki Hyuk (Pak Ki-hyŏk) et al. *A Study of Land Tenure System in Korea.* Seoul, Korea Land Economics Research Center, 1966.

Park, Jin Hwan (Pak Chin-hwan). *An Economic Analysis of Land Development Activities in Korea.* Department of Agricultural Economics, Seoul National University, 1969.

Pitts, Forrest R. "A Factor Ecology of Taegu City," in M. G. Lee and H. R. Barringer, eds., *A City in Transition: Urbanization in Taegu, Korea.* Seoul, Hollym Corporation, 1971.

Renaud, Bertrand S. "Conflicts Between National Growth and Regional Income Inequality in a Rapidly Growing Economy: The Case of Korea," *Economic Development and Cultural Change,* April 1973.

Roll, C. R. "The Distribution of Rural Income in China: A Comparison of the 1930's and the 1950's." Unpublished PhD dissertation, Harvard University, 1974.

Sedjo, Roger A. "Korean Historical Experience and the Labor Surplus Model," *Journal of Developing Areas,* June 1976.

——. "The Turning Point for the Korean Economy," in Sung-hwan Jo (Sŏng-hwan Cho) and Seong-Yeung Park (Sŏng-yong Pak), *Basic Documents and Selected Papers of Korea's Third Five-Year Economic Plan.* Seoul, Soyang University Press, 1972.

Shimpo, Mitsuru. *Three Decades in Shiwa: Economic Development and Social Change in a Japanese Farming Community.* Vancouver, University of British Columbia Press, 1976.

Smith, Robert J. *Kurusu: The Price of Progress in a Japanese Village, 1951–1975.* Stanford, Stanford University Press, forthcoming.

Smith, Theodore R. *East Asian Agrarian Reform: Japan, Republic of Korea, Taiwan, and the Philippines.* Hartford, John C. Lincoln Institute, n.d.

Stouffer, S. A. "Intervening Opportunities and Competing Migrants," *Journal of Regional Science,* February 1960.

Suh, Sang-Chul (Sŏ Sang-ch'ŏl). *Growth and Structural Changes in the Korean Economy, 1910–1940.* Cambridge, Council on East Asian Studies, Harvard University, 1978.

United States Operations Mission to Korea (USOM/K). *Rural Development Program Evaluation Report, Korea, 1967.* Seoul, 1967.

Vogel, Ezra F. "Kinship Structure, Migration to the City, and Modernization," in *Aspects of Social Change in Modern Japan.* Princeton, Princeton University Press, 1967.

Wery, R., G. B. Rodgers, and M. D. Hopkins. *BACHUE-2: Version I: A Population and Employment Model for the Philippines.* Geneva, International Labor Organization, 1974.

White, James W. "Internal Migration in Prewar Japan," *Journal of Japanese Studies,* winter 1978.

Yap, Lorene Y. L. "Internal Migration in Less Developed Countries: A Survey of the Literature." World Bank Staff Paper No. 215. Washington, D.C., 1975.

Yi Han-sun (Lee, Han-Soon). "Uri nara in'gu ŭi chiyŏkkan idong e kwanhan yŏn'gu" (A study of migration of population in Korea). *In'gu munje nonjip* (Journal of population studies). In'gu Munje Yŏn'guso (The Institute of Population Problems), August 1969.

Yoon, Jong Ju (Yun Chong-ju). *A Study of Fertility and Out-migration in a Rural Area.* Seoul, Seoul Women's College, 1971.

——. *A Study on Rural Population.* Seoul, Seoul Women's College, 1975.

——. *A Study of the Population of Seoul.* Seoul, Seoul Women's College, 1975.

Yun, D. J. (Yun Tŏk-chin). S. J. (Sang-jae Yi) and other as compiled by O. R. Moon (Ong-nyun Mun) and J. W. Hong (Chae-ung Hong) in "Health Services Outcome Data: A Survey of Data and Research Findings of the Provision of Health Services in Korea." Unpublished paper, Seoul National University, 1975.

Zachariah, K. C. "A Note on the Census Survival Ratio of Estimating Net Migration," *Journal of the American Statistical Association,* 1962.

Index

Africa, 16

Agricultural Censuses 366, 394, 404–405

Agricultural Cooperatives Law, 168–169, 213

Agricultural Extension, Office of, 167, 169. *See also* Extension services, agricultural

Agricultural Finance Bonds, 212, 215–216, 219, 230

Agricultural Land Mortgage Law (1966), 227

Agricultural Product Price-Stabilization Fund, 256

Agricultural Production Loan, 209

Agricultural products: markets for, 5, 116; imports, 27–29; exports, 27–28, 175; growth in, 39–50; demand for, 85; production under Japanese, 163; price regulation of, 168, 234–259; specialized areas for, 177; investment in, 179; distribution of, 394; inventory, 431. *See also* Agriculture; and specific commodities

Agricultural statistics, 393–396

Agriculture: share in GNP, 11, 13; share in GDP, 15; transfers out of, 17–24, 316–382; regional differences in, 112–159; shifts from, 126; work force in, 134, 316–382; budget for, 166; investment in, 167–174, 178–182, 192–195, 200, 218, 231, 275

Agriculture Bank Ltd. (1956), Agriculture Bank (1958), 167–169, 205–212, 406

Agriculture and Fisheries (Forestry), Ministry of, 169, 205, 226, 262, 263, 270, 273, 279, 393, 395, 400–404

Aid, 10

Aid, U.S.: and fertilizer imports, 6, 9, 101, 386, 388; and grain imports, 29, 51, 239, 384; and land reclamation, 80, 83, 388; and road-building, 148, 154; and economic rehabilitation, 165–166; as source of funding, 218; and irrigation, 220; and ORD, 270; and fertilizer plants, 288; and institution-building, 290; and land survey, 395. *See also* Public Law 480

Alcohol, 23

Asan Bay, 189

Bank of Korea (1950, formerly Bank of Chosun), 203–204, 206, 214, 219, 226, 230, 232

Banks: Japanese, 197–199, 201; Korean, 201–212; after WWII, 235; and NACF, 264

Barley, 3; imports, 29, 167, 251; production of, 40–41, 44, 46, 50, 130–132, 407, 414; on new upland, 86; and chemical fertilizer, 108; double-cropping of, 114, by region, 115, 123; prices for, 129–132, 243, 246–251, 253; gov't purchases of, 128–132, 238; terms of trade for, 139, 243; farm income in, 141; research on, 170, 176, 191; two-price system for, 246–250; handling costs for, 247; consumption of, 247–250, 400; forward-pricing for, 258; yield data on, 396, 398, 399. *See also* Foodgrains; Grains

Basic Agricultural Law (1967), 175

Basic Agricultural Statistics (MAF), 393

455

Japan: agricultural trends in, 7, 13-17, 370; imports of silk by, 48; fertilizer consumption in, 106; demand for commodities in, 163, 164; land reform in, 284; Korean migrants to, 319-320; rural-urban migration in, 353, 375, 381; women's wages in, 372-373

Japanese colonial period: agricultural output in, 4, 163-164; inheritances from, 14, 277, 389; use of chemical fertilizers in, 6, 163; reforestation in, 94, 97; land productivity in, 136, 164; rural credit in, 197-201; grain market control in, 235, 320; administration during, 261; land tenure in, 283-284, 384; size of farms in, 297; class structure in, 308; education in, 311; and Korean migrants, 319-320; rural non-farm employment in, 373; and rural dislocation, 377

Japanese Government General, 4, 27, 99, 100, 165, 269

Johnson, Edwin C., 205

Kangwŏn province: characteristics of, 114, 116-118; paddy land in, 119; irrigation in, 120-121; fertilizer consumption in, 121; output from, 122-124, 127, 137-139; terms of trade for, 129; population of, 135-136, 350-353; farm income in, 140, 145; electrification in, 145; roads in, 155; size of farms in (1937), 294; farm households in, 324, 352; off-farm migration in, 325-327, 329

Keidel, Albert, 110

Kimjang, 255

Korea Food Corporation, 202

Korea Development Institute Migration Data, 371

Korea Reconstruction Bank (Korea Development Bank), 203, 204, 206, 208

Korean Agricultural Association (1926), 164

Korean Agricultural Sector Study—Population, 335

Korean Farming Society, 202

Korean Federation of Financial Associations (1933), 198-199

Korean Federation of Irrigation

Associations, 203

Korean Financial Debentures, 199

Korean Industrial Bank (1917), 198, 199, 202-203

Korean War: and agriculture, 35, 36, 38, 47-48, 51, 291; and agricultural labor, 53, 64, 134; and agricultural capital, 54, 56; and productivity growth, 60, 62; and road damages, 148; and economic development, 166; rehabilitation from, 238-239; and land reform, 290, and disease, 311; refugees, 320, 352; casualties, 320; and population displacement, 352, 377

Kŭm River, 188, 195

Kuomintang, 284

Kuro Industrial Export Zone, 374

Kuznets, Simon, 316

Kwangju, 157

Kwon, Tai Hwan, 320

Kye (financing clubs), 22, 197

Kyŏnggi province: characteristics of, 115, 116-118, 349; paddy land in, 119-120; irrigation in, 120-121; fertilizer consumption in, 121; output from, 122-125, 137, 137-139; terms of trade for, 129; rice prices in, 133; population of, 134-136, 350; farm income in, 140; electrification in, 145; roads in, 155; standard of living in, 158-159; size of farms in (1937), 294; farm population in, 323, farm households in, 324, off-farm migration in, 326-327, 329, 348; village studies in, 368

Kyongju area, 195

Labor: rural-urban migration of, 5, 6, 23-24, 176, 318-319; rural "surplus," 5, 11, 71-72, 76, 79; supply of, 6, 7, 52-53, 385-386; farm, 15, 53-54, 357-359; 385, 410-411, 418; urban-rural migration of, 53, 64, 319; productivity, 62-65, 79, 422; hours used, 72, 74; seasonal shortages of, 74, 190; female, 74, 338-340; and mechanization, 72-79, 190; for large farms, 293; interregional mobility of, 317; farm to non-farm shifts of, 316-382; forced, under Japanese, 320; migrant, in Seoul, 271; as productivity

Harvard East Asian Monographs

21. Kwang-Ching Liu, ed., *American Missionaries in China: Papers from Harvard Seminars*

22. George Moseley, *A Sino-Soviet Cultural Frontier: The Ili Kazakh Autonomous Chou*

23. Carl F. Nathan, *Plague Prevention and Politics in Manchuria, 1910–1931*

24. Adrian Arthur Bennett, *John Fryer: The Introduction of Western Science and Technology into Nineteenth-Century China*

25. Donald J. Friedman, *The Road from Isolation: The Campaign of the American Committee for Non-Participation in Japanese Aggression, 1938–1941*

26. Edward Le Fevour, *Western Enterprise in Late Ch'ing China: A Selective Survey of Jardine, Matheson and Company's Operations, 1842–1895*

27. Charles Neuhauser, *Third World Politics: China and the Afro-Asian People's Solidarity Organization, 1957–1967*

28. Kungtu C. Sun, assisted by Ralph W. Huenemann, *The Economic Development of Manchuria in the First Half of the Twentieth Century*

29. Shahid Javed Burki, *A Study of Chinese Communes, 1965*

30. John Carter Vincent, *The Extraterritorial System in China: Final Phase*

31. Madeleine Chi, *China Diplomacy, 1914–1918*

32. Clifton Jackson Phillips, *Protestant America and the Pagan World: The First Half Century of the American Board of Commissioners for Foreign Missions, 1810–1860*

33. James Pusey, *Wu Han: Attacking the Present through the Past*

34. Ying-wan Cheng, *Postal Communication in China and Its Modernization, 1860–1896*

35. Tuvia Blumenthal, *Saving in Postwar Japan*

36. Peter Frost, *The Bakumatsu Currency Crisis*

37. Stephen C. Lockwood, *Augustine Heard and Company, 1858–1862*

38. Robert R. Campbell, *James Duncan Campbell: A Memoir by His Son*

39. Jerome Alan Cohen, ed., *The Dynamics of China's Foreign Relations*

40. V. V. Vishnyakova-Akimova, *Two Years in Revolutionary China, 1925–1927*, tr. Steven I. Levine

41. Meron Medzini, *French Policy in Japan during the Closing Years of the Tokugawa Regime*

42. *The Cultural Revolution in the Provinces*

43. Sidney A. Forsythe, *An American Missionary Community in China, 1895–1905*

44. Benjamin I. Schwartz, ed., *Reflections on the May Fourth Movement: A Symposium*

45. Ching Young Choe, *The Rule of the Taewŏn'gun, 1864–1873: Restoration in Yi Korea*

46. W. P. J. Hall, *A Bibliographical Guide to Japanese Research on the Chinese Economy, 1958–1970*

47. Jack J. Gerson, *Horatio Nelson Lay and Sino-British Relations, 1854–1864*

48. Paul Richard Bohr, *Famine and the Missionary: Timothy Richard as Relief Administrator and Advocate of National Reform*

49. Endymion Wilkinson, *The History of Imperial China: A Research Guide*

50. Britten Dean, *China and Great Britain: The Diplomacy of Commerical Relations, 1860–1864*

51. Ellsworth C. Carlson, *The Foochow Missionaries, 1847–1880*

52. Yeh-chien Wang, *An Estimate of the Land-Tax Collection in China, 1753 and 1908*

53. Richard M. Pfeffer, *Understanding Business Contracts in China, 1949–1963*

54. Han-sheng Chuan and Richard Kraus, *Mid-Ch'ing Rice Markets and Trade, An Essay in Price History*

55. Ranbir Vohra, *Lao She and the Chinese Revolution*

56. Liang-lin Hsiao, *China's Foreign Trade Statistics, 1864–1949*

57. Lee-hsia Hsu Ting, *Government Control of the Press in Modern China, 1900–1949*

58. Edward W. Wagner, *The Literati Purges: Political Conflict in Early Yi Korea*

59. Joungwon A. Kim, *Divided Korea: The Politics of Development, 1945–1972*

60. Noriko Kamachi, John K. Fairbank, and Chūzō Ichiko, *Japanese Studies of Modern China Since 1953: A Bibliographical Guide to Historical and Social-Science Research on the Nineteenth and Twentieth Centuries, Supplementary Volume for 1953–1969*

61. Donald A. Gibbs and Yun-chen Li, *A Bibliography of Studies and Translations of Modern Chinese Literature, 1918–1942*

62. Robert H. Silin, *Leadership and Values: The Organization of Large-Scale Taiwanese Enterprises*

63. David Pong, *A Critical Guide to the Kwangtung Provincial Archives Deposited at the Public Record Office of London*

64. Fred W. Drake, *China Charts the World: Hsu Chi-yü and His Geography of 1848*

65. William A. Brown and Urgunge Onon, translators and annotators, *History of the Mongolian People's Republic*

66. Edward L. Farmer, *Early Ming Government: The Evolution of Dual Capitals*

67. Ralph C. Croizier, *Koxinga and Chinese Nationalism: History, Myth, and the Hero*

68. William J. Tyler, tr., *The Psychological World of Natsumi Sōseki*, by Doi Takeo